International Climate Protection

Michael Palocz-Andresen · Dóra Szalay ·
András Gosztom · László Sípos ·
Tímea Taligás
Editors

International Climate Protection

Springer

Editors
Michael Palocz-Andresen
Institute of Sustainable and Environmental Chemistry
Leuphana University of Lüneburg
Lüneburg, Germany

Department of Building Services and Process Engineering
Budapest University of Technology and Economics
Budapest, Hungary

András Gosztom
Institute of Applied Arts
University of Sopron
Sopron, Hungary

Tímea Taligás
Plant Protection Institute, Centre for Agricultural Researches
Hungarian Academy of Sciences
Martonvásár, Hungary

Dóra Szalay
Institute of Forest- and Environmental Techniques
University of Sopron
Sopron, Hungary

László Sípos
István Széchenyi Management and Organisation Doctoral School
University of Sopron
Sopron, Hungary

ISBN 978-3-030-03815-1 ISBN 978-3-030-03816-8 (eBook)
https://doi.org/10.1007/978-3-030-03816-8

Library of Congress Control Number: 2018963042

This Springer imprint is published by the registered company Springer Nature Switzerland AG.
The registered company address is: Gewerbestrasse 11, 6330 Cham, Switzerland

The Alexander von Humboldt Foundation Bonn has sponsored both the Humboldt Kolleg 2015 in Budapest and the current book.

The participants of the Kolleg and the editors of this book thank the foundation for its support.

Alexander von Humboldt
Stiftung / Foundation

Dedicated to Professor Dr. Hartmut Grassl

Preface

Abstract The concentration of climate-active gases has been intensively affected by the industrial production in the past decades. This tendency leads to higher infrared absorption in the atmosphere and causes the global warming, which is a natural phenomenon and cannot be the object of any discussion. The consequences of this change are higher temperatures, and these lead to tropical diseases also in temperate zones, decreasing the diversity of the biosphere in many regions in the world. Furthermore, among the consequences, we can find a reduction of safety in the water supply and a more complicated situation at using agricultural arable land, on the one hand, and the protection for flood, on the other hand.

Human and security aspects of global migration have caused an important problem in the last years. Financial and social impacts of climate change are influencing the global economy more and more. Crises and disturbances caused by the negative effects of climate change along with the circumstances arising from social tensions, such as warfare, destruction of the environment, soil degradation, agricultural production failures, etc., are creating a highly explosive potential in many regions of the world.

Urgent technical assistance and service are needed to solve the problem of salinization and drought in large regions of the world. Environment and climate protection are gaining a leading role also in international politics, according to the resolution of COP 21.

The cooperation between political and civilian organizations is one of the basic elements of future development. However, not only legislation but also technology is key to protecting the climate. In the energy legislation, environmental-friendly fossil and renewable energy sources are decisive factors not only for the generation of electrical energy but also for the production of fuels.

Using alternative energy generated by wind, solar, geothermal, and hydrogen sources demands new and highly smart transportation systems for the produced energy, such as transmission grids, pipelines, trucks, trains, ships, etc. Besides, transportation and large storage systems for renewable energy are needed as soon as possible.

The complexity of mixed fossil and renewable energy sources as well as the issue of decentralized energy production, transportation, storing, and using systems requires a high degree of attention for security aspects.

The main task is to further develop a liveable environment and a stabilized climate by increasing welfare and by creating healthy green cities in the future.

Hamburg, Germany Michael Palocz-Andresen

Proposer

Abstract The Alexander von Humboldt Foundation supports with its "Humboldt Kollegs" initiatives of Humboldt Alumni Associations and individual Humboldtians—as we call our alumni—in order to strengthen regional and interdisciplinary networking among them. Alexander von Humboldt himself was well aware of the importance of cross-cultural exchange and of the opportunity to learn from each other.

The Humboldt Kolleg "Symposium on International Climate Protection" addressed a topical and interdisciplinary issue and, hence, was perfectly in line with Humboldt's own research as he was one of the first scientists doing profound analyses on the human influence on the environment. Furthermore, he published on the influence of deforestation on erosion and drought in South America and warned against the damage from early factories on natural resources and human health.

From a current point of view, the global challenge of climate change can only be met by cross-border international collaborations and cooperation across the disciplines. In this spirit, the Humboldt Kolleg "Symposium on International Climate Protection" demonstrated that the concept of climate protection can be approached from totally different angles: the causes of climate change have to be addressed as well as its consequences. How can legal regulations help us to reach mitigation of emissions? How can sustainable adaptation strategies for agriculture and forestry be formulated and implemented? And how could a climate-friendly answer to a growing demand for transport and mobility look like? Those were only a few questions which were impressively addressed during the conference.

On the other hand, this Humboldt Kolleg also contributed to the Humboldt Foundation's mission to build and foster networks—scientific networks as networks of trust that go far beyond research purposes. By granting research fellowships and awards, the Humboldt Foundation brings more than 700 excellent researchers from all over the world to Germany each year. Our worldwide network now comprises more than 27,000 researchers in over 140 countries and our alumni continuously support our work by spreading the spirit of Alexander von Humboldt to new generations of researchers.

The Humboldt Foundation would like to thank our Humboldtians Prof. Beatrix Farkas and Prof. Michael Palocz-Andresen for their dedication and the initiative to organize the Humboldt Kolleg "Symposium on International Climate Protection".

Bonn, Germany Anke Hoffmann-Pantha

Contents

Part I
Climate Research and Climate Modelling

Introduction

Zoltán Kern

Advance in our understanding of recent changes in the climate system results from the combination of observations, studies of feedback processes, and model simulations [1]. Instrumental observations and climate modelling provide a comprehensive overview of the variability and long-term changes in the atmosphere, the oceans, the cryosphere, and the land surface.

Observations of the various systems of the climate come from surface-based measurements and satellite-based remote sensing observations. However, instrumental records for the key meteorological variables, such as surface air temperature and precipitation amount are available for a period extending back only to the 17th century. The earliest instrumental data date back to 1654 for temperature [2] and to 1697 for precipitation [3]. Global-scale observations from the instrumental era began around the mid-19th century for temperature and other variables with more comprehensive and diverse set of observations available mainly for the second half of the 20th century. Palaeoclimatological reconstructions, however, can in some cases extend our knowledge back over hundreds to millions of years.

A global climate model (GCM) is a complex mathematical representation of the major climate system components (atmosphere, land surface, ocean, and sea ice), and their interactions. The Earth's energy balance between the four components is the key to long-term climate prediction [4]. Each of the components (atmosphere, land surface, ocean, and sea ice) has its own equations calculated on a global grid for a set of climate variables. Model components have not only computed how they change over time, and how the different parts exchange fluxes of heat, water, and momentum, but model components also interact with one another as a coupled system [4].

A recent study investigated the long-term response of the Earth's global climate system in the future using periods in climate history that were warmer than our recent past [5]. The study showed that marine and terrestrial ecosystems will spatially shift

Z. Kern (✉)
Institute for Geological and Geochemical Research, Research Centre for Astronomy and Earth Sciences, Hungarian Academy of Sciences, Budapest, Hungary
e-mail: kern.zoltan@csfk.mta.hu

M. Palocz-Andresen et al. (eds.), *International Climate Protection*,
https://doi.org/10.1007/978-3-030-03816-8_1

and sea level will rise by several meters over the next few thousand years even under strict mitigation scenarios as foreseen in the Paris Agreement [6]. This evidence stresses the need for climate models to include such long-term effects to forecast the full spectrum of future Earth System changes.

This chapter collects seven papers dealing with different aspects of climate research. Georg Feulner briefly reviews the current state of the climate system and discusses certain related questions in light of the Paris Agreement. József Pálfy and his co-authors summarize state-of-art knowledge about a possible analogue for current conditions at ~200 million years ago, when the Earth's atmosphere was characterised by very high atmospheric CO_2 concentrations, climate change and ocean acidification. They conclude that the reconstruction and understanding of such past environmental and biotic crises may allow us to take a crucial step towards an understanding of processes operating in the Earth's climatic system at times of extreme change. Beside greenhouse gases of the Earth's atmosphere incoming solar energy is also a crucial external factor driving the global climate. Zoltán Mitre discusses the effects of changes in solar radiation on scales ranging from the multimillennial to sub-decadal. Judit Bartholy and Rita Pongrácz report the projected climate change for the Carpathian Region. The authors keep the focus of the contribution on the regional temperature and precipitation extremes. Árpád Barsi presents a short introduction to geographic information systems and attempts to highlight their potential in application to climate protection efforts. In the next paper, János Mika gives an overview of contributions from space-borne observations to climate science. A successful and a less successful example for validation of climate models against satellite-based observations are presented in this contribution. Finally, as a case study of satellite derived records Stamatia Doniki presents the calculation of the ozone longwave radiative effect from satellite observation. This is an important issue because ozone is a radiatively active component and its radiative forcing was not well understood until recently.

References

1. IPCC.: Climate Change 2013: The Physical Science Basis. Contribution of Working Group I to the Fifth Assessment Report of the Intergovernmental Panel on Climate Change [Stocker, T.F., et al. (eds.)]. Cambridge University Press, Cambridge, United Kingdom and New York, NY, USA, 1535 pp (2013)
2. Camuffo, D., Bertolin, C.: The earliest temperature observations in the world: the Medici Network (1654–1670). Clim. Change **111**(2), 335–363 (2012)
3. Wales-Smith, G.B. Monthly and annual totals of rainfall representative of Kew, Surrey, for 1697 to 1970. Meteorol. Mag. **100**, 345–362 (1971)
4. GFDL.: Climate Modeling. https://www.gfdl.noaa.gov/climate-modeling/. Access date 03 July 2018
5. Fischer, H., Meissner, K.J., Mix, A.C., Abram, N.J., Austermann, J., Brovkin, V., et al.: Palaeoclimate constraints on the impact of 2 °C anthropogenic warming and beyond. Nat. Geosci. **11**, 474–485. https://doi.org/10.1038/s41561-018-0146-0
6. United Nations.: Paris Agreement (2015). http://unfccc.int/files/essential_background/convention/application/pdf/english_paris_agreement.pdf

The Future of Earth's Climate After Paris

Georg Feulner

1 Current State of the Climate

Over the last few centuries, humanity has become a significant driver of changes in Earth's climate. The primary cause of the current (anthropogenic) climate change are emissions of the greenhouse gas carbon dioxide, which is released by the combustion of fossil fuels like coal, petroleum or natural gas powering a major part of industrial production, electric energy generation, domestic heating and transportation in modern civilisations.

As a result of human activities, the atmospheric concentration of carbon dioxide has steadily increased since the beginning of the industrial era. While pre-industrial concentrations of carbon dioxide in Earth's atmosphere were around 280 ppm (parts per million), they have recently surpassed the level of 400 ppm. The *daily* mean concentration as measured since the late 1950s on the remote volcano Mauna Loa on Hawaii exceeded this symbolic level for the first time in May 2013 [1]. The latest Mauna Loa data are shown in Fig. 1. *Annual* mean concentrations were above 400 ppm for the first time in 2015 and may now remain above this threshold for the foreseeable future [2]. It is also obvious from this iconic 'Keeling Curve'[1] that the atmospheric concentration of carbon dioxide continues to increase to the present day—and the increase is even accelerating.

Since the pioneering works of Joseph Fourier (1768–1830), John Tyndall (1820–1893) and Svante Arrhenius (1859–1927) it is well established that carbon dioxide acts as a greenhouse gas: it absorbs infrared radiation in the atmosphere and thus helps to warm the climate [3]. The observed increase in atmospheric carbon

[1] The Keeling Curve is named after Charles David Keeling (1928–2005) who started the measurements of carbon dioxide at Mauna Loa.

G. Feulner (✉)
Potsdam Institute for Climate Impact Research, Potsdam, Germany
e-mail: feulner@pik-potsdam.de

M. Palocz-Andresen et al. (eds.), *International Climate Protection*,
https://doi.org/10.1007/978-3-030-03816-8_2

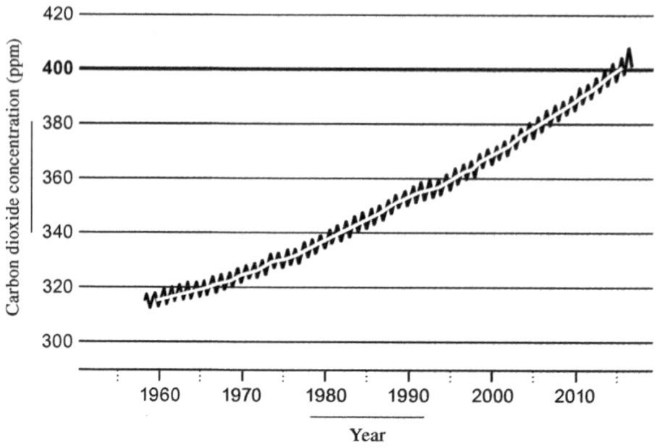

Fig. 1 Monthly (grey) and annual mean (black line) values for the atmospheric carbon dioxide concentration as measured on Mauna Loa (Hawaii) from March 1958 until September 2016 [10]. The dashed line indicates the 400-ppm level

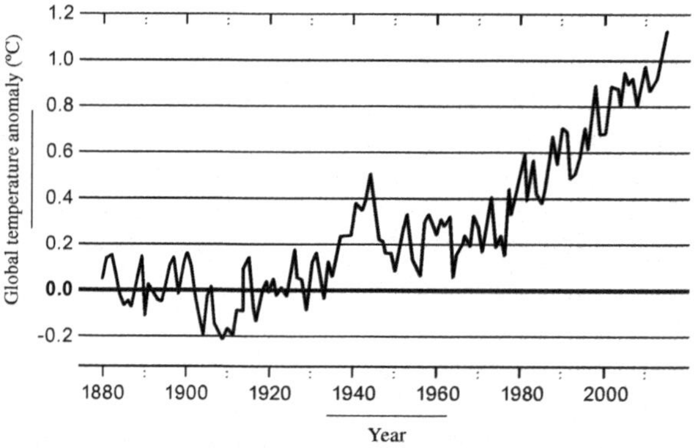

Fig. 2 Anomalies of the annual global mean temperature [11] since 1880 relative to the average over the period 1880–1909 (black)

dioxide since the beginning of the industrialisation thus has to result in increasing temperatures.

Indeed, global warming resulting from anthropogenic greenhouse-gas emissions is now well established. It is not only indicated by observable phenomena like melting glaciers, the retreat of Arctic sea-ice or changes in the vegetation period, the warming is also documented through direct temperature measurements at thousands of weather stations around the world [4]. The global temperature curve derived from these measurements, another iconic graph in climate science, is shown in Fig. 2.

The global temperature time series in Fig. 2 clearly shows that the Earth's climate has warmed by about 1 °C since the end of the 19th century. The year 2015 was the warmest year since the beginning of instrumental measurements, and most of these years of record warmth occurred within the last few decades. In 2015, the global mean temperature exceeded a value of 1 °C above pre-industrial levels for the first time. It is thus evident that global warming caused by anthropogenic greenhouse gas emissions continues unabated.

2 Future Warming and Impacts of Climate Change

How will our climate change in the future? The answer to this essential question depends, of course, on the amount of greenhouse gases that will be emitted over the coming decades. For the latest assessment report [5] of the Intergovernmental Panel on Climate Change (IPCC), climate modelling groups around the world provided projections of future warming based on different scenarios for future emissions (see the left panel of Fig. 3).

If we continue to emit greenhouse gases at the current rate, the expected additional temperature rise over the 21st century will amount to about 4 °C (or roughly 5 °C above pre-industrial temperatures), see the high-emission scenario in the left panel of Fig. 3. For a climate-mitigation scenario, the additional warming until 2100 will be around 1 °C (corresponding to about 2 °C warming compared to the pre-industrial period). It is humanity's choice which of these paths it wishes to pursue. Of course, the expected ecological, social and economical impacts for these different trajectories are of paramount importance for this decision.

An ever-increasing number of scientific studies of climate impacts for different world regions and sectors clearly demonstrate that the negative impacts of climate change increase with rising temperatures [6]. This is illustrated in the right panel of Fig. 3 where climate-change related risks at different warming levels are summarised for the 5th IPCC Assessment Report. This assessment of current and future climate impacts shows that global warming should not exceed 2 °C above pre-industrial levels (temperature scale to the right of the graph) to avoid unmanageable climate risks.

Although below this "2 °C guardrail" the globally aggregated impacts are still comparatively moderate, it should be pointed out that there are already severe impacts for certain systems and sectors even below a temperature increase of 2 °C above pre-industrial levels. This concerns in particular the increase of extreme weather events and threats to unique ecosystems (like coral reefs or habitats in the rapidly warming Arctic) for which this level of warming can be viewed as too high. The 2 °C guardrail derived from these assessments should therefore be considered as a minimum target for international climate policy. To achieve this minimum target, however, global greenhouse-gas emissions have to be significantly reduced as soon as possible.

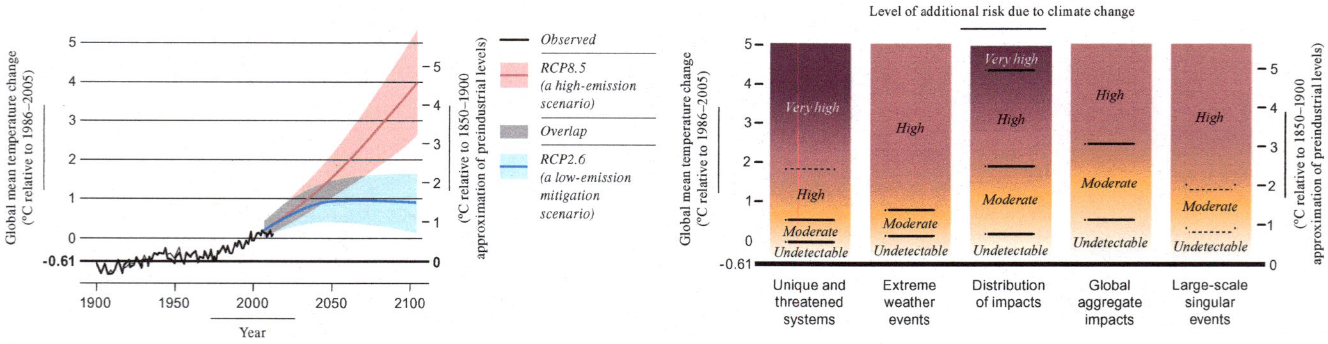

Fig. 3 Left: historical and future warming for a high-emission scenarios (red) and a climate-mitigation scenario (blue). Right: global aggregation of climate impacts at different warming levels for different sectors. Reproduced from Box TS.5 Fig. 1 of the Technical Summary in [6] with permission

3 The Run-up to Paris

The sense of urgency concerning quick and decisive emission reductions generated by the scientific findings outlined above resulted in high expectations in the Paris climate summit, particularly among concerned scientists, climate protesters and environmental politicians. It was hoped that COP21, i.e. the 21st Conference of the Parties of the United Nations Framework Convention on Climate Change (UNFCCC), would finally mark a breakthrough in international climate policy.

The sense of optimism in the run-up to Paris was fuelled by a number of factors and events. First, the Intended Nationally Determined Contributions (INDCs), non-binding declarations of national emission reduction targets declared by countries before the Paris meeting, marked a progress compared to earlier climate targets, although they are by no means sufficient for meeting the 2 °C guardrail. Second, the G7 Summit in Germany in 2015 endorsed the 2 °C guardrail, declared support for marked reductions in greenhouse-gas emissions until the mid of the century and set out the aim of decarbonising the world economy by 2100 [7]. Third, declarations by various religious organisations on the urgency of climate mitigation had a powerful impact, in particular Pope Francis' encyclical on the environment [8]. Finally, an increasing sense of awareness among members of civil society, demonstrated for example through enhanced climate protests in the time before the Paris summit, should not be neglected in this context.

All these factors and declarations did not guarantee a positive outcome of the Paris climate summit, but certainly paved the way for the Paris Agreement [9], a milestone in the history of international climate protection.

4 The Paris Agreement

Following the disappointment after earlier climate summits, the agreement signed in Paris was greeted with widespread approval since some of the expectations outlined above had even been exceeded. The Paris Agreement [9] not only establishes the 2 °C guardrail as the focal point of international climate policy, it recommends to "pursue efforts to limit the temperature increase to 1.5 °C above pre-industrial levels". This is a rather ambitious goal, of course, since the current warming already amounts to about 1 °C as discussed above. To achieve these goals, global emissions will essentially have to be reduced to zero by the mid of the 21st century. In this sense, the Paris Agreement is a powerful signal that the international community is finally serious about climate mitigation. This recognition of the objective of limiting global warming to less than 2 °C sends a clear signal to the financial markets and enterprises. Investments in fossil fuels are already declining today, while investment in renewable energies and research into sustainable technologies will continue to grow. The signal from Paris will further strengthen this development and thus hopefully make an important contribution to limiting climate change to a level still manageable for humanity.

At the core of the Paris agreement are the self-commitment of the states to reduce their greenhouse gas emissions, the so-called Nationally Determined Contributions (NDCs). The agreement establishes a transparent reporting mechanism for verifying compliance with these commitments every five years, starting in 2020. The reporting obligation and regular monitoring are intended to exert political and moral pressure on those countries that do not meet their targets or do not provide sufficiently ambitious self-commitments. Responsibility for the implementation of measures to meet the respective obligations rests with the national states. Therefore, the hope of the Paris agreement is not based on binding objectives. Rather, a transparent system for the reporting of national emission reductions in combination with their regular review is hoped to create increasing momentum towards decarbonated societies. It will have to be seen whether this hope is justified.

The result of the UN climate change summit in Paris raises important questions for climate research: How could transformation pathways to a carbon-free (or at least low-carbon) society look like? What measures could be used to limit global warming to below 2 °C (or even 1.5 °C) above pre-industrial levels? And how would the adaptation measures necessary for this warming have to be designed and financed in the different world regions and sectors?

However, human societies face the main challenges resulting from the Paris Agreement. As impressive as the diplomatic negotiation success of Paris may be, it remains unclear how the objectives set out in the agreement can be achieved. In particular, it is obvious that the national emission reduction commitments announced by the national states so far will not be sufficient to limit global warming to no more than 2 °C against pre-industrial levels, let alone to 1.5 °C.

In any case, the declarations of intent of Paris are to be followed quickly within the next few years. To this end, the state's climate protection commitments must be implemented rapidly and gradually strengthened in the course of the regular reviews in order to meet the 2 °C guardrail. As pointed out above, the phasing-out of fossil fuels and the rapid establishment of renewable energy sources to achieve a decarbonisation of the global economy are inevitable in order to safeguard our planet. The historical diplomatic breakthrough of the United Nations climate-change conference in Paris certainly does not mark the end, but only the beginning of this journey into a sustainable future.

References

1. Showstack, R.: Carbon dioxide tops 400 ppm at Mauna Loa, Hawaii. EOS Trans. AGU **94**(21), 192 (2013)
2. Betts, R.A., Jones, C.D., Knight, J.R., Keeling, R.F., Kennedy, J.J.: El Niño and a record CO_2 rise. Nat. Clim. Change **6**, 806–810 (2016)
3. Lacis, A.A., Schmidt, G.A., Rind, D., Ruedy, R.A.: Atmospheric CO_2: principal control knob governing earth's temperature. Science **330**(6002), 356–359 (2010)

4. Hansen, J., Ruedy, R., Sato, M., Lo, K.: Global surface temperature change. Rev. Geophys. **48**, RG4004 (2010)
5. Stocker, T.F., Qin, D., Plattner, G.-K., Tignor, M., Allen, S.K., Boschung, J., Nauels, A., Xia, Y., Bex, V., Midgley, P.M. (eds.): Climate Change 2013: The Physical Science Basis. Contribution of Working Group I to the Fifth Assessment Report of the Intergovernmental Panel on Climate Change. Cambridge University Press, Cambridge, United Kingdom and New York, NY, USA (2013)
6. Field, C.B., Barros, V.R., Dokken, D.J., Mach, K.J., Mastrandrea, M.D., Bilir, T.E., Chatterjee, M., Ebi, K.L., Estrada, Y.O., Genova, R.C., Girma, B., Kissel, E.S., Levy, A.N., MacCracken, S., Mastrandrea, P.R., White, L.L. (eds.): Climate Change 2014: Impacts, Adaptation, and Vulnerability. Part A: Global and Sectoral Aspects. Contribution of Working Group II to the Fifth Assessment Report of the Intergovernmental Panel on Climate Change. Cambridge University Press, Cambridge, United Kingdom and New York, NY, USA (2014)
7. Leaders' Declaration G7 Summit: Think Ahead. Act Together/An morgen denken. Gemeinsam handeln, https://www.g7germany.de/Content/EN/_Anlagen/G7/2015-06-08-g7-abschluss-eng_en.pdf?__blob=publicationFile&v=3 (2015). 7–8 June 2015
8. Pope Francis: Encyclical Letter Laudato si' of the Holy Father Francis on Care for Our Common Home, Vatican (2015)
9. United Nations: Paris Agreement http://unfccc.int/files/essential_background/convention/application/pdf/english_paris_agreement.pdf (2015)
10. ftp://aftp.cmdl.noaa.gov/products/trends/co2/co2_mm_mlo.txtftp://aftp.cmdl.noaa.gov/products/trends/co2/co2_annmean_mlo.txt (date of access 25 October 2016)
11. http://data.giss.nasa.gov/gistemp/graphs/graph_data/mixedGLB.Ts.ERSSTV4.GHCN.CL.PA.csv (date of access 2 November 2016)

Lessons from Earth's Deep Past: Climate Change and Ocean Acidification 200 Million Years Ago

József Pálfy, T. Ádám Kocsis, Zsófia Kovács and Szabina Karancz

1 Past and Present Climate Change, Ocean Acidification and Biodiversity Crises

The Earth system is composed of four major interconnected subsystems, the lithosphere, hydrosphere, atmosphere and biosphere. Geologists use the rock record to understand processes and past changes in climate and environment, both on land and in the ocean, whereas paleontologists study the fossil record to reveal the history and past diversity of life on Earth.

The Holocene, when our geologically recent past merges with human history, is a more than 10,000-year-long epoch with relatively stable interglacial climate which was an important natural background for development of human societies. The climate record of the past millennium reveals that this stability was terminated by a rapid rise in temperature since the 20th century.

Ongoing global warming is primarily driven by the anthropogenic increase of atmospheric CO_2, mainly from fossil fuel burning. Adverse effects of global warming are exacerbated by increasing uptake of CO_2 by the ocean which in itself restrains temperature increase, but leads to a decrease of seawater pH, known as ocean acidification. This process is detrimental to corals and other marine organisms which

J. Pálfy (✉) · Z. Kovács · S. Karancz
Department of Geology, Eötvös Loránd University, Budapest, Hungary
e-mail: palfy@elte.hu

J. Pálfy · T. Á. Kocsis
Research Group for Paleontology, Hungarian Academy of Sciences-Hungarian Natural History Museum-Eötvös Loránd University, Budapest, Hungary

T. Á. Kocsis
Friedrich-Alexander-Universität Erlangen-Nürnberg, GeoZentrum Nordbayern, Erlangen, Germany

Z. Kovács
Karl-Franzens-Universität Graz, Institute of Earth Sciences, Graz, Austria

M. Palocz-Andresen et al. (eds.), *International Climate Protection*,
https://doi.org/10.1007/978-3-030-03816-8_3

secrete calcium-carbonate skeleton, endangering reef ecosystems and driving many species to extinction.

Such changes unfolding in our human time scale are not unprecedented in Earth history. Five major mass extinctions punctuate the history of life on Earth, including the one at the Triassic-Jurassic boundary ~200 million years ago. The modern rate of species loss is comparable with or exceeds of those measured at the "Big Five" events [9], suggesting that mankind is unintentionally triggering the Sixth Extinction. Geological evidence is mounting to show that drivers of past biotic crises include rapid climatic and environmental changes, with many natural parallels to the modern world. The end-Triassic event has been subject to both detailed studies and recent reviews [4, 8] and offers a prominent case of Earth system's response to perturbation in deep time.

2 The Earth 200 Million Years Ago

Planet Earth looked substantially different 200 million years ago. All landmasses formed the supercontinent Pangea which then started to break up by voluminous volcanism of the giant Central Atlantic Magmatic Province (CAMP) (Fig. 1). Carbon-dioxide emission and other consequences of CAMP volcanism are key in driving environmental change at the end of the Triassic period.

Radiometric dating of zircon crystals in volcanic ashes from marine sedimentary rocks in western Canada and Peru allows to pinpoint the age of the Triassic-Jurassic boundary at ~201 Ma. Dating of volcanic rocks from CAMP yielded the same dates, proving the synchrony of large-scale volcanism and extinction, helping to establish their cause-and-effect relationship.

2.1 Biodiversity Crisis and Global Warming at the Triassic-Jurassic Boundary

Dramatic extinctions of marine organisms happened at the Triassic-Jurassic boundary. For instance, assemblages of radiolarians, marine siliceous microfossils, are markedly different across the Triassic-Jurassic boundary, but are remarkably similar on both sides of the vast Panthalassa superocean, now preserved in western Canada and Japan. This fossil group clearly shows the severe extinction at the end of Triassic, coeval with the CAMP volcanism.

Although previous analyses of radiolarian turnover rates indicated otherwise, our new global analyses at higher stratigraphic resolution indicated that radiolarians were just as affected as bivalves, corals, or brachiopods [5]. This indicates that both global warming and ocean acidification were important extinction agents.

Few proxies are available to reconstruct the concentration of atmospheric CO_2 in the deep geological past. Leaves of fossil plants offer valuable clues as the density

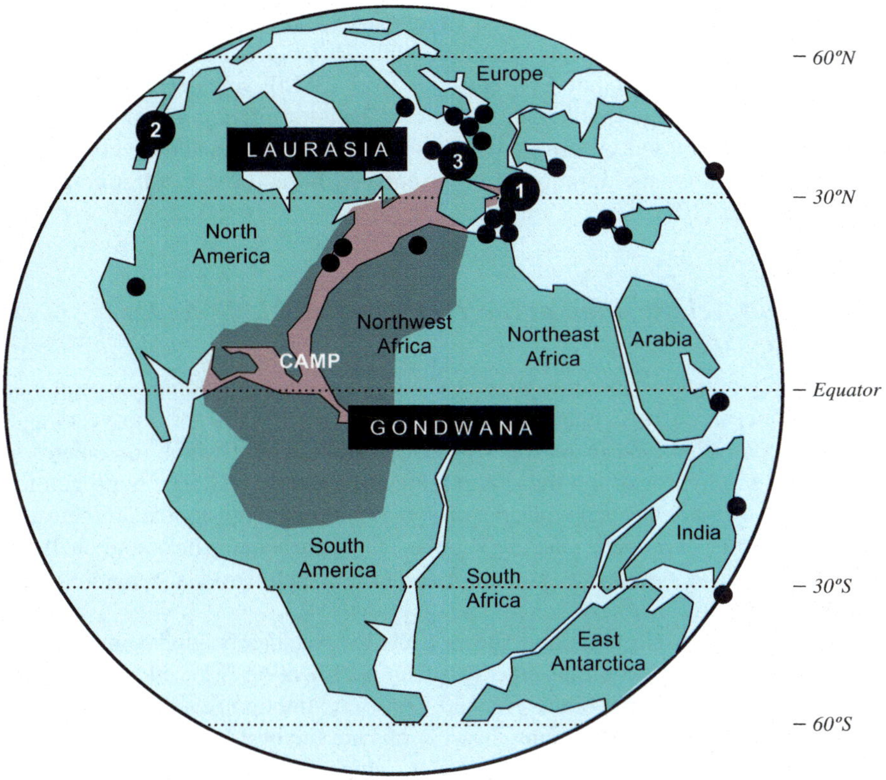

Fig. 1 Paleogeographic map showing key stratigraphic sections with $\delta^{13}C$ data across the Triassic-Jurassic boundary. 1—Csővár, Hungary; 2—Kennecott Point, Canada; 3—St. Audrie's Bay, UK. Modified from Pálfy and Kocsis [8]

of their stomata, allowing the gas exchange, is proportional to pCO_2. Paleobotanical data from different localities indicate a massive increase in the CO_2 level at the Triassic-Jurassic boundary, leading to a major episode of global warming [11].

2.2 Carbon Cycle Perturbation at the Triassic-Jurassic Boundary

If CO_2 were a key factor of climate and environmental change, then a perturbation of the global carbon cycle could be expected. Our best means to trace changes in the carbon cycle in the geological record is to measure the carbon isotopic ratios preserved in carbonate or organic carbon. So far nearly 50 Triassic-Jurassic boundary sections have been studied for stable isotope geochemistry worldwide. The first three of them, from Hungary [7], Canada [12] and England [3], recognized a negative carbon isotope excursion in both organic matter and carbonate, before the first appearance of the earliest Jurassic ammonite genus *Psiloceras*. Possible causes of the sharp 2–4‰ excursion include CO_2 emission from CAMP volcanism, followed by methane release from gas hydrate dissociation. The latter process could be induced by gradual warming and in turn may have led to runaway greenhouse conditions. The best estimate for the duration of the excursion, based on cyclostratigraphy, is 20–40,000 years [10].

2.3 Ocean Acidification at the Triassic-Jurassic Boundary

As a corollary of the rise of atmospheric CO_2 which drove high global temperatures, changes in ocean chemistry also characterized the Earth ~200 million years ago. Carbon dioxide dissolved in seawater produce carbonic acid which dissociates to bicarbonate and hydrogen ion thus decreasing the seawater pH. The hydrogen ion reacts with the carbonate, and when the dissolved carbonate ion content of the water is depleted, more carbonate ions are supplied from carbonate rocks and shells or skeletons of marine organisms through dissolution, which leaves a signature in the stratigraphic record.

The Triassic-Jurassic transition shows a distinctive, nearly global interruption of carbonate sedimentation that hints to ocean acidification [2], which also had a severe effect on the marine biota. The end-Triassic extinction severely affected the carbonate secreting groups. Among these, corals are the best known, and represent one of the most ecologically sensitive groups, which suffered a major extinction at the end-Triassic, resulting in a reef gap in the earliest Jurassic.

Past global warming and acidification events are analogous in many ways to the ongoing global change. Even though the source of the CO_2 is different, the alteration of the atmospheric composition and ocean chemistry is comparable. The continuous uptake of CO_2 by the oceans results in increasing acidification. The ocean pH has already dropped 0.1 unit since the industrial revolution, and a further 0.3–0.4 unit fall is expected in the next century [6].

3 The Triassic-Jurassic Boundary in Hungary

Sedimentary strata spanning the Triassic-Jurassic boundary crop out in different parts of Hungary and were deposited in different environments. Marine rocks occur in the Transdanubian Range, whereas a coal-bearing terrestrial sequence is known from the Mecsek Mountains. A globally significant, continuous marine sedimentary section across the Triassic-Jurassic boundary is exposed near the village Csővár. Simultaneously with the worldwide documented excursion in the carbon isotope record, the marine and the terrestrial palynomorph assemblages also show spikes in their abundance (Fig. 2). The proliferation of stress-tolerant green algae, the prasinophytes,

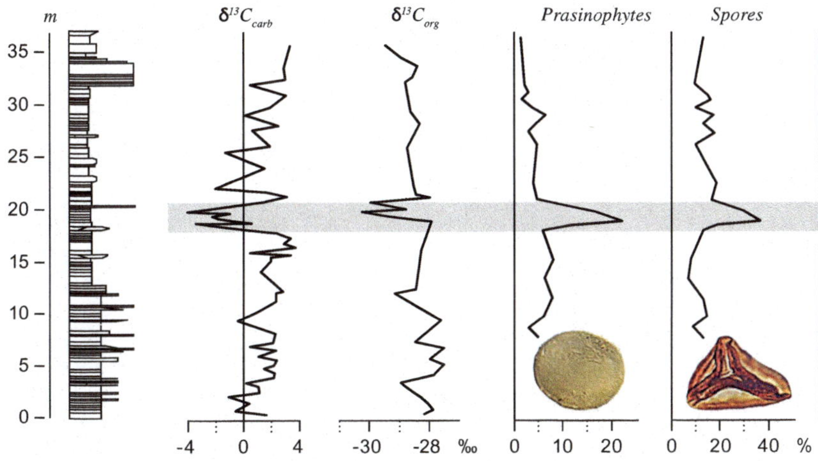

Fig. 2 Carbon isotope excursion and synchronous spikes in abundance of fern spores and prasinophyte algae mark synchronous changes in both terrestrial and marine ecosystems and perturbation of the carbon cycle at the Triassic-Jurassic boundary in the Csővár section, Hungary. After Götz et al. [1]

point to the disturbance of the marine ecosystem, and is interpreted as a response to changes of ocean chemistry and temperature [1].

An abandoned quarry at the Kálvária Hill in Tata provides a spectacular exposure of the Triassic-Jurassic boundary developed in shallow marine environment (Fig. 3). Here, as in several sections elsewhere in the Transdanubian Range, light grey platform limestone layers are abruptly terminated and are overlain by pink lowermost Jurassic limestone. The gap separating the two formations represents the boundary and is likely related to the cumulative effects of ocean acidification and extinction of marine organisms.

Studies of Triassic-Jurassic boundary sections in Hungary in the past 15 years has provided much new data and significant new insights. Ongoing research continues to focus on the paleontological, sedimentological and geochemical record to advance our understanding of the boundary events.

4 Conclusions

At the Triassic-Jurassic boundary, global warming and ocean acidification, with attendant geochemical and sedimentological signatures were part of a cascade of environmental changes that led to a major mass extinction event (Fig. 4). Triggered by volcanic eruptions and CO_2 emission of the Central Atlantic Magmatic Province, this crisis some 200 million years ago offers valuable lessons for understanding human-induced changes of today. The modern, anthropogenic surge in greenhouse

Fig. 3 The Triassic-Jurassic boundary, marked by an abrupt change in sedimentation, in the abandoned quarry at Tata, Hungary

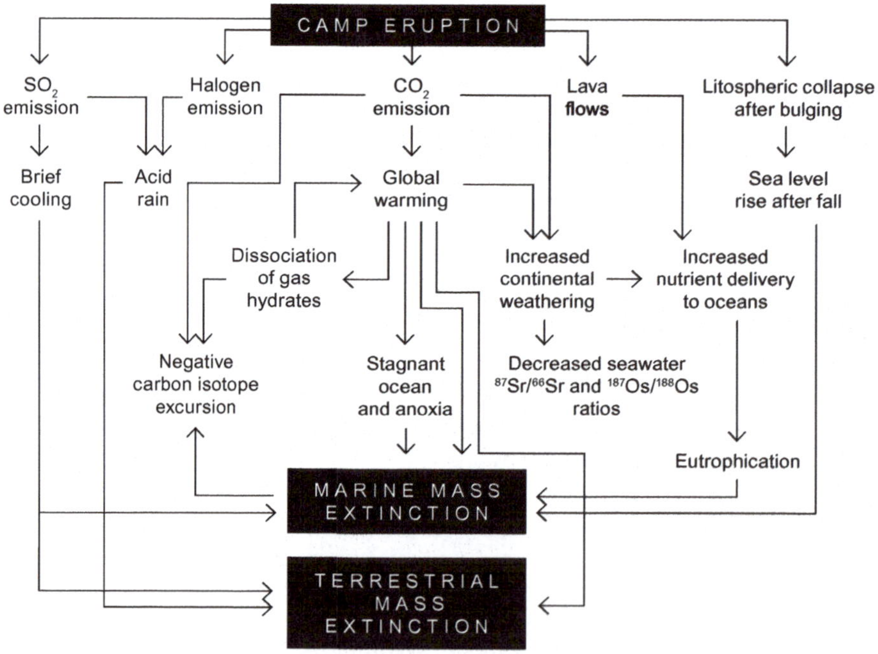

Fig. 4 Schematic diagram summarizing the cause-and-effect relationship of various events taking part in the global change at the Triassic-Jurassic boundary. Adapted from Wignall [13]

gas emission has natural analogs in Earth's deep past. The many similarities make reconstruction of past environmental and biotic crises, such as that at the end of the Triassic, a crucial step in understanding Earth system processes at times of extreme change. The Triassic-Jurassic boundary sections in Hungary preserve important records of these changes.

References

1. Götz, A.E., Ruckwied, K., Pálfy, J., Haas, J.: Palynological evidence of synchronous changes within the terrestrial and marine realm at the Triassic/Jurassic boundary (Csővár section, Hungary). Rev. Palaeobot. Palynol. **156**(3–4), 401–409 (2009)
2. Greene, S.E., Martindale, R.C., Ritterbush, K.A., Bottjer, D.J., Corsetti, F.A., Berelson, W.M.: Recognising ocean acidification in deep time: an evaluation of the evidence for acidification across the Triassic-Jurassic boundary. Earth Sci. Rev. **113**(1–2), 72–93 (2012)
3. Hesselbo, S.P., Robinson, S.A., Surlyk, F., Piasecki, S.: Terrestrial and marine mass extinction at the Triassic-Jurassic boundary synchronized with major carbon-cycle perturbation: a link to initiation of massive volcanism? Geology **30**(3), 251–254 (2002)
4. Hesselbo, S.P., McRoberts, C.A., Pálfy, J.: Triassic-Jurassic boundary events: problems, progress, possibilities. Palaeogeogr. Palaeoclimatol. Palaeoecol. **244**(1–4), 1–10 (2007)
5. Kocsis, Á.T., Kiessling, W., Pálfy, J.: Radiolarian biodiversity dynamics through the Triassic and Jurassic: implications for proximate causes of the end-Triassic mass extinction. Paleobiology **40**(4), 625–639 (2014)
6. Orr, J.C., Fabry, V.J., Aumont, O., Bopp, L., Doney, S.C., Feely, R.A., Gnanadesikan, A., Gruber, N., Ishida, A., Joos, F.: Anthropogenic ocean acidification over the twenty-first century and its impact on calcifying organisms. Nature **437**(7059), 681–686 (2005)
7. Pálfy, J., Demény, A., Haas, J., Hetényi, M., Orchard, M., Vető, I.: Carbon isotope anomaly and other geochemical changes at the Triassic-Jurassic boundary from a marine section in Hungary. Geology **29**(11), 1047–1050 (2001)
8. Pálfy, J., Kocsis, T.Á.: Volcanism of the Central Atlantic Magmatic Province as the trigger of environmental and biotic changes around the Triassic-Jurassic boundary. In: Keller, G., Kerr, A.C. (eds.) Volcanism, Impacts and Mass Extinctions: Causes and Effects, vol. 505. Geological Society of America Special Paper. Geological Society of America, Boulder, CO, pp. 245–261 (2014)
9. Raup, D.M., Sepkoski Jr., J.J.: Mass extinctions in the marine fossil record. Science **215**, 1501–1503 (1982)
10. Ruhl, M., Deenen, M.H.L., Abels, H.A., Bonis, N.R., Krijgsman, W., Kürschner, W.M.: Astronomical constraints on the duration of the early Jurassic Hettangian stage and recovery rates following the end-Triassic mass extinction (St Audrie's Bay/East Quantoxhead, UK). Earth Planet. Sci. Lett. **295**(1–2), 262–276 (2010)
11. Steinthorsdottir, M., Jeram, A.J., McElwain, J.C.: Extremely elevated CO_2 concentrations at the Triassic/Jurassic boundary. Palaeogeogr. Palaeoclimatol. Palaeoecol. **308**(3–4), 418–432 (2011)
12. Ward, P.D., Haggart, J.W., Carter, E.S., Wilbur, D., Tipper, H.W., Evans, T.: Sudden productivity collapse associated with the Triassic-Jurassic boundary mass extinction. Science **292**, 1148–1151 (2001)
13. Wignall, P.B.: Large igneous provinces and mass extinctions. Earth Sci. Rev. **53**(1–2), 1–33 (2001)

Global and Regional Climate Change, Extreme Events

Judit Bartholy and Rita Pongrácz

1 Global Warming

Most of the regional environmental issues can clearly be related to global and regional climate change. The Intergovernmental Panel on Climate Change (IPCC) Assessment Reports (AR) summarize the global warming issues, detected changes, possible causes, and global model projections. The detected average global warming is 0.85 °C over the period 1880–2012 [1]. The linear trend of temperature increase includes two major warming periods, (i) from the beginning of the 20th century until about the early 1940s, (ii) from the 1970s to the present. The second period is clearly due to the enhanced anthropogenic emissions of greenhouse gases. In order to estimate the future climate changes, physically-based climate models are used. Because of the uncertainty of several socio-economic factors, the predictions of climate change are based on scenarios, which specify these conditions, e.g., global population, energy sources, greenhouse gas emissions, industrial developments at different parts of the world, etc. Before the IPCC AR5 [1] SRES scenarios [2] were widely used, in the IPCC AR5 the newly developed radiative forcing-based scenarios [3] are introduced. On the basis of the estimated changes of the radiative forcing, the two groups of scenarios can be compared. The new RCP4.5, RCP6.0 and RCP8.5 scenarios are similar to the previously used SRES B1, A1B, and A2 scenarios, respectively. However, there is one completely new scenario, i.e. RCP2.6, to which any SRES scenario cannot be assigned. This assumes strong mitigation actions in the coming decades.

According to the latest simulation results of global climate models the greatest further warming is projected to the northern polar region, moreover, the predicted warming is greater over land than over ocean. The estimated global average tem-

J. Bartholy (✉) · R. Pongrácz
Department of Meteorology, Eötvös Loránd University, Budapest, Hungary
e-mail: bartholy@caesar.elte.hu

J. Bartholy · R. Pongrácz
Faculty of Science, Excellence Center, Eötvös Loránd University, Martonvásár, Hungary

M. Palocz-Andresen et al. (eds.), *International Climate Protection*,
https://doi.org/10.1007/978-3-030-03816-8_4

Table 1 The projected global average increase of annual mean temperature values for the target period 2081–2100 relative to the reference period 1986–2005 (Stocker et al. [1])

Scenario	Predicted global warming (°C)
RCP2.6	0.3–1.7
RCP4.5	1.1–2.6
RCP6.0	1.4–3.1
RCP8.5	2.6–4.8

perature increases by 2081–2100 relative to 1986–2005 are summarised in Table 1. The projected global warming ranges from 0.3 to 1.7 °C in the case of the smallest increase of radiative forcing (RCP2.6) to 2.6–4.8 °C in the case of the greatest increase of radiative forcing (RCP8.5). The projected warming patterns are similar for the different scenarios. For instance, within Europe the greatest and the smallest temperature increases in annual mean values are projected in the northeastern and southwestern parts, respectively.

2 Regional Climate Change

For a region with the size of the Carpathian Basin, global climate models may provide a relatively good estimation for temperature change. However, their spatial resolutions (typically 100–200 km) are too coarse for the appropriate prediction of precipitation change. Therefore, global climate simulation results must be downscaled for regional analysis, for which physically-based regional climate models (RCMs) are essential. RCMs are limited area models nested in global climate models, i.e. the initial and lateral boundary conditions are provided by the outputs of global climate models' simulations.

The projected annual and seasonal mean temperature and precipitation changes for Hungary are summarised in Figs. 1 and 2, respectively.

The estimations presented here are based on the RCM outputs of the research program ENSEMBLES [4] using an intermediate scenario (SRES A1B, for which the estimated global CO_2 concentration levels by 2050 and 2100 are 532 and 717 ppm, respectively), and discussed in details in Pongrácz et al. [5]. As a regional consequence of global warming, clear temperature increase is predicted in the target region for the coming decades. The multi-model average estimated warming on annual scale is 1.6 and 3.4 °C by the mid- and late-century, respectively. The greatest regional warming is projected in summer by the end of the 21st century (when the multi-model average temperature increase is close to 4 °C), which will consequently result in a substantial increase of heat stress.

The precipitation projections include more uncertainty due to the different model parameterisations, as well as the high spatial and temporal variability of the hydrological cycle. The average annual totals are not likely to change substantially in the

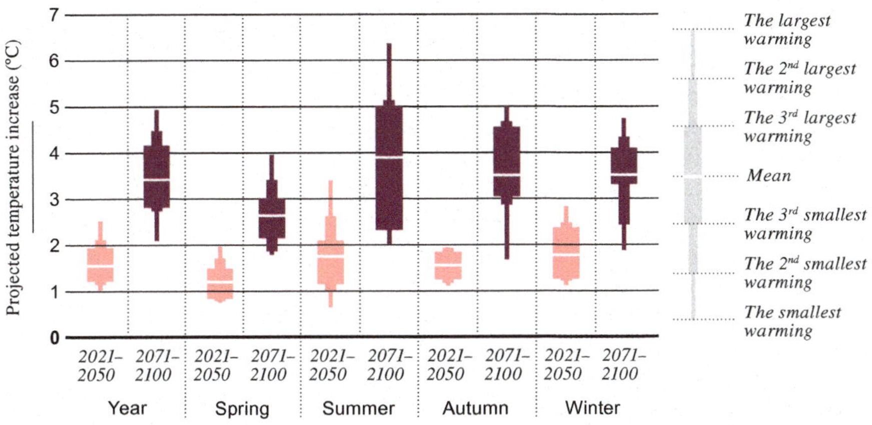

Fig. 1 The projected annual and seasonal mean temperature changes for Hungary using SRES A1B scenario, reference period: 1961–1990 (based on 11 RCM simulations)

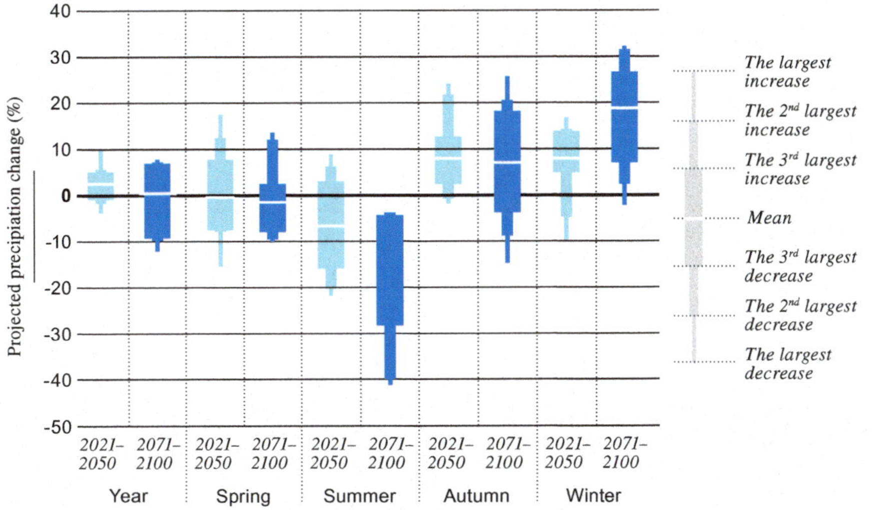

Fig. 2 The projected annual and seasonal mean changes of precipitation totals for Hungary using SRES A1B scenario, reference period: 1961–1990 (based on 11 RCM simulations)

coming decades. However, summer is projected to become drier, especially by the end of the 21st century when the multi-model average precipitation decrease is about 20%. Furthermore, opposite changes are predicted for winter, i.e. wetter conditions are likely to occur on average compared to the 1961–1990 reference period, the multi-model average precipitation increase is about 20%.

3 Regional Temperature Extremes

Simulation results of RCMs can also be used to project future changes in extreme climatological conditions. First, extreme temperature conditions are analysed using simulated daily minimum and maximum temperature values. For this purpose, the simulation outputs of model PRECIS are used (the overall analysis of simulations for various scenarios can be found in Pieczka et al. [6]). Figure 3 shows the predicted shifts of temperature distribution, which is a clear consequence of regional warming. The average occurrences of negative extremes are projected to decrease, whereas the frequencies of positive extremes tend to increase in the future.

Since the positive temperature extremes tend to cause more problems than the negative temperature extremes in the continental Carpathian region, we focus on the estimated changes of heat stress in this paper. For this purpose, the three warning levels within the Heat Health Watch Warning System developed on the basis of a retrospective analysis of mortality and meteorological data from Hungary [7] are used. The definitions of warning levels are associated to the different thresholds (25 and 27 °C) of the daily mean temperature and the duration (at least 1 day, at least 3 days) of the event as summarised in Table 2.

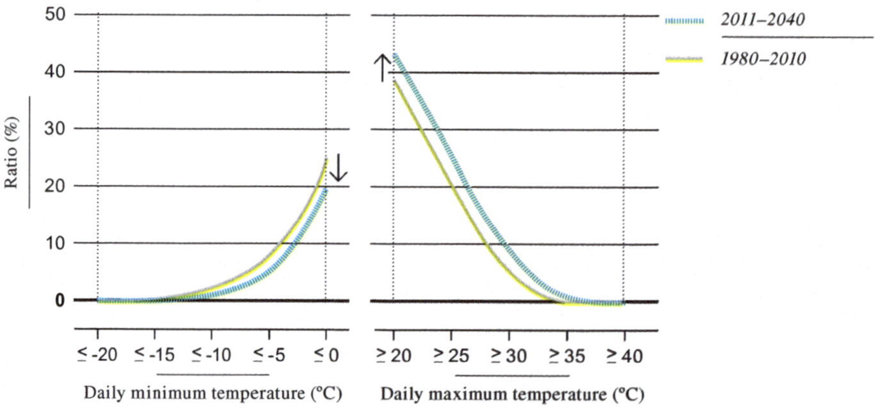

Fig. 3 The projected changes in the exceedances of temperature thresholds for Hungary by the target period 2011–2040 relative to the reference period 1981–2010

Table 2 The definition of the warning levels within the Heat Health Watch Warning System in Hungary

Warning level	Definition
Level 1	The daily mean temperature exceeds 25 °C for at least 1 day
Level 2	The daily mean temperature exceeds 25 °C for at least 3 days
Level 3	The daily mean temperature exceeds 27 °C for at least 3 days

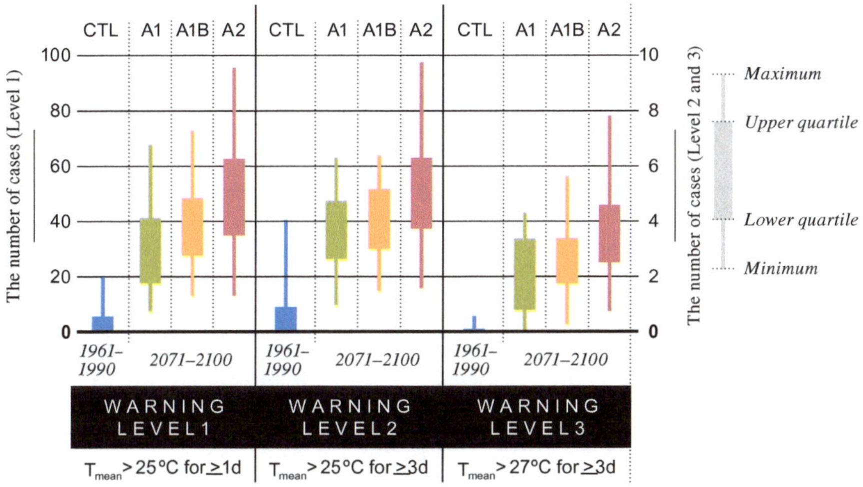

Fig. 4 The simulated occurrences of the warning levels within the Heat Health Watch Warning System in 2071–2100 compared to the reference period 1961–1990 for Hungary. The small rectangles indicate the middle half (between the quartiles) of the simulated annual occurrences, and the vertical lines are drawn from the minimum to the maximum simulated annual occurrences

For instance, at the grid cell representing Budapest heat wave warning level conditions occurred 7 times (Level 1) in a year, once per year (Level 2), and only once per decade (Level 3) on average in the reference period (1961–1990). During the last three decades of the 21st century heat wave warning level 1 conditions are projected to occur 39 times, 39 times, and 56 times in a year on average [8] taking into account B2, A1B, and A2 scenarios, respectively. Naturally, level 2 and 3 conditions are likely to occur fewer times in the future than level 1 conditions, however, similarly to level 1, significantly more frequent warning events are projected by 2071–2100 than in 1961–1990. The future annual average level 2 occurrences are 4, 5, and 5 in case of B2, A1B, and A2 scenarios, respectively. The future annual average level 3 occurrences are 2, 3, and 4 in case of B2, A1B, and A2 scenarios, respectively. The 30-year variances are shown by the Box-Whisker plot diagram of Fig. 4.

4 Regional Precipitation Extremes

In the case of precipitation the asymmetric distribution of daily precipitation implies that excessive rainfall events last shorter in the midlatitude continental climate of the Carpathian region [9], however, the lack of precipitation might cause severe economical problems if it occurs for a long time.

Due to the summer drying tendency predicted for the region, we focus on the dry conditions. For this purpose, daily precipitation data are used, which are available from the RCM simulations of research program ENSEMBLES [4]. The detailed analysis of various precipitation-related indices can be found in Pongrácz et al. [10].

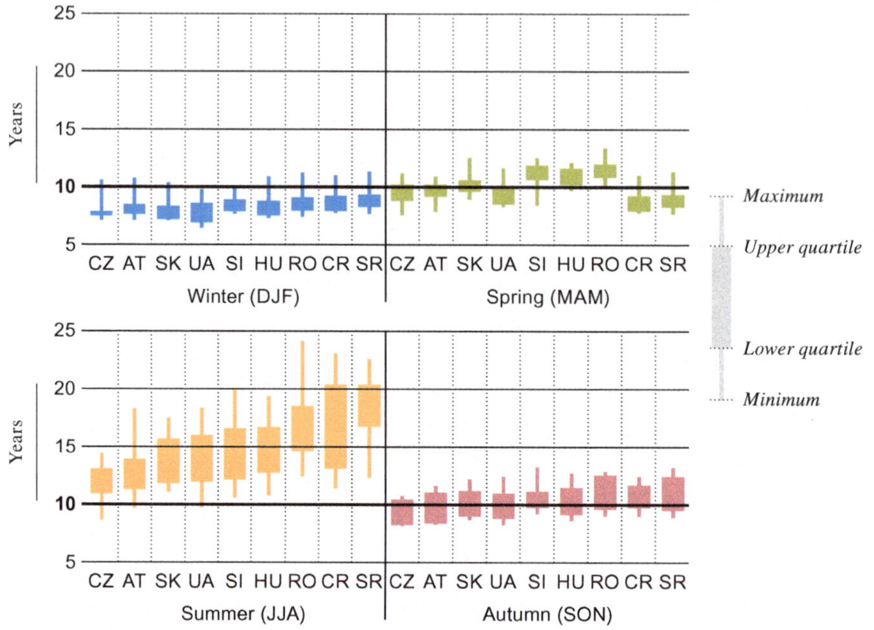

Fig. 5 The Box-Whisker diagrams to indicate the distribution of the spatially averaged 10-year seasonal return period of daily precipitation for 9 subregions within the Central/Eastern European region (CZ: southeastern Czech Republic, AT: eastern Austria, SK: Slovakia, UA: southwestern Ukraine, SI: Slovenia, HU: Hungary, RO: Romania, CR: Croatia and SR: northern Serbia). Target period: 2071–2100, reference period: 1961–1990

Here, the predicted changes in the 10-year return period are analysed. Due to the opposite signs of the seasonal precipitation changes, this statistical measure is also evaluated on a seasonal scale. In our analysis nine subregions are defined in the entire Central/Eastern European region. First, the 10-year return values of simulated precipitation data are calculated for each grid cell using historical simulations (1961–1990) of 11 RCM. Then, the empirical distributions of the simulated daily precipitation for the future target period (2071–2100) are used to determine the return periods of these values. Hence, the result refers to the change of the 10-year return period from 1961–1990 to 2071–2100. The multi-model composite of the seasonal changes are shown in Fig. 5 for all the 9 subregions. The seasonal return period is mostly projected to decrease in winter in the entire domain. A zonal-like spatial pattern can be recognised in the projected changes both in spring and autumn; most of the RCM simulations predict a decrease of the 10-year return period in the northern and northwestern parts of the domain, while an increase is projected in the south (i.e. in Serbia and Romania). Due to the projected summer drying, the 10-year return period tends to increase by the late 21st century. A larger increase (i.e. more intense drying) is projected in the south compared to the northern part of the selected domain (Fig. 6). The predicted multi-model average return period is 12–20 years for 2071–2100.

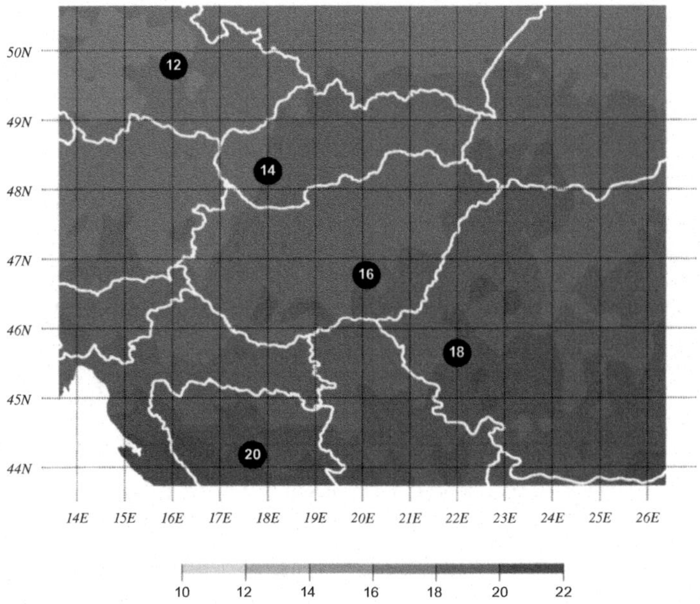

Fig. 6 The composite map of 11 RCM simulations indicating the projected mean increase of the 10-year return period of daily precipitation in summer. Target period: 2071–2100, reference period: 1961–1990

5 Conclusions

On the basis of the results presented in this paper, similarly to the global tendencies, the regional warming is projected to continue in the Carpathian region. As a consequence, the heat stress is likely to increase since the greatest warming is estimated during summer, and the frequency and duration of heat waves are projected to increase during the 21st century. Precipitation totals are projected to increase in winter and decrease in summer. The projected summer drying results in a substantial increase of the 10-year return period of the daily precipitation.

Acknowledgements Research leading to this paper has been supported by the following sources: the Ministry of National Development of the Hungarian Government via the AGRÁRKLIMA2 project (VKSZ_12-1-2013-0034), the Széchenyi 2020 programme, the European Regional Development Fund and the Hungarian Government via the AgroMo project (GINOP-2.3.2-15-2016-0028), the Hungarian Scientific Research Fund under grants K-129162 and K-120605. The authors wish to thank the Hadley Centre of the UK MetOffice to provide the model PRECIS for regional climate change analysis. The ENSEMBLES data used in this work was funded by the EU FP6 Integrated Project ENSEMBLES (Contract number 505539) whose support is gratefully acknowledged.

References

1. Stocker, T.F., Qin, D., Plattner, G.K., Tignor, M., Allen, S.K., Boschung, J., Nauels, A., Xia, Y., Bex, V., Midgley, P.M. (eds.) Climate Change 2013: The Physical Science Basis. Contribution of Working Group I to the Fifth Assessment Report of the Intergovernmental Panel on Climate Change. Cambridge University Press, Cambridge, UK (2013)
2. Nakicenovic, N., Swart, R. (eds.): Emissions Scenarios. A Special Report of IPCC Working Group III. Cambridge University Press, Cambridge, UK (2000)
3. van Vuuren, D.P., Edmonds, J.A., Kainuma, M., Riahi, K., Thomson, A.M., Hibbard, K., Hurtt, G.C., Kram, T., Krey, V., Lamarque, J.-F., Masui, T., Meinshausen, M., Nakicenovic, N., Smith, S.J., Rose, S.: The representative concentration pathways: an overview. Clim. Change **109**, 5–31 (2011)
4. van der Linden, P., Mitchell, J.F.B. (eds.): ENSEMBLES: Climate Change and Its Impacts: Summary of Research and Results from the ENSEMBLES Project. UK Met Office Hadley Centre, Exeter, UK (2009)
5. Pongrácz, R., Bartholy, J., Miklós, E.: Analysis of projected climate change for Hungary using ENSEMBLES simulations. Appl. Ecol. Environ. Res. **9**, 387–398 (2011)
6. Pieczka, I., Pongrácz, R., Bartholy, J.: Comparison of simulated trends of regional climate change in the Carpathian Basin for the 21st century using three different emission scenarios. Acta Silvatica et Lignaria Hungarica **7**, 9–22 (2011)
7. Páldy, A., Bobvos, J., Vámos, A., Kovats, R.S., Hajat, S.: The effect of temperature and heat waves on daily mortality in Budapest, Hungary, 1970-2000. In: Kirch, W., Menne, B., Bertollini, R. (eds.) Extreme Weather Events and Public Health Responses. WHO, Springer, pp. 99–108 (2005)
8. Pongrácz, R., Bartholy, J., Bartha, E.B.: Analysis of projected changes in the occurrence of heat waves in Hungary. Adv. Geosci. **35**, 115–122 (2013)
9. Bartholy, J., Pongrácz, R., Kis, A.: Projected changes of extreme precipitation using multi-model approach. Időjárás Q. J. Hung. Meteorol. Serv. **119**, 129–142 (2015)
10. Pongrácz, R., Bartholy, J., Kis, A.: Estimation of future precipitation conditions for Hungary with special focus on dry periods. Időjárás Q. J. Hung. Meteorol. Serv. **118**, 305–321 (2014)

Environmental Effect of a Solar Eclipse: What Happens, When the Solar Radiation Changes?

Zoltán Mitre

1 When Solar Radiation Changes

The climate of the Earth is highly influenced by the radiation of the Sun. In a short period—some million years—it has no change, just small variations occur. Any little change in the solar radiation has an effect on the Earth. In this section we will study a couple of examples about this problem, due to natural circumstances in the celestial space, higher atmosphere (volcanic aerosols) or anthropogenic processes [1–4].

1.1 Ice Ages, Glacial, Interglacial Periods

We have evidence of climate changes in the history of the Earth, about "ice ages" and warmer periods. In the present we live in an interglacial period of an ice age, started 33.5 million years ago. Interglacial period means a warmer period within an ice age, glacial period means a cooler period. We know 6–7 great ice ages from the history of Earth, with several glacial and interglacial periods [5].

There are several—sometimes not fully understood—reasons why ice ages and glacial, interglacial periods occur, like atmospheric composition, tectonic plate motion, volcanism, orbit and obliquity of the Earth, large meteorites, etc. One of them is Milutin Milanković's theory, presumed that glacial periods are influenced by astronomical reasons. These changes influence the solar radiation touching the Earth's surface at a given geographical latitude [1, 6]. Periods based on this theory (approximately 100,000 years) were clearly detectable at 140 million-year-old (Jurassic period) limestone layers from sea animals, whose life was influenced by the climate change [7].

Z. Mitre (✉)
Faculty of Natural Sciences, Eszterházy Károly University, Eger, Hungary
e-mail: mitre.zoltan@sek.elte.hu

M. Palocz-Andresen et al. (eds.), *International Climate Protection*,
https://doi.org/10.1007/978-3-030-03816-8_5

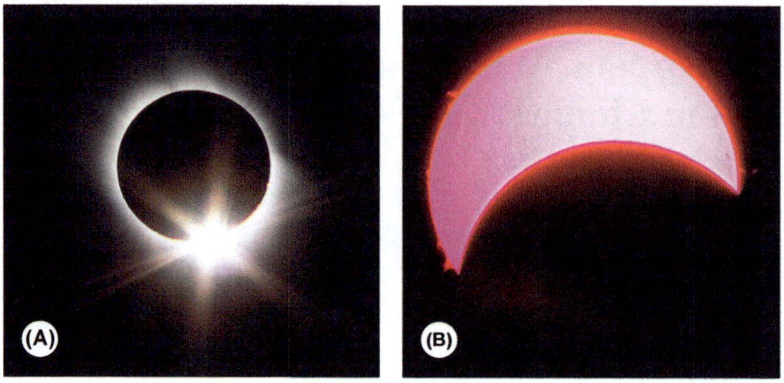

Fig. 1 **a** Total solar eclipse with diamond ring effect from Turkey, 29th March, 2006 (Photo by Z. Mitre). **b** Partial solar eclipse through hydrogen-alpha telescope from Hungary, 20th March, 2015. We can recognize some solar prominences (from sunspots) all around the edge of the Sun (Photo by Z. Mitre)

1.2 Sunspot Cycles, Clouds, Climate

Sunspots are dark, temporary, colder areas on the photosphere of the Sun, with intense magnetic activity. Most solar flares, prominences and coronal mass ejections originate from around sunspot groups, these ejections cause geomagnetic storms on the Earth (Fig. 1b). Their number on the Sun varies in a period of 11 years, it is called solar cycle or solar magnetic activity cycle.

There is also a longer-period trend in the activity of the Sun. These are the great minimum and maximum periods, which last for decades. Of course these long periods also contain the 11-years cycles. The average number of sunspots is lower during the minimum periods, and higher during the maximum periods [4].

There is a hypothesis, which says the solar cycles have an effect to the Earth's climate through the cloud formation [2, 4]. Cosmic rays from the universe help the chemical reactions to produce aerosol for cloud formation. Stronger magnetic activity of the Sun during sunspot maximum reflects more cosmic rays than in sunspot minimum. A decrease in cosmic radiation reaching the Earth results in less clouds in the atmosphere, and thus more solar radiation reaches the surface [8].

From the 13th century till the 18th century there was a long colder period ("Little Ice Age"), when the solar activity was lower, than now. During this period more cosmic rays reached the Earth, more clouds formed, less solar radiation touched the ground. There was a period during the 10th century with intensive solar activity, less cosmic rays reached the Earth. Climate of the Earth was warmer, for example the Vikings reached Greenland this time, which was really green, with more vegetation diversity and less ice than now [9].

1.3 Volcanic Aerosols in the Atmosphere

It seems the Little Ice Age has another reason too, related to the volcanic activity, and influenced the climate more than the sunspot minimum. When a volcano erupts, a huge amount of ash rises to the higher atmosphere and covers the whole Earth. The ash cloud blocks the solar radiation, temperature decreases [10].

This is not the only one period when volcanic eruption influenced the temperature decrease. During the 6th century, around 535/536 AD there was a significant global temperature decrease. There are descriptions about low temperature, snow during the summer, crop failures, dense dry fog, drought, epidemics. Byzantine historian Procopius described that the Sun gave its light without brightness, like *the Sun in eclipse*. Many scientific evidence validate the volcano eruption theory, based on tree ring analysis, ice cores from Greenland, Antarctica with higher sulfate amount than usual [11].

1.4 Global Dimming

The global dimming is a problem of anthropogenic origin, similar to the natural phenomena we studied above. We produce artificial dust from engines, power plants, it spreads in the higher atmosphere, reduces the amount of solar light reaching the ground, but helps the cloud formation. Global dimming has cooling effects [12].

Air traffic also adds to the global dimming problem, producing contrails, which become clouds and appear many times over territories where naturally no cloud formation would occur. After the terror attack against the United States of America in 2001 September, all of the aircrafts were grounded for 3 days. Climatologist recognized, that, as there were no contrails above the USA, there was an observable increase in diurnal temperature range during this period [13].

Global dimming influences the weather as well. Due to the reduced sunlight, the monsoon has reduced force. There are ongoing research about the effects on the agriculture [12, 14, 15].

2 Solar Eclipse and Environment

The solar eclipse is an ideal phenomenon to study and analyze, what happens when the solar radiation decreases. As the Moon moves in front of the solar disc, the lunar shadow causes changes in the environment, including but not limited to the temperature and light decrease (Fig. 1a).

Diameter of the shadow depends on the distance of the Moon from the Earth, as it has elliptical orbit. Partial eclipse is, when the Moon covers a certain percent of the solar disc and total coverage does not happen. This is a more frequent phenomena,

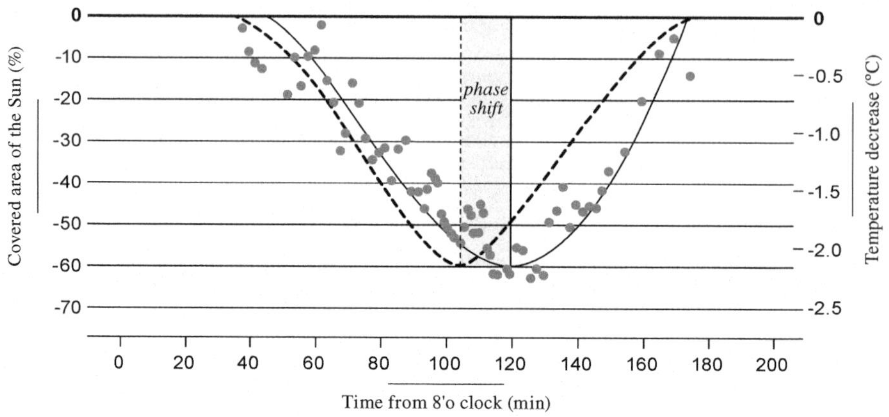

Fig. 2 Visible area of the Sun in percent and temperature decrease from professional measurement as a function of time during 03.20.2015. Data from measurements of students with home thermometers are represented by dots (Edited by Z. Mitre)

because it is observable from several parts of the Earth, while the diameter of the track of totality usually 100–200 km of a total solar eclipse.

In the last 15 years we observed several solar eclipses, both partial and total ones, from several parts of the world. All examinations showed the same results, like light and temperature decrease. In 2009, we recorded data from the total solar eclipse in Shanghai with different heights above the ground. We experienced that the higher we measure the temperature, the less decrease we observe. Based on multiple observations, it seems, temperature decrease is observable up to a height of 20–25 m above the ground. In 1999, during the "great" total solar eclipse in Hungary, we made observations 48, 82, 112 m above the ground, and no temperature decrease was experienced.

All of the temperature decreases at both total and partial solar eclipses has a shift in time from the phase of solar eclipse, due to the convection in the air. This delay is influenced by the local conditions. For example in 2006, Turkey, the temperature still decreased for 15 min after the end of the total phase of the eclipse, then it began to increase again [16–18].

Our latest examination shows that different land covers influence the degree of temperature decrease and the delay from the phase of the eclipse. On the 20th March 2015. we executed a widespread observation project, as there was a partial solar eclipse with a maximum coverage of 60% above Hungary. We had five measurement stations in areas not so far from each other, with different vegetation. Areas with dense vegetation (bushes, trees, park) has only 10–11 min delay and less decrease in temperature (–1.1 °C). Highest temperature decrease (–2.1 °C) happened in a city parking zone with dark asphalt coverage with a shift of 14–16 min [19] (Fig. 2).

High school students were involved into the experimental measure above. Parallel with our professional temperature measurements, they measured the temperature at all of the observation stations with their own home thermometers, and made description of their subjective point of view about the environmental changes (what they feel on their own skin, birds, animal's behavior, etc.).

The inquiry-based learning is a good way to demonstrate the students the effects of the solar radiation decrease included but not limited to the microclimate. After the phenomena we discussed the data experienced and kept lessons about the usage of their observation regarding climate change as well. This is a good way of motivation for students to collect more information about the climate change problem and discuss about the possible solutions [19].

Latest researches try to apply a more precise theoretical model, based on meteorological and physical theories, to the solar radiation decrease and temperature change during a solar eclipse. In the last three years we have also improved our former analysis method, included but not limited to the theoretical model of the solar radiation change [19]. It seems a small coverage already induces temperature decrease, so a little decrease of the solar light has observable effects. That is why it is important to make detailed observations of partial eclipses with several percent of maximum coverages, to analyze the environment reactions to the solar radiation change with different amounts.

3 Conclusions

Solar eclipses are good "reality experiments" about the environmental reaction of the solar radiation change. It can decrease by natural processes, like sunspot cycles or volcanic aerosols, etc., and anthropogenic influences, like air pollution. Measurements show, that land coverage has an influence on the microclimate. These examinations can help the civil engineering of urban territories to design cities with good microclimate.

Global dimming could be a problem for the future, as, apparently, the fossil fuel emission, air traffic, etc. have an observable role in it. We saw, that small amount of solar radiation decrease has an effect on the environment both in light and temperature, which reduce the efficiency of solar energy production as well.

With inquiry-based learning, we can start discussions about the climate change problem in schools, involve high school students in experiments related to solar eclipses. The changes we experience in light, temperature, vegetation, fauna, etc. have useful information for several type of lessons to discuss about this problem.

Acknowledgements Supported by the ÚNKP-18-2 new national excellence program of the Ministry of Human Capacities.

References

1. Érdi, B.: A naprendszer dinamikája. ELTE Eötvös Kiadó, Budapest (2001)
2. Friis-Christensen, E., Svensmark, H.: What do we really know about the Sunclimate connection? Adv. Space Res. **20**(4–5), 913–921 (1997)
3. Laskar, J., Joutel, F., Robutel, P.: Stabilization of Earth's obliquity by the Moon. Nature **361**, 615–617 (1993)
4. Lassen, K., Friis-Christensen, E.: Variability of the solar cycle length during the past five centuries and the apparent association with terrestrial climate. J. Atmos. Terr. Phys. **57**(8), 835–845 (1995)
5. Gradstein, F.M., Ogg, J.G., Schmitz, M.D., Ogg, G.M. (eds.): The Geologic Time Scale 2012, 1st edn, 1176 p. Elsevier (2012)
6. Bacsák, G.: A pliocén és a pleisztocén az égi mechanika megvilágításában. In: Földtani Közlöny LXXXV, vol. 1, pp. 70–101. Budapest, Hungary (1955)
7. Haas, J., Kovács, L.Ó., Tardi-Filácz, E.: Orbitally forced cyclical changes in the quantity of calcareous and siliceous microfossils in an upper jurassic to lower cretaceous pelagic basin succession, Bakony mountains. Sedimentology **41**, 643–653 (1994)
8. National Research Council: Radiative Forcing of Climate Change: Expanding the Concept and Addressing Uncertainties, 222 p. The National Academy Press, Washington, D.C. (2005)
9. Mann, M.E.: Little ice age. In: Munn, T. (ed.) The Earth system: physical and chemical dimensions of global environmental change, pp. 504–509 (2002)
10. Miller, G.H., Geirsdóttir, Á., Zhong, Y., Larsen, D.J., Otto-Bliesner, B.L., Holland, M.M., Bailey, D.A., Refsnider, K.A., Lehman, S.J., Southon, J.R., Anderson, C., Björnsson, H., Thordarson, T.: Abrupt onset of the little ice age triggered by volcanism and sustained by sea-ice/ocean feedbacks. Geophys. Res. Lett. **39**(2), L02708 (2012)
11. Larsen, L.B., Vinther, B.M., Briffa, K.R., Melvin, T.M., Clausen, H.B., Jones, P.D., Siggaard-Andersen, M.-L., Hammer, C.U., Eronen, M., Grudd, H., Gunnarson, B.E., Hantemirov, R.M., Naurzbaev, M.M., Nicolussi, K.: New ice core evidence for a volcanic cause of the A.D. 536 dust veil. Geophys. Res. Lett. **35**(4), L04708 (2008)
12. Stanhill, G., Cohen, S.: Global dimming: a review of the evidence for a widespread and significant reduction in global radiation with discussion of its probable causes and possible agricultural consequences. Agric. Forest Meteorol **107**, 255–278 (2001)
13. Travis, D.J., Carleton, A.M., Lauritsen, R.G.: Contrails reduce daily temperature range. Nature **418**, 601 (2002)
14. Soni, V.K., Pandithurai, G., Pai, D.S.: Is there a transition of solar radiation from dimming to brightening over India? Atmos Res **169**, 209–224 (2016)
15. Yang, X., Asseng, S., Fook Wong, M.T., Yu, Q., Li, J., Liu, E.: Quantifying the interactive impacts of global dimming and warming on wheat yield and water use in China. Agric. Forest Meteorol. 182–183, 342–351 (2013)
16. Peñaloza-Murillo, M.A., Pasachoff, J.M.: Air-cooling mathematical analysis as inferred from the air-temperature observation during the 1st total occultation of the Sun of the 21st century at Lusaka, Zambia. J. Atmos. Solar-Terr. Phys. **125–126**, 59–77 (2015)
17. Pintér, T.P., Péntek, K., Mitre, Z.: Mathematical analysis of temperature results of the total solar eclipse on 29 March 2006. In: 19. Celoštátny Slnečný Seminár Proceedings, pp. 106–113 (2008)
18. Pintér, T.P., Péntek, K., Mitre, Z.: Mathematical analysis of temperature results of the total solar eclipse in China on the 22nd July, 2009. In: 20. Celoštátny Slnečný Seminár Proceedings, pp. 144–154 (2010)
19. Finta, Z.S., Mitre, Z.: A kutatás alapú tanulás alkalmazása a 2015. március 20-i napfogyatkozás során végzett hőmérséklet mérésre. XIV. In: Természet-, Műszaki- és Gazdaságtudományok Alkalmazása Nemzetközi Konferencia előadások, pp. 57–64 (2015)

How Can GIS Support the Climate Protection?

Árpád Barsi

1 Introduction: What Is GIS?

The 21st century is said as the era of informatics, which is meant as the information must be collected, managed and analyzed by computer systems. A possible definition of such information system is given by Wikipedia: "*an organized system for the collection, organization, storage and communication of information*". Furthermore it has been stated, that "*it is the study of complementary networks that people and organizations use to collect, filter, process, create and distribute data*" [7].

The information systems have different goals; their application fields vary from enterprise systems (e.g. customer relationship management at a telecommunication company) to decision support systems (e.g. supporting business or medical diagnosis). Under this wide variety one can find the geographic information system (abbreviated generally as GIS). These systems are dedicated the handle geo-related data, i.e. they have to capture, store, manipulate, analyze and present all types of spatial data. In this sense GIS has huge role in information science, because "*80% of the world's data includes some kind of spatial aspect, 80% of data has a location component, 80% of data possesses a geographic reference, 80% of transactional data has a location component*" [5].

This citation clearly states that GIS is an excellent tool for managing of geographically related data. Having the global climate protection in our mind, GIS seems to be the most suitable solution.

Let's have a closer look on it! As an information system, GIS also has three essential components: hardware, software and data. Some authors in the literature extends this list by a fourth item: the users. A common agreement of the components' value is 1:10:100, as hardware, software and data. In our paper we will focus on the software and data aspects, on the two most valuable parts (Fig. 1).

Á. Barsi (✉)
Faculty of Civil Engineering, Department of Photogrammetry and Geoinformatics, Budapest University of Technology and Economics, Budapest, Hungary
e-mail: barsi.arpad@epito.bme.hu

M. Palocz-Andresen et al. (eds.), *International Climate Protection*,
https://doi.org/10.1007/978-3-030-03816-8_6

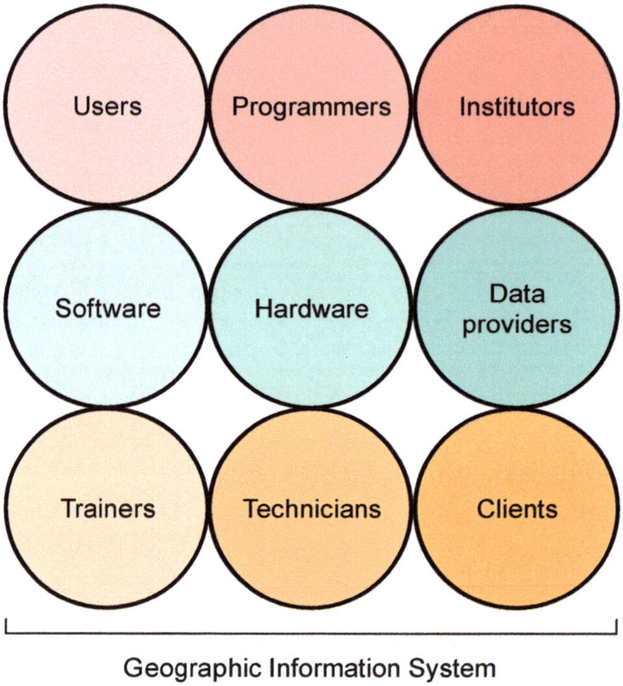

Fig. 1 The GIS and its user community [3]

Taken the technology in a closer look, GIS can be presented as a system of data capture (or input), data management (or storage), data analysis and a data presentation (or visualization) functionality. We try to show all these function groups in the later chapters.

Several practical realizations of GIS solutions can be found with independent names, like

- cadastral information systems,
- multipurpose cadastral systems,
- land data systems,
- land information systems,
- urban information systems.

This nomenclature has the focus on the parcels, but in geoscience these variations belong to the first group having a three level categorization by the data scale. The big picture contains the following scales: local (which was demonstrated by the prior systems), regional (e.g. country-wide focus) and global (for the whole globe). One can formulate it as high, medium and low scale systems. Although we are concentrating on the global aspects, the tools can successfully adapt for the other two categories, like e.g. in micro-climatic projects.

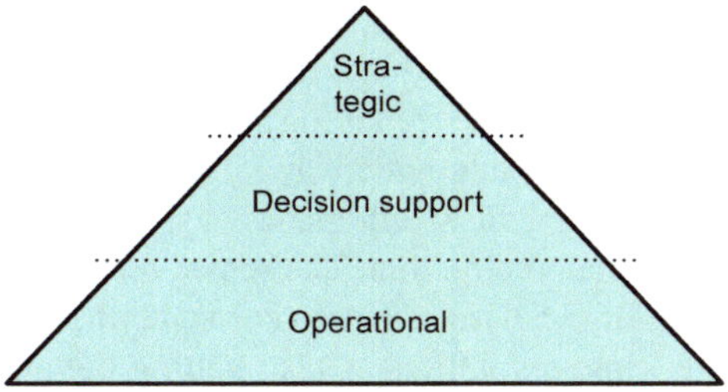

Fig. 2 Decision levels as GIS application categories

The application fields of GIS are also quite wide: facility management (~30%), municipality information systems (~25%), environmental systems (~10%), topographic-cartographic systems and land management (~30%) [2]. The percent values represent the ratio of the realizations compared to all GIS applications. The applications can be categorized also in three further bins, as the next figure shows (Fig. 2).

The climate protection as GIS application rather belongs to the strategic, but this highest level task requires huge amount of captured and processed data (lowest level) and sophisticated analysis engines to support the expert in an efficient way (medium level).

2 Some New GIS Data Capture and Storage Methods

As it was presented in the introduction, the very first function group of geographic information systems is the data capturing. The data to be collected has again a wide range of variety. Lodwick and Feuchtwanger have categorized the groups as follows [6]:

Group A: Environmental and natural resource data (e.g. geologic, hydrologic, climatologic, biologic data)
Group B: Socio-economic data (e.g. economic, financial, demography data)
Group C: Infrastructure data (e.g. transportation, facility, service data)

The most relevant category for our study is the first one, although the others can have extreme impact on the Globe's climate. The mentioned literature lists the climatologic data as examples like precipitation, temperature, air quality, wind.

The data to be collected for GIS use have two important features: geometry and attributes. There is large amount of data capturing techniques for obtaining these features: surveying, GPS measurements, photogrammetry, remote sensing—as the most known geometric data capturing disciplines. For the attributes the GIS users involve other experts from meteorology, climatology, ecology, biology and so on. The obtained attributes are smoothly imported into the applied GIS data base.

Today the data capture methodologies are coupled by the storage of the metadata, where the system archives beyond the essential data at least their origin, accuracy, precision, consistency, actuality and completeness. These data quality measures are documented in adequate reports.

Nowadays the technology is developed enough to capture data by an automatic or at least semi-automatic way. The satellites orbiting around the Earth are shutting the imagery in a fully automatic way, where the data transmission works also without any human interaction. The sensor on the space platforms have increasing resolution (smaller pixel size, better color depth, more spectral channel) and the repetition rate (time resolution) is also getting better. These satellite based earth observation techniques ensures enough amount and quality data for the climatological studies.

The next interesting technique is the Internet-of-Things (IoT) sensors. This technique means an ensemble of similar sensor nodes having communication among them, so the observed data can be passed through their network till they are stored in a database. The IoT solution is a direct data capture measuring the features of prior specification. These data can excellently be combined by the satellite based ones, so the expert can work with enough detailed data warehouses.

The rapidly increasing data amount can be stored in different formats: vector formats (like points, lines, polygons) or raster formats (like images or data matrices). The modern relational data base management systems (RDBMS) are capable to serve the hybrid (which means both) data sources. The distributed and several times open access data bases means redundant and high speed data access required by the specialists. Thanks to the distributed storage, the potential and size of archives has been practically unlimited. The data access has been standardized: the Structured Query Language (SQL) is the common language also in the geoscience; the newest GIS software packages handle it as native tool.

3 GIS Data Analysis for Climate Protection

The most general GIS analysis questions are the following [4]:

a. What can be found in the location of …?
b. Where is …?
c. What has been changed?
d. What is the phenomena in …?
e. What will happen, if …?
f. Which is the most optimal place/way …?

If these questions are interpreted in climate protection, all of them can have importance. Otherwise it must be underlined, that GIS differs from a simple database management system, differs from computer aided drawing (CAD) and desktop mapping systems, because it is capable to derive value added outputs thanks to the analysis functionality.

The analysis can be understood as Bernhardsen published [1]:

- data selection by geometric and/or attribute feature(s),

- performing analysis operation(s) (e.g. logical, arithmetical, geometric, statistical),
- presentation of the results.

The most popular analysis operation block can be detailed as follows:

- measurement, counting, computations,
- overlaying surface, buffer generation,
- statistical computations, data mining operations,
- network analysis functions, optimization tasks,
- modelling and simulation,
- using digital elevation/terrain/surface models,
- others.

The modelling and simulation item has a common agreement in the GIS community to involve experts of other scientific and technical disciplines. For example, if a contamination propagation project is running, chemists, meteorologist, water engineers, biologists and other professionals are involved to achieve to project's goal. GIS behaves as a support base for the work of these experts.

We have to underline, that the geographic information systems of nowadays have convenient tools to integrate the data sources of several disciplines and the usage of such software tools, GIS offers fresh research instruments!

4 GIS Visualizations

The old style hard copy maps are moved nowadays into museums, or is kept only for safety reasons. The today's GIS databases are visualized on screens of mobile equipment (tablet, mobile phone) or smaller or bigger desktop monitors. The visualization therefore is not limited for static content: the screens can show changing legends, database content by varying the outlook, but also animation (displaying maps series, results of simulations) is also possible. This is extended by the new Virtual and Augmented Reality (VR/AR) facilities, known and taken from game industry. The visualization is of course not limited to 2D, but using the last graphic engines and monitor technique, real 3D presentation is available.

The GIS output is not only the maps and similar drawings, but also reports, graphs, diagrams or even automatic electronic notification (SMS, e-mail, tweet etc.) can be produced. One can imagine the usefulness of it in case of emergency situations.

5 Conclusion

The paper summarizes the essence of geographic information systems having a soft focus on the climate protection. The current systems capture adequate data sets, handle them and perform excellent analysis steps in very efficient way. The result of such work is visualized by the newest computer graphics techniques. The author of the paper is very sure that geographic information systems can help professionals in protecting our planet.

References

1. Bernhardsen, T.: Geographic Information System. Arendal (1992)
2. Bill, R., Fritsch, D.: Grundlage der Geo-Informations-Systeme. Wichmann Verlag, Karlsruhe (1991)
3. DeMers, M.N.: GIS for Dummies. Wiley (2009)
4. Detrekői, Á., Szabó, G.: Introduction to Geoinformatics [in Hungarian]. Nemzeti Tankönyvkiadó, Budapest (1995)
5. https://www.gislounge.com/80-percent-data-is-geographic/
6. Lodwick, G.D., Feuchtwanger, M.: Land-Related Information Systems. University Calgary (1987)
7. Wikipedia: https://en.wikipedia.org/wiki/Main_Page

Contribution of Satellite Observations to Climate Science

János Mika

1 Introduction

Changes of climate can always be detected during the earth's history. But, the changes of the distant Past were relatively slow and they were of natural origin. In the recent century the situation has very likely been changing. Besides the natural forces, human activity has been added to the climate determining factors. In a few decades it can bring about changes of the present climate of such extent and rate that has not been experienced in the past one hundred thousand years.

The modern climate change problem has been surveyed in detail by the IPCC AR5 WG-I [2]. The present chapter provides an insight into the results of climate science obtained by satellite methods.

Satellite technology is based on electromagnetic radiation observations. Satellite images have fairly high spatial and temporal resolution. This technology allows us to measure locations of the Earth system impossible or difficult to access, mainly by the all-weather day-and-night capability for microwave sensing. Climate applications of the satellites mostly rely on fair spatial resolution (few kilometres) of the images rather than their good time resolution (15 min).

The Chapter continues with satellite observation of the climate forcing factors (Sect. 2). The observed changes are divided into two parts, according to the content of Sect. 3. So, they illustrate the possibilities of satellite technology in detection of the atmospheric variables, as well, as those in the oceans and the cryosphere. Validation of the present climate models is another important issue, which is tackled in further two aspects by Sect. 4. One of them is simulation of the recent climate changes by the models, the second one is of the model simulations' coherence in projection of the future. The latter is possible mostly indirectly, via analysis of the individual feedback mechanisms.

J. Mika (✉)
Eszterházy Károly University, Eger, Hungary
e-mail: mika.janos@uni-eszterhazy.hu

M. Palocz-Andresen et al. (eds.), *International Climate Protection*,
https://doi.org/10.1007/978-3-030-03816-8_7

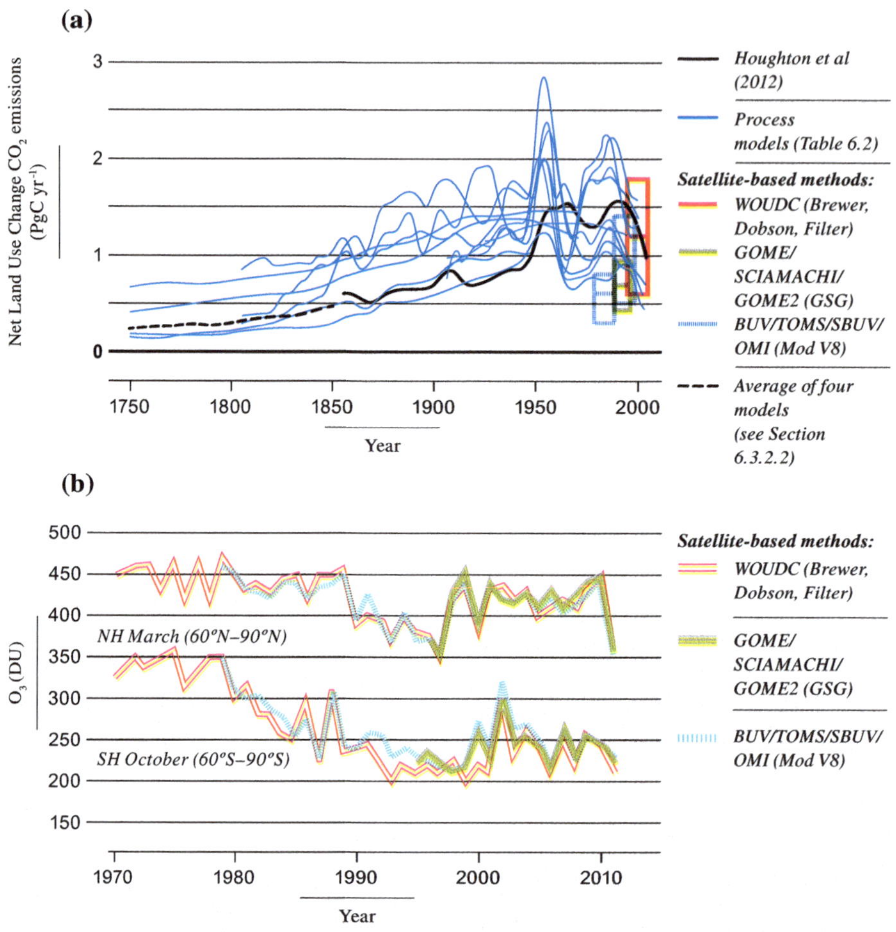

Fig. 1 **a** Land use change CO_2 emissions (PgC yr^{-1}) according to two empirical, one model-based and one satellite estimates. The satellite-based methods are available for the tropics (IPCC [2]: Fig. 6.10). **b** Zonally averaged, annual mean total column ozone in Dobson Units (1 DU = 2.69 × 1016 O_3/cm^2) from ground-based spectrometer data, WOUDC, GOME/SCIAMACHY/GOME2 and satellite BUV/TOMS/SBUV/OMI data for March (60–90°N) and October (60°–90°S) (redrawn from IPCC [2]: Fig. 2.6)

2 Climate Forcing Factors

Right part of Fig. 1 contributes to estimation of CO_2 emission due to the land-use changes. This Figure also demonstrates that results by the various satellite based methodologies differ from each other by tens of percent. At the same time, the uncertainties among the surface-based estimations are of similar magnitude.

Left part of Fig. 1 demonstrates that even the vertical ozone content can already be reconstructed since the beginning of the 1980s. Moreover, these estimates coincide the more accurate but point-wise surface-based observations. The differences between the estimates, spreading from a few to 10% are distributed randomly in time, so they could not strongly influence the observed tendencies.

Spatial distribution of the greenhouse gases, that extremely likely cause the recent warming, is rather even over the Globe. Therefore, observation of these gases is rarely in the focus of satellite development projects. The components of atmospheric aerosols have modified the atmospheric radiation balance in the opposite direction, namely decreasing the warming. The direct effect of aerosol components mainly means the backscattering of solar radiation. The direct effect of aerosols could be described three ways. The optical thickness of aerosol, τaer shows the ration Sun radiation does not reach the bottom of atmosphere as negative exponent of e natural number. The α albedo of aerosol shows the ration of radiation reflected back to the space in the given wavelength. Finally, DRE, the common effect of natural and anthropogenic aerosols, shows how much plus energy will leave the Earth-atmosphere system comparing with the no aerosol at all system.

3 Changes in Climate Series

3.1 Atmospheric Variables

Let us first have a look at air temperatures near the surface and at higher altitudes. The latter values can be observed by radio-sounds and satellite technologies. The latter approach is possible at the microwaves which are not absorbed by cloudiness, so not only bright subsamples can be climatologically analysed as it would be the case in the long-wave part of the spectrum.

Left part of Fig. 2 indicates that the lower troposphere had been continuously warming from the 1960s to the end of the 20th century. Since then, warming has slowed down significantly. The lower stratosphere exhibits opposite tendencies with unequivocal cooling and stopping of it before the turn of the century. This cooling tendency can be explained be the stronger lapse rate that leads to colder stratosphere, as the surface warms. Most likely, the cooling is also supported by increasing absorption by the more and more greenhouse gases, but diminishing of the cooling takes place, despite the further increase of the greenhouse effect.

Stratospheric water vapour content is also often investigated by satellite observations. In the right part of Fig. 2 they are compared with the USA NCAR radio-sounding observations. Clear coincidence between the two observations is seen, especially in their inter-annual fluctuations. Their longer tendencies diverge a little, but it may be a consequence of the relatively short existence of observations.

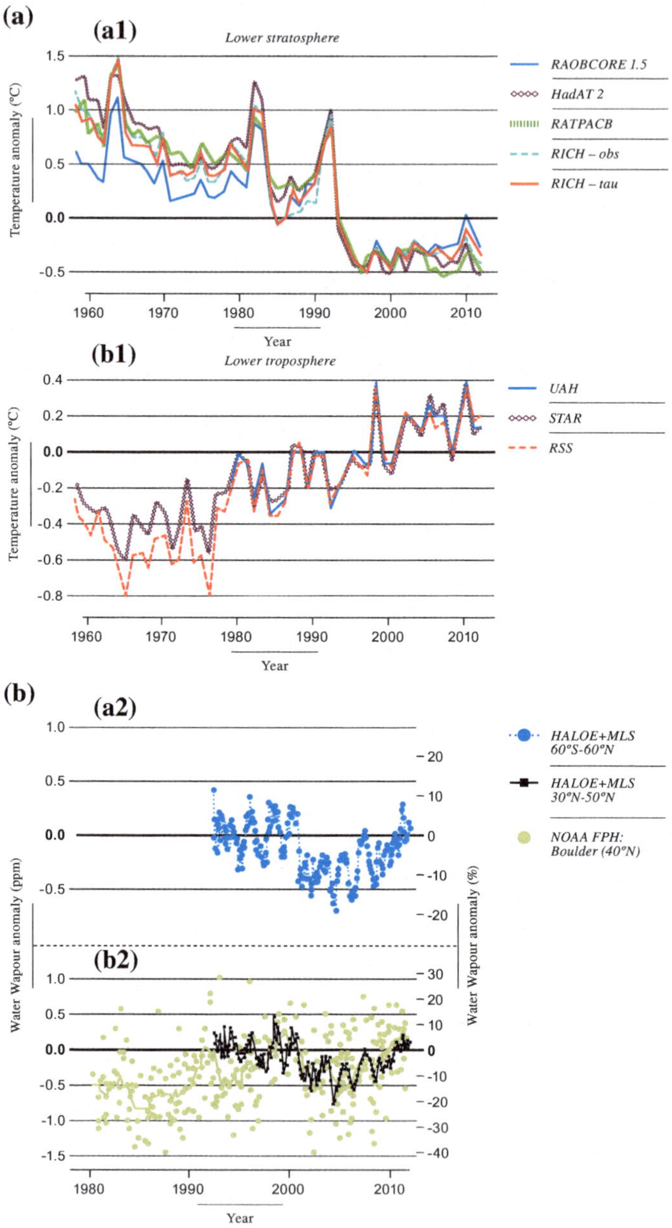

Fig. 2 **a** Global annual mean temperature anomalies in the lower stratosphere (**a1**) and lower troposphere (**b1**), relative to 1981–2010, from different data sets. Note that the y-axis resolution differs between the two panels (**a1** & **b1**). The curves contain five surface-based and three satellite-based methods, listed in the **a1** and the **b1** panels, respectively. (IPCC [2]: Fig. 2.24). **b** Water vapour in the stratosphere from satellites and in situ measurements, compared to 2000–2011. **a2** Monthly mean at 83 hPa for 60°S–60°N. **b2** Monthly balloon-born water vapour from Boulder, at 40°N (green) in 16–18 km, and monthly means as in (**a2**), over 30°–50°N (black) (redrawn from IPCC [2]: Fig. 2.5)

3.2 Oceans, Cryosphere

Turning to the oceans, the first task is observation of sea-surface temperatures (left part of Fig. 3). It is seen that satellite reconstructions fairly coincide with those derived from the surface based observations. In the last decade the more intensive warming has slowed down here since the turn of the century.

The following parameters detect changes in the cryosphere. Snow cover data are drawn in the right part of Fig. 3. The space-based and the earth-based technologies show fair coincidence here, too, although they refer to different parts o the year.

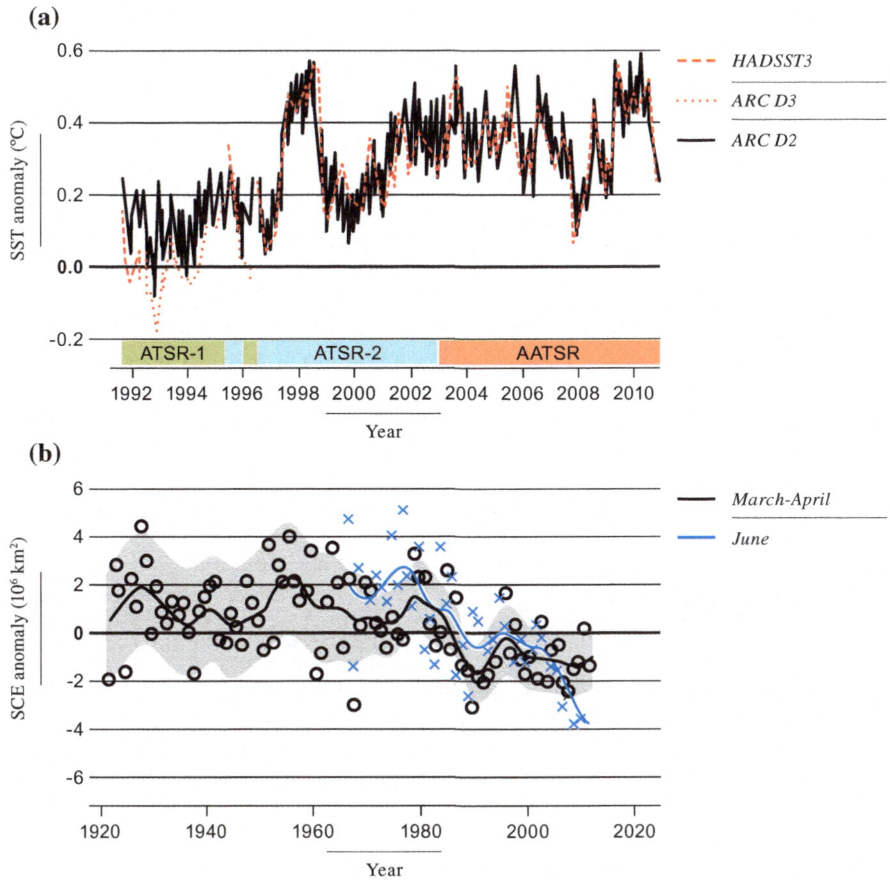

Fig. 3 **a** Global sea-surface temperature relative to 1961–1990 from satellites and in situ records (HadSST3). Black lines: satellite-based night-time subsurface temperature at 0.2 m depth estimates from the ARC project (IPCC [2]: Fig. 2.17). **b** In situ March–April NH snow cover extent (SCE, circles) with shading for the 95% confidence interval; and June (crosses, satellite data), both calculated relative to the 1971–2000 means (redrawn from IPCC [2]: Fig. 4.19)

Considering the annual cycle, the absolute changes in June are even stronger in their relative values, expressed in proportion of the snow-covered area.

Tendencies of the sea-ice show rather unique patterns in Fig. 4. The two graphs of the Figure can be characterised by completely opposite tendencies. Shrinking of sea ice over the period, maybe with slow-down in the recent decade, can well be explained by the overall warming of the climate system. But, increase of sea ice in the southern hemisphere, near the Antarctic is no way coherent with the overall warming! This problem, and the actual missing of correct explanation, has also been admired by the IPCC WG-I Report [2].

According to our view, derived from the various figures and statements by the IPCC AR4 WG-I [2], this contradiction can be explained by the experienced stronger heat uptake by the deep oceans, which is the likely reason of stagnation of the global warming. Since 2014 the warming continues by an even stronger pace than before the ca. 2002–2013 period of temporary stagnation (e.g. NOAA [3]).

Near the end of this section, two more variables are used to assess the climate change. Sea-level is seen in the left part of Fig. 5 in four different approaches. They are satellite-born and sea-shore direct sea-level indications, as well, as water volume changes connected with melting of ice and also with thermal expansion. The previous variables are characterised by unequivocally increasing tendency. The latter components of sea level rise are of nearly equal importance.

Right part of Fig. 5 indicates ice-sheet changes in Greenland and Antarctic. According to the upper maps, ice is shrinking over whole Greenland. At the same time, ice cover of Antarctic indicates thicker ice in larger areas of the island, than those with thinner ice. (In the black-and-white Figure, left parts of the continents are the thinning areas.) This latter part refers to the 2003–2012 period, only. Considering the recent two decades, both islands have lost from its ice volumes.

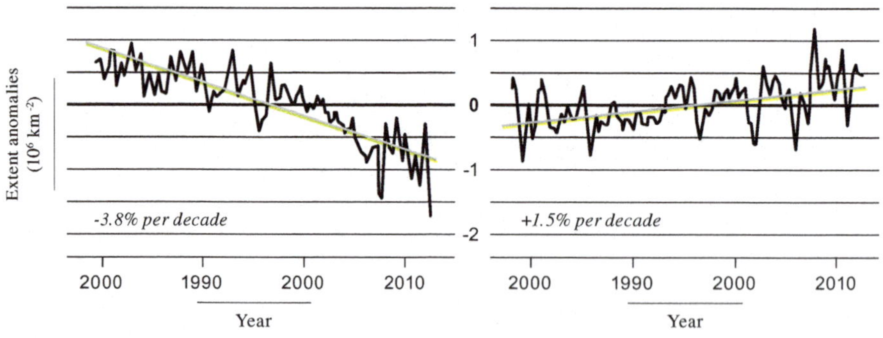

Fig. 4 Decadal trends (in km^2, expressed in %) of annual sea-ice anomalies, after removal of the seasonal cycle in the Arctic and Antarctic oceans. The average sea ice cover for the period 1979 through 2012 (redrawn from IPCC [2]: FAQ 4.1 Fig. 1)

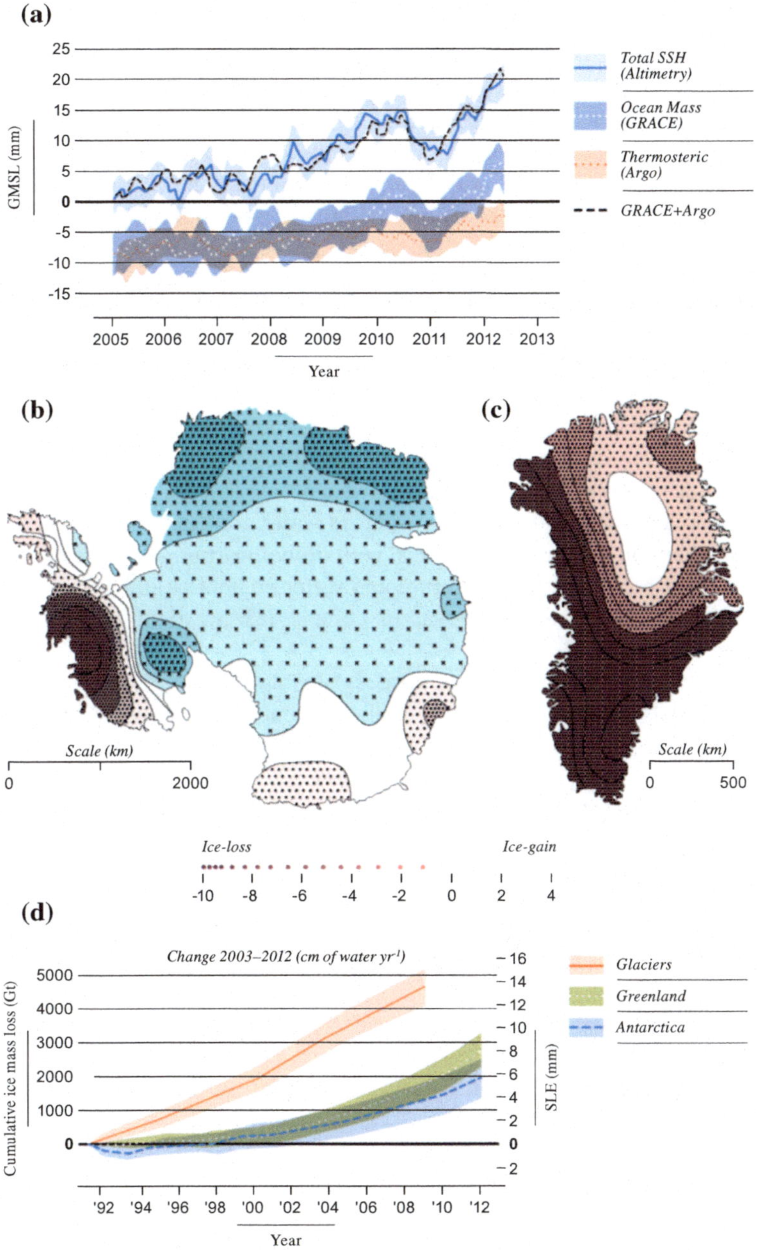

Fig. 5 **a** Global mean sea level from altimetry, 2005–2012 (upper solid line). Ocean mass changes (middle solid line, by Gravity Recovery and Climate Experiment, GRACE) and thermosteric sea level changes (lower solid line, by the Argo Project). The dashed line shows sum of ocean mass and thermosteric contributions (IPCC [2]: Fig. 13.6). Distribution of ice loss determined from the GRACE for **b** Antarctica and **c** Greenland, shown in cm yr^{-1} for the period 2003–2012. **d** The assessment of the total loss of ice from glaciers and ice sheets in sea level equivalent (mm) (redrawn from IPCC [2]: Fig. TS.3)

4 Validation of Climate Models

Climate system is one of the most complex and non-linear systems. The important space scales spread from sub-millimetres of cloud processes to the length of the Equator. Time scales are important from the seconds of micro-turbulence to the hundreds of years for the ocean circulation. None of the present climate models can correctly tackle all scales. Hence, testing of climate models is very important.

4.1 Simulation of the Recent Climate Variations

The first aspect of climate model validation answers the question if the model simulations of the past decades are correct. A positive example of this validation is shown in Fig. 6. It demonstrates that the water content of atmosphere and its changes was estimated relatively well by the model and was fitted to the reality via sea surface temperature as lower boundary condition. We can state that the dynamical processes of the atmosphere can handle the atmospheric water content. The increasing trend of water content during this two decades, with global warming behind, points at the positive inter-relatedness of temperature and water content at global scales: Warming climate initiates increased water vapour content, leading to further warming, also mentioned in Sect. 4.2.

As a negative example, i.e. limited success of simulation, Fig. 7 indicates to what extent the models were capable to mimic the tendencies of sea-ice cover in the two

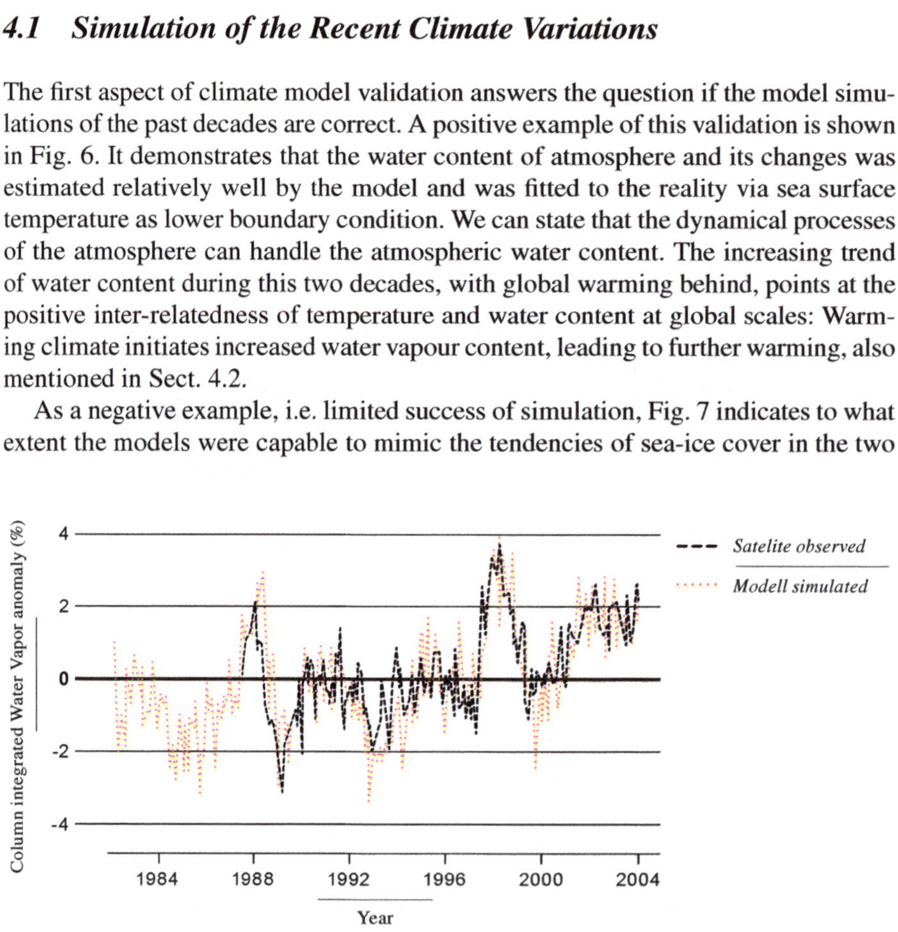

Fig. 6 The anomaly of vertically integrated water vapour content above the ocean, expressed in percent of 1987–2000 period average. The values are simulated by the general circulation model of Geophysical Fluid Laboratory, Princeton and observed by the SSM/I satellite. The model was driven by observed sea-surface temperature, as lower boundary condition, otherwise by external climate forcing. The model well reproduces the slow increase of water vapour content in connection with warming, and the inter-annual fluctuation in relation to the El Nino/La Nina oscillation (redrawn from IPCC [1]: Fig. 9.17)

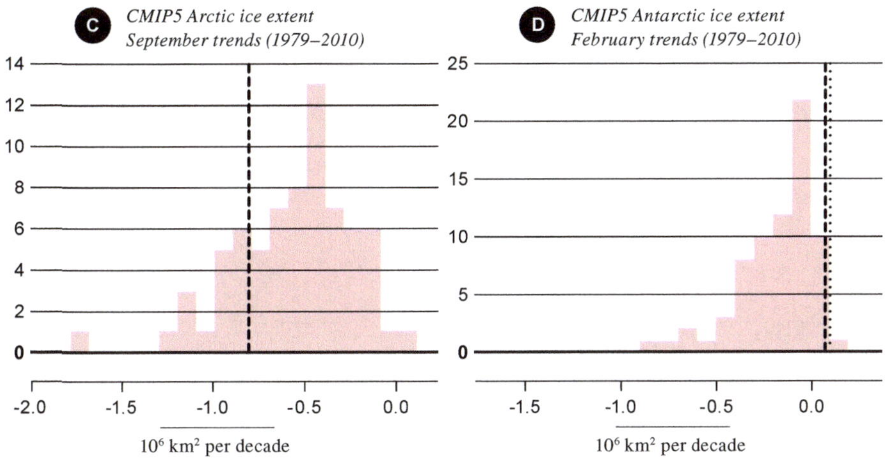

Fig. 7 CMIP5 sea ice extent trends over 1979–2010 for **a** the Arctic in September and **d** the Antarctic in February. 66 realizations of 26 models are compared to observation-based trends (redrawn from IPCC [2]: Fig. 9.24)

Hemispheres during the 1979–2010 period. One must admire that not too much! Majority of the models underestimated melting in the Northern Hemisphere, but strongly overestimated the retardation of sea-ice around the Antarctic. (CMIP5 is the abbreviation for the Coupled Model Intercomparison Project Phase 5 to obtain a wide set of new global climate model simulations for the IPCC AR5 Report.)

4.2 Simulation of Future Climates

Final aim of climate modelling is to project the future climate in response to changes in the external forcing factors. These external factors and their uncertainty are influenced by several circumstances, e.g. the world population, the structure of energy industry, development difference between the regions, etc.

The other uncertainty factor is how correctly we simulate the sensitivity of climate system, namely the expected temperature in response to given changes of the external factors. We are not really able to estimate the first uncertainty source, due to its complexity, but we can validate the climate sensitivity simulations through testing certain particular processes. Sensitivity of the climate system to the forcing factors is determined by feedback mechanisms in the climate system.

Left part of Fig. 8 demonstrates them according to our present knowledge, derived from the global models. The strongest, negative feedback is the long-wave irradiation, while the positive feedback mechanisms became slightly stronger since the previous IPCC Report, especially concerning cloudiness and water vapour. This means that the models became more sensitive to the external forcing factors. Right part of Fig. 8 indicates that model-based snow-albedo feedback estimations spread evenly around the satellite observation estimate. This figure indicates that though the average feedback over all models fit the reality, the spread between the models is rather wide.

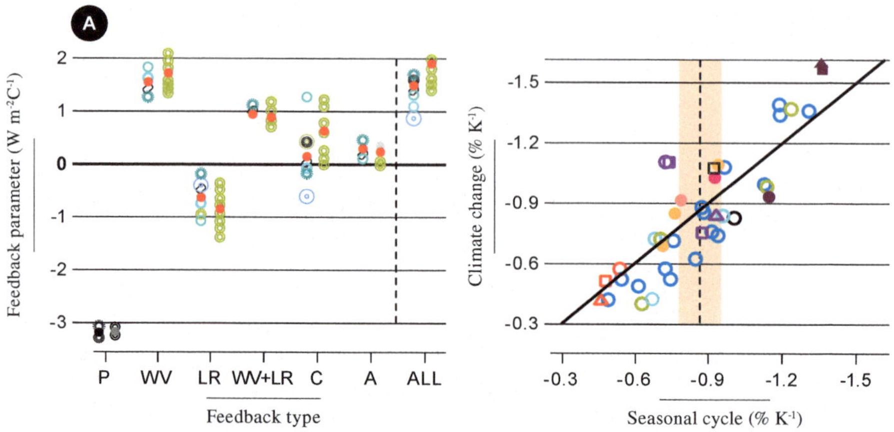

Fig. 8 Left panel: strength of individual feedbacks for CMIP3 and CMIP5 models (left and right columns of symbols) for Planck (P), water vapour (WV), lapse rate (LR), combination of latter two (WV+LR) clouds (C), albedo (A), and sum of all feedbacks except Planck (ALL) (IPCC [2]: Fig. 9.43). Right panel: Scatter-plot of simulated springtime snow–albedo feedback ($\Delta\alpha s/\Delta Ts$) values in climate change (y-axis) versus simulated springtime values in the seasonal cycle (x-axis) in climate change simulations from 17 CMIP3 and 24 CMIP5 models (αs and Ts are surface albedo and air temperature). The vertical shaded band shows observation-based estimate of snow-albedo feedback (redrawn from IPCC [2]: Fig. 9.45)

5 Conclusion

Modern satellite technology contributes to climate science with increasing weight to answer the still existing questions. They are not focusing on what the main reasons of the changes are, but on why and to what extent do the inter-annual and even inter-decadal fluctuations perturb the monotonically warming tendencies of our global climate. Probably, this is the strongest challenge, the recent IPCC WGI [2] Report delivers for the climate scientists.

References

1. IPCC.: Climate Change (2007): The Physical Science Basis. Contribution of WG I to the 4th Assessment Report of the Intergovernmental Panel on Climate Change, 2007 [Solomon, S. et al., (eds.)] Cambridge University Press, Cambridge UK and NY, USA (2007)
2. IPCC AR5 WG-I.: Climate Change 2013: The Physical Science Basis. Contribution of Working Group I to the Fifth Assessment Report of the Intergovernmental Panel on Climate Change [Stocker, T.F., et al. (eds.)]. Cambridge University Press, Cambridge, United Kingdom and New York, NY, USA, 1535 pp (2013)
3. NOAA.: Global Analysis—September 2016 Top 15 Monthly Global Land and Ocean Temperature Departures from Average, 29 Oct 2016, https://www.ncdc.noaa.gov/sotc/global/2016/9/supplemental/page-1

Calculating the O_3 Instantaneous Longwave Radiative Impact from Satellite Observations

Stamatia Doniki

Through the years, there have been a number of studies on ozone and its impact on climate, which is summarized in the Intergovernmental Panel on Climate Change Assessment Report (IPCC AR). The latter uses the concept of radiative forcing (RF) as a parameter to quantify the change in the radiation budget, introduced by a change in a surface or atmospheric variable, such as for instance the concentration of atmospheric species [26]. Through these studies, ozone emerged as a key parameter to RF estimations. Tropospheric ozone was found to be the third most important GHG in terms of radiative forcing, after the well-mixed GHGs [23]. Ozone is distinguished from the other GHGs due to the high spatial and temporal variability it presents, owed to a relatively short lifetime.

A lot of studies have identified the drivers of the ozone RF, the most important of which are its own vertical distribution, the vertical profile of temperature, the temperature of the surface, humidity and clouds (e.g., Lacis et al. [21], Forster and Shine [15], Gauss et al. [16], Worden et al. [29, 30], Bowman et al. [4]). Considering the non-linear relations between ozone and these variables, it is well understood that the representation of the ozone RF is non-trivial, and requires a good knowledge of its distribution, both horizontally and vertically, as well as its long-term trends. Up until recently, the ozone RF calculations and changes over time were entirely model-based, mainly because there are no available records of ozone for the pre-industrial era. However, model calculations depend on assumptions and own radiative transfer codes, leading to intermodel differences, which are then presented in the latest IPCC assessment report [23]. The anthropogenic tropospheric ozone contribution to RF is given as +40 Wm^{-2} (−0.20 to +0.60 Wm^{-2}) for the time period of 1750–2011, for the ensemble of model calculations. But different models produce results that vary around this reported value. For example, Søvde et al. [28] report a value of + 0.45 Wm^{-2} for the same time period, while Skeie et al. [27] report +0.41 Wm^{-2}.

S. Doniki (✉)
Spectroscopie de l'Atmosphère, Chimie Quantique et Photophysique, Université Libre de Bruxelles, Brussels, Belgium
e-mail: sdoniki@ulb.ac.be

M. Palocz-Andresen et al. (eds.), *International Climate Protection*,
https://doi.org/10.1007/978-3-030-03816-8_8

The situation is similar for the stratospheric ozone as well, although for a RF much smaller: here the contribution is -0.05 ± 0.10 Wm^{-2} [23] for the model ensemble, but Conley et al. [10] report a value of -0.02 Wm^{-2} (-0.09 to +0.05 Wm^{-2}). It is quite obvious that the large confidence intervals of the ensemble values raise concerns on model reliability. These estimates have to be tested against observations on large scales.

Satellite measurements provide global coverage within a few hours or days, and many of them giving coarse vertical distribution of species. This is the case for ozone, for which the vertical profiles are available from satellite observations since a decade. However, only a few studies have made use of them to quantify the radiative effect of ozone directly from the observations (e.g. Worden et al. [29], Joiner et al. [20]). Of particular interest towards such studies are the infrared (IR) nadir sounders, which have sufficient vertical sensitivity to be able to distinguish between tropospheric and stratospheric ozone, but also provide direct measurements of the top-of-the-atmosphere (TOA) radiance in the IR ozone band globally, with high sampling.

A first effort to benefit from the IR sounders for ozone RF assessment is described in Worden et al. [30], where they use the measurements of TES-Aura (Tropospheric Emission Spectrometer) and propose a method to calculate the longwave radiative effect (LWRE) due to ozone with respect to the TOA radiative flux. The LWRE is different form the RF in the IPCC assessment reports, as this is not calculated at the TOA, neither refers to changes in the ozone concentration between the pre-industrial and present time periods. Despite this, the LWRE is a very useful parameter which can be directly compared to climate model output data. The method of Worden et al. [30] consists of two steps in the calculation: first the calculation of a so-called instantaneous radiative kernel (IRK) of each measurement, which represents the TOA radiative flux sensitivity with respect to the ozone profile, and then the calculation of the LWRE. Based on this study, Aghedo et al. [2] used the IRKs from TES to evaluate the radiative impact of ozone biases for the chemistry-climate models that participate in the Atmospheric Chemistry and Climate Model Intercomparison Project (ACCMIP). Taking the study a step further, Bowman et al. [4] revealed a correlation between model outgoing longwave radiation (OLR) bias and RF in the ACCMIP models. Their results were included in the IPCC assessment report of 2013 [23].

Another instrument suitable for calculating the O_3 LWRE, which already has proven its capabilities on atmospheric chemistry and numerical weather prediction, is the Infrared Atmospheric Sounding Interferometer (IASI) on board the MetOp platforms. IASI is a Michelson interferometer which measures the radiation emitted by the Earth and the atmosphere in the thermal infrared (TIR) spectral range [9]. The MetOp-A and -B platforms—launched in October 2007 and September 2012 respectively, with a third one scheduled for 2018—are polar-orbiting, sun-synchronous satellites, crossing the Equator at around 09.30 and 21.30 LT. IASI uses nadir geometry, complemented by off-nadir measurements up to 48.3° at each side of the satellite track (swath of 2,200 km approximately), which allows the instrument to cover the globe twice per day, providing a total of about 1.3 million observations. Each field of view of IASI comprises 2×2 circular pixels, with each pixel having a

12 km diameter footprint on the ground at nadir. The spectral range of IASI covers the band between 645 and 2,760 cm^{-1}, with a 0.25 cm^{-1} sampling (0.5 cm^{-1} apodized resolution for the Level 1C data [13]) and radiometric noise between 0.25 and 0.3 K, at around 1000 cm^{-1} and a reference temperature of 280 K. For more information on IASI and its potential, readers can refer to Clerbaux et al. [8], Coheur et al. [9], Clarisse et al. [7] and Hilton et al. [18], as well as EUMETSAT [13] and August et al. [3] for technical characteristics and operational data availability.

As IASI is a very capable instrument, covering a wide range of atmospheric research applications, there has been a few algorithms [12] dedicated to handle the vast amount of data it produces, and retrieve different species and parameters. Concerning the radiative impact discussed here, there is only one algorithm at the moment that is able to perform the demanding calculations. The Fast Optimal Retrieval on Layers for IASI (FORLI) is a radiative transfer and retrieval software, built at the Université Libre de Bruxelles (ULB), dedicated to serving the atmospheric composition objectives of the IASI mission. At the moment FORLI is running at ULB in near-real time, providing total columns and vertical profiles of O_3, CO and HNO_3 for each IASI observation. The retrieval algorithm is based on the maximum a posteriori solution, as described in Rodgers [24]. A full description of FORLI, the radiative transfer, other methods, input parameters and approximations can be found in Hurtmans et al. [19]. The ozone product of FORLI (FORLI-O_3) has been extensively validated against ground-based, sonde and satellite data [5, 6, 12, 17, 25], revealing a positive bias of 3–6% for the total column on global scale, and specifically around 10–0 15% for the 10–25 km. The most recent validation by Boynard et al. [6] showed that, when IASI is compared to GOME-2 (UV satellite instrument), IASI presents a positive bias of 5.5–7.1% for the total columns, depending on latitude and season. The highest bias is observed over the SH Polar region, and can reach up to 30% during the austral summer and fall. When IASI is compared to the Brewer-Dobson network (UV ground-based instruments), it is found biased around $5 \pm 1\%$ in the NH and $6 \pm 1\%$ in the SH, with differences over 20% in the Antarctic. Regarding the ozone vertical profiles, which were compared against ozonesondes around the globe, IASI generally underestimates ozone in the troposphere, up to 12% in the mid-latitudes, and overestimates it in the stratosphere, with up to 35% in the tropics. The largest biases are found in the upper troposphere—lower stratosphere (UTLS), especially around the tropics.

The following quantities are output of FORLI-O_3, and we will provide a couple of definitions, to avoid any misunderstanding between the parameters in use.

The radiance Jacobians: analytic Jacobians of the last iteration during retrieval, representing the sensitivity of IASI TOA spectral radiances to the ozone abundance at each vertical level; they are measurement-angle and wavenumber dependent (in W cm^{-2} sr^{-1} cm ppb^{-1}).

The Instantaneous Radiative Kernel (IRK); representing the sensitivity of the TOA flux, F_{TOA} (in Wm^{-2}), to a change in the vertical distribution of an atmospheric parameter, here in ozone, also referred to as climate sensitivity (in W m^{-2} ppb^{-1}).

The LongWave Radiative Effect (LWRE), which represents the reduction in outgoing longwave radiation due to ozone absorption with respect to each atmospheric state at each vertical level (in Wm^{-2}). It can refer either to a 100% change in ozone, when only the satellite observation is considered, and corresponds to the instantaneous longwave radiative impact of the observed O_3, or to the difference between two measurements, e.g. model simulated O_3 compared to satellite observed O_3.

The calculation of the LWRE is completed in two steps, following Worden et al. [30]. First, the calculation of the IRKs and then that of the LWRE. To calculate the IRKs, a double integration is performed on the Jacobians, over the observation angle and the wavenumber. There are two possible ways to calculate the IRK, depending on the approximation for the angular integration: (a) the Anisotropy approximation and (b) the Direct Integration method. Both methods are based on simulations over different angles, provided by the Gaussian Quadrature for the angular integration of moments [1, 11, 22]. However, as already proven in Doniki et al. [11], the Anisotropy approximation can under- or over-estimate the IRKs from −25 to +12% and the LWRE total or tropospheric columns from −25% at 0° viewing angle, to +20% at the maximum viewing angle of IASI. Based on these results, the Direct Integration should be preferred over the Anisotropy approximation. To complete the IRK calculation, a simple integration over the wavenumber is performed. The IRK of IASI peaks at the middle to higher troposphere, which is a common feature among IR sounders [11, 30].

The LWRE is obtained by multiplying the IRK with the O_3 profile (in ppb) for the same measurement. As the parameters are instantaneous, it is obvious that the triplet O_3 profile—IRK—LWRE refers to a single observation each time. For a single day, IASI will provide the vertical distribution of LWRE as shown in Fig. 1. The advantage of the method just described lies exactly on the fact that we can obtain a detailed distribution of the LWRE, and determine the climate sensitivity of the instrument to ozone, but also determine other regions of interest. Since we can obtain the vertical distribution of the LWRE, we can also calculate tropospheric, total or other columns, by summing up the appropriate vertical layers, e.g. for the tropospheric LWRE we sum up from the ground to the tropopause. In Fig. 1 we can spot that the highest LWRE values are found where the highest O_3 is expected, that is the stratosphere. However we can also spot high values between 30° and 60°N, linked to the dominating presence of land. In Figs. 2 and 3, where we present the tropospheric and total LWRE respectively, we find higher values within the tropical belt (30°S–30°N), linked with high temperatures, but also between 30° and 60°N, in the form of hotspots, during the day overpass, which disappear during the evening overpass, again associated with high temperatures. For more information, readers are kindly referred to Doniki et al. [11].

Finally, both the IRKs and the LWRE can be directly compared to chemistry-climate models, as these can also compute the same parameters. In such a case, the chemistry-climate models can be evaluated for the first time, not only on the ozone content they simulate, but also on the radiative content, and how biases in ozone can propagate in biases in the radiative effect and therefore to the radiative forcing.

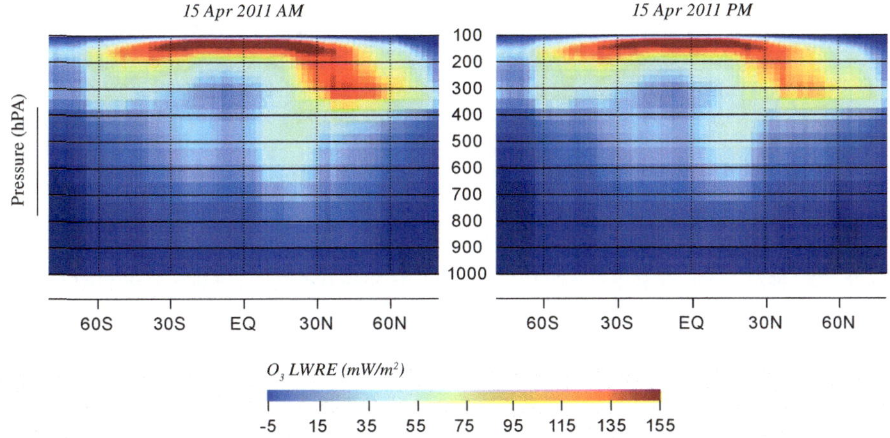

Fig. 1 Single-day clear-sky zonal average O_3 LWRE vertical distribution for local morning (left) and evening (right) IASI overpasses, on 15 April 2011. Adapted from Doniki et al. [11]

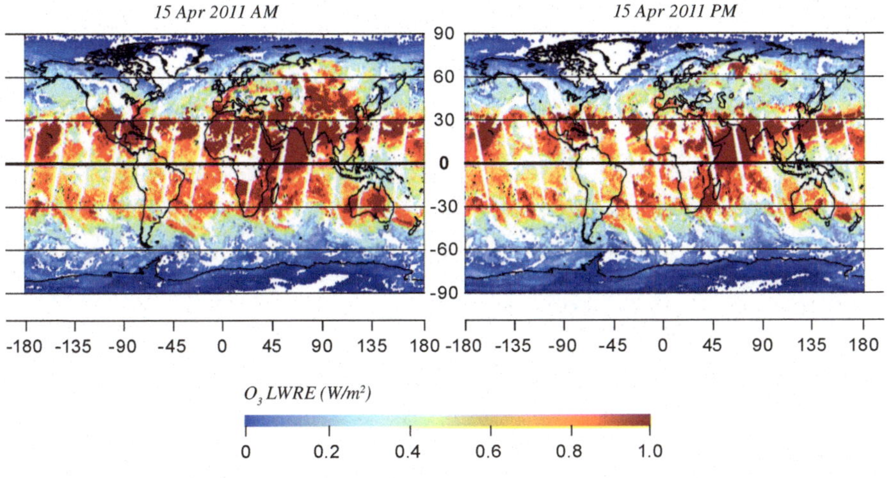

Fig. 2 Global distribution of the tropospheric O_3 LWRE for morning (left) and evening (right) IASI overpasses, on 15 April 2011. Adapted from Doniki et al. [11]

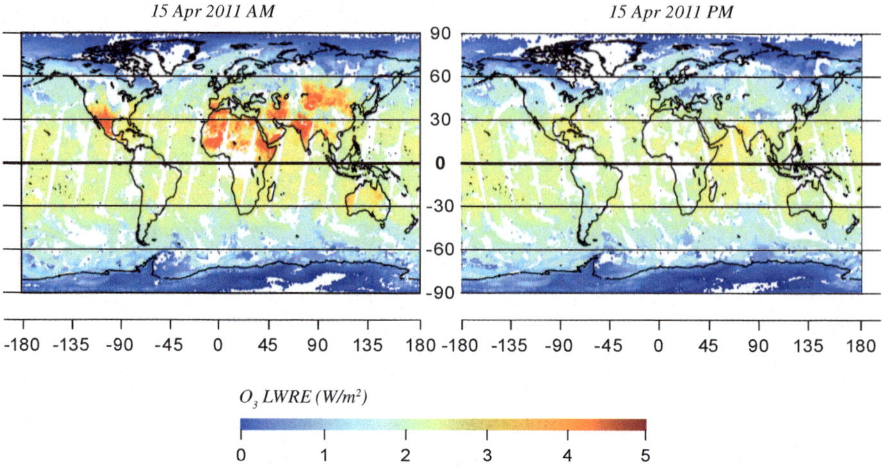

Fig. 3 Global distribution of the total O_3 LWRE for morning (left) and evening (right) IASI overpasses, on 15 April 2011. Adapted from Doniki et al. [11]

Acknowledgments The author would like to acknowledge the contribution of D. Hurtmans, L. Clarisse and P.-F. Coheur [Université Libre de Bruxelles, Belgium (ULB)], C. Clerbaux (LATMOS/IPSL, UPMC Univ. Paris 06 Sorbonne Universités, UVSQ, CNRS, France and ULB), H. M. Worden (Atmospheric Chemistry Observations & Modeling Laboratory, NCAR, USA) and K. W. Bowman (Jet Propulsion Laboratory, USA).

References

1. Abramowitz, M., Stegun, I.: Handbook of Mathematical Functions with Formulas, Graphs, and Mathematical Tables, National Bureau of Standards, Applied Mathematics Series-55, Dover, Mineola, NY, USA (1972)
2. Aghedo, A.M., Bowman, K.W., Worden, H.M., Kulawik, S.S., Shindell, D.T., Lamarque, J.F., Faluvegi, G., Parrington, M., Jones, D.B.A., Rast, S.: The vertical distribution of ozone instantaneous radiative forcing from satellite and chemistry climate models. J. Geophys. Res. **11**, D01305 (2011). https://doi.org/10.1029/2010jd014243
3. August, T., Klaes, D., Schlüssel, P., Hultberg, T., Crapeau, M., Arriaga, A., O'Carroll, A., Coppens, D., Munro, R., Calbet, X.: IASI on and Metop-A: operational level 2 retrievals after five years in orbit. J. Quant. Spectrosc. Radiat. **113**, 1340–1371 (2012). https://doi.org/10.1029/2010jd014243
4. Bowman, K.W., Shindell, D.T., Worden, H.M., Lamarque, J.F., Young, P.J., Stevenson, D.S., Qu, Z., de la Torre, M., Bergmann, D., Cameron-Smith, P.J., Collins, W.J., Doherty, R., Dalsøren, S.B., Faluvegi, G., Folberth, G., Horowitz, L.W., Josse, B.M., Lee, Y.H., MacKenzie, I.A., Myhre, G., Nagashima, T., Naik, V., Plummer, D.A., Rumbold, S.T., Skeie, R.B., Strode, S.A., Sudo, K., Szopa, S., Voulgarakis, A., Zeng, G., Kulawik, S.S., Aghedo, A.M., Worden, J.R.: Evaluation of ACCMIP outgoing longwave radiation from tropospheric ozone using TES satellite observations. Atmos. Chem. Phys. **13**, 4057–4072 (2013). https://doi.org/10.5194/acp-13-4057-2013

5. Boynard, A., Clerbaux, C., Coheur, P.-F., Hurtmans, D., Turquety, S., George, M., Hadji-Lazaro, J., Keim, C., Meyer-Arnek, J.: Measurements of total and tropospheric ozone from IASI: comparison with correlative satellite, ground-based and ozonesonde observations. Atmos. Chem. Phys. **9**, 6255–6271 (2009). https://doi.org/10.5194/acp-9-6255-2009
6. Boynard, A., Hurtmans, D., Koukouli, M.E., Goutail, F., Bureau, J., Safieddine, S., Lerot, C., Hadji-Lazaro, J., Pommereau, J.-P., Pazmino, A., Zyrichidou, I., Balis, D., Barbe, A., Mikhailenko, S.N., Loyola, D., Valks, P., Van Roozendael, M., Coheur, P.-F., Clerbaux, C.: Seven years of IASI ozone retrievals from FORLI: validation with independent total column and vertical profile measurements. Atmos. Meas. Tech. Discuss. (2016). https://doi.org/10.5194/amt-2016-11 (in review)
7. Clarisse, L., R'Honi, Y., Coheur, P.-F., Hurtmans, D., Clerbaux, C.: Thermal infrared nadir observations of 24 atmospheric gases. Geophys. Res. Lett. **3**, L10802 (2011). https://doi.org/10.1029/2011GL047271
8. Clerbaux, C., Boynard, A., Clarisse, L., George, M., Hadji-Lazaro, J., Herbin, H., Hurtmans, D., Pommier, M., Razavi, A., Turquety, S., Wespes, C., Coheur, P.-F.: Monitoring of atmospheric composition using the thermal infrared IASI/MetOp sounder. Atmos. Chem. Phys. **9**, 6041–6054 (2009). https://doi.org/10.5194/acp-9-6041-2009
9. Coheur, P.-F., Clarisse, L., Turquety, S., Hurtmans, D., Clerbaux, C.: IASI measurements of reactive trace species in biomass burning plumes. Atmos. Chem. Phys. **9**, 5655–5667 (2009). https://doi.org/10.5194/acp-9-5655-2009
10. Conley, A.J., Lamarque, J.-F., Vitt, F., Collins, W.D., Kiehl, J.: PORT, a CESM tool for the diagnosis of radiative forcing. Geosci. Model Dev. **6**, 469–476 (2013). https://doi.org/10.5194/gmd-6-469-2013
11. Doniki, S., Hurtmans, D., Clarisse, L., Clerbaux, C., Worden, H.M., Bowman, K.W., Coheur, P.-F.: Instantaneous longwave radiative impact of ozone: an application on IASI/MetOp observations. Atmos. Chem. Phys. **15**, 12971–12987 (2015). https://doi.org/10.5194/acp-15-12971-2015
12. Dufour, G., Eremenko, M., Griesfeller, A., Barret, B., LeFlochmoën, E., Clerbaux, C., Hadji-Lazaro, J., Coheur, P.-F., Hurtmans, D.: Validation of three different scientific ozone products retrieved from IASI spectra using ozonesondes. Atmos. Meas. Tech. **5**, 611–630 (2012). https://doi.org/10.5194/amt-5-611-2012
13. EUMETSAT.: IASI Level 1: Product Guide, EUM/OPSEPS/MAN/04/0032 v4C (2014). Available at: http://www.eumetsat.int/website/home/Data/Products/Level1Data/index.html. Last access 3 Nov 2015
14. Forster, P., Ramaswamy, V., Artaxo, P., Berntsen, T., Betts, R., Fahey, D.W., Haywood, J., Lean, J., Lowe, D.C., Myhre, G., Nganga, J., Prinn, R., Raga, G., Schulz, M., Dorland, R.V.: Changes in atmospheric constituents and in radiative forcing. In: Climate Change 2007: The Physical Science Basis, Contribution of Working Group I to the Fourth Assessment Report of the Intergovernmental Panel on Climate Change. Cambridge University Press, Cambridge, UK and New York, NY, USA (2007)
15. Forster, P.M.D., Shine, K.P.: Radiative forcing and temperature trends from stratospheric ozone changes. J. Geophys. Res. **102**, 10841–10855 (1997). https://doi.org/10.1029/96JD03510
16. Gauss, M., Myhre, G., Pitari, G., Prather, M.J., Isaksen, I.S.A., Berntsen, T.K., Brasseur, G.P., Dentener, F.J., Derwent, R.G., Hauglustaine, D.A., Horowitz, L.W., Jacob, D.J., Johnson, M., Law, K.S., Mickley, L.J., Müller, J.-F., Plantevin, P.-H., Pyle, J.A., Rogers, H.L., Stevenson, D.S., Sundet, J.K., van Weele, M., Wild, O.: Radiative forcing in the 21st century due to ozone changes in the troposphere and the lower stratosphere, J. Geophys. Res. **108**, 4292 (2003). https://doi.org/10.1029/2002jd002624
17. Gazeaux, J., Clerbaux, C., George, M., Hadji-Lazaro, J., Kuttippurath, J., Coheur, P.-F., Hurtmans, D., Deshler, T., Kovilakam, M., Campbell, P., Guidard, V., Rabier, F., Thépaut, J.-N.: Intercomparison of polar ozone profiles by IASI/MetOp sounder with 2010 Concordiasi ozonesonde observations. Atmos. Meas. Tech. **6**, 613–620 (2013). https://doi.org/10.5194/amt-6-613-2013

18. Hilton, F., Armante, R., August, T., Barnet, C., Bouchard, A., Camy-Peyret, C., Capelle, V., Clarisse, L., Clerbaux, C., Coheur, P.-F., Collard, A., Crevoisier, C., Dufour, G., Edwards, D., Faijan, F., Fourrié, N., Gambacorta, A., Goldberg, M., Guidard, V., Hurtmans, D., Illingworth, S., Jacquinet-Husson, N., Kerzenmacher, T., Klaes, D., Lavanant, L., Masiello, G., Matricardi, M., McNally, A., Newman, S., Pavelin, E., Payan, S., Péquignot, E., Peyridieu, S., Phulpin, T., Remedios, J., Schlüssel, P., Serio, C., Strow, L., Stubenrauch, C., Taylor, J., Tobin, D., Wolf, W., Zhou, D.: Hyperspectral earth observation from IASI: five years of accomplishments. Am. Meteorol. Soc. **93**, 347–370 (2011). https://doi.org/10.1175/BAMS-D-11-00027.1
19. Hurtmans, D., Coheur, P.-F., Wespes, C., Clarisse, L., Scharf, O., Clerbaux, C., Hadji-Lazaro, J., George, M., Turquety, S.: FORLI radiative transfer and retrieval code for IASI. J. Quant. Spectrosc. Rad. Transf. **113**, 1391–1408 (2012). https://doi.org/10.1016/j.jqsrt.2012.02.036
20. Joiner, J., Schoeberl, M.R., Vasilkov, A.P., Oreopoulos, L., Platnick, S., Livesey, N.J., Levelt, P.F.: Accurate satellite derived estimates of the tropospheric ozone impact on the global radiation budget. Atmos. Chem. Phys. **9**, 4447–4465 (2009). https://doi.org/10.5194/acp-9-4447-2009
21. Lacis, A.A., Wuebbles, D.J., Logan, J.A.: Radiative forcing of climate by changes in the vertical distribution of ozone. J. Geophys. Res. **95**, 9971–9981 (1990). https://doi.org/10.1029/JD095iD07p09971
22. Li, J.: Gaussian quadrature and its application to infrared radiation. J. Atmos. Sci. **57**, 753–765 (2000)
23. Myhre, G., Shindell, D., Bréon, F.-M., Collins, W., Fuglestvedt, J., Huang, J., Koch, D., Lamarque, J.-F., Lee, D., Mendoza, B., Nakajima, T., Robock, A., Stephens, G., Takemura, T., Zhang, H.: Anthropogenic and natural radiative forcing. In: Climate Change 2013: The Physical Science Basis, Contribution of Working Group I to the Fifth Assessment Report of the Intergovernmental Panel on Climate Change. Cambridge University Press, Cambridge, UK and New York, NY, USA (2013)
24. Rodgers, C.D.: Inverse Methods for Atmospheric Sounding: Theory and Practice, vol. 2. World Scientific Publishing Co. Pte. Ltd., Signapore (2000)
25. Scannell, C., Hurtmans, D., Boynard, A., Hadji-Lazaro, J., George, M., Delcloo, A., Tuinder, O., Coheur, P.-F., Clerbaux, C.: Antarctic ozone hole as observed by IASI/MetOp for 2008–2010. Atmos. Meas. Tech. **5**, 123–139 (2012). https://doi.org/10.5194/amt-5-123-2012
26. Shine, K.P., Derwent, R.G., Wuebbles, D.J., Morcrette, J.-J.: Radiative forcing of climate in climate change: The IPCC scientific assessment, report prepared for the intergovernmental panel on climate change by working group 1. Cambridge University Press, Cambridge, NY and Melbourne, Sydney (1990)
27. Skeie, R.B., Berntsen, T.K., Myhre, G., Tanaka, K., Kvalevåg, M.M., Hoyle, C.R.: Anthropogenic radiative forcing time series from pre-industrial times until 2010. Atmos. Chem. Phys. **11**, 11827–11857 (2011). https://doi.org/10.5194/acp-11-11827-2011
28. Søvde, O.A., Hoyle, C.R., Myhre, G., Isaksen, I.S.A.: The HNO_3 forming branch of the HO_2 CNO reaction: pre-industrial-to-present trends in atmospheric species and radiative forcings. Atmos. Chem. Phys. **11**, 8929–8943 (2011). https://doi.org/10.5194/acp-11-8929-2011
29. Worden, H.M., Bowman, K.W., Worden, J.R., Eldering, A., Beer, R.: Satellite measurements of the clear-sky greenhouse effect from tropospheric ozone. Nat. Geosci. **1**, 305–308 (2008). https://doi.org/10.1038/ngeo182
30. Worden, H.M., Bowman, K.W., Kulawik, S.S., Aghedo, A.M.: Sensitivity of outgoing longwave radiative flux to the global vertical distribution of ozone characterized by instantaneous radiative kernels from Aura-TES. J. Geophys. Res. **116**, D03309 (2011). https://doi.org/10.1029/2006JD007258

Part II
Adaptation in Agriculture, Forestry and Water Resources Economy

Introduction

Borbála Gálos

Recent climate change has already widespread observable effects on the environment. The WMO Statement on state of the global climate [25] reports that the average global temperature for 2013–2017 is the highest five-year average on record. The global mean temperatures in 2017 were 1.1 °C ± 0.1 °C above pre-industrial levels. It is more than half way towards the maximum limit of temperature increase of 2 °C sought through the Paris Agreement and we are two thirds of the way to a 1.5 °C world. Based on the IPCC AR5 it is very likely that anthropogenic influence, particularly greenhouse gases and stratospheric ozone depletion, has led to a detectable observed pattern of tropospheric warming and it is likely that human influences have affected the global water cycle since 1960 [10]. There has been an overall decrease in the number of cold extremes, an overall increase in the number of warm extremes and an intensification of the hydrological cycle [9].

Extreme events have greater impacts on sectors with closer links to climate, such as water, agriculture and food security, forestry, health, and tourism [9]. These sectors are sensitive and vulnerable to the changes of the frequency and severity of climate extremes rather than to the relative slow changes of the temperature and precipitation means. Recurrent droughts and long lasting heatwaves of the last decades have strongly affected the distribution, health status and production of the vulnerable forest tree species (e.g. [1, 13, 16]). In the Carpathian Basin the vitality and survival of forests is limited by the available water. The recurrent, 3–4-year long drought periods induced damage chains: the vitality loss of Beech, Oak and Black Pine [2, 8, 17, 20, 21] was followed by mortality due to the increasing occurrence of pests and diseases [4, 15]. Weather extremes are also a major driver of the interannual variability in agricultural production.

A changing climate leads to changes in the frequency, intensity, spatial extent, duration and timing of extreme weather and climate events [10, 11, 23]. It is virtually certain that increases in the frequency and magnitude of warm temperature extremes

B. Gálos (✉)
Faculty of Forestry, University of Sopron, Sopron, Hungary
e-mail: galos.borbala@uni-sopron.hu

M. Palocz-Andresen et al. (eds.), *International Climate Protection*,
https://doi.org/10.1007/978-3-030-03816-8_9

will occur in the 21st century at the global scale. Furthermore, it is very likely that the length, frequency, and/or intensity of warm spells or heat waves will increase over most land areas. There is medium confidence that droughts will intensify in the 21st century in some seasons and areas, due to reduced precipitation and/or increased evapotranspiration [9]. The World Climate Research Program (WRCP) listed the weather and climate extremes as one of the Grand Challenges that requires urgent attention in order to better identify the underlying factors and mechanisms. This high-priority research and information is needed in the near-term (from a season to a year) to mitigate risks to society and ecosystems, and in the longer term (from a decade to centuries) for effective adaptation planning. Even under a 2 °C global warming, in most regions of Europe, the projected regional warming is more pronounced than the global mean temperature increase and an European-wide increase in the frequency of extreme events is expected [12, 14, 24]. This stresses the need of adaptation strategies even under the 1.5 or 2 °C threshold.

Forest ecosystems can be especially threatened by the projected changes, since forest tree species have low migration rate and extremely long life cycle. The key challenges and issues regarding to climate change are the expected drought tendencies and risks [7], their impacts on the water balance of forests [3], appearance of new biotic damages [5], changes of the distribution and vitality of trees [19], increasing mortality and its effects on the carbon sequestration and climate change mitigation potential of forests [22], as well as the development of decision support systems for the selection of adaptive tree species [6, 18] in order to maintain the ecosystem services of forests.

In Chap. 2, impacts of climate change on the water balance of forests as well as on the agricultural systems are discussed. Methods, technologies and management options are introduced that can be applied to reduce the negative effects of extreme events on crop production. Furthermore, the role of agro-forestry in adaptation to climate change is investigated. All of these recent studies point out the already observable and the expected impacts of climate change on regional and local scale. They underline the importance and the urgent need of the development of adaptation strategies in the affected sectors. Reaching the +2 °C target as slowly as possible will allow for a longer time period to implement the adaptation measures. To meet the challenge of climate change in agriculture, forestry and water management, multi- and interdisciplinary approaches, international collaborations, clear communication of the related scientific results and the continuously improvement of the public awareness are required.

References

1. Allen, C.D., Breshears, D.D., McDowell, N.G.: On underestimation of global vulnerability to tree mortality and forest die-off from hotter drought in the Anthropocene. Ecosphere **6**(8), 129 (2015). https://doi.org/10.1890/ES15-00203.1

2. Berki, I., Rasztovits, E., Móricz, N., Kolozs, L.: The role of tree mortality in vitality assessment of Sessile Oak Forests. South-East Eur. For. **7**(2), 91–97 (2016). https://doi.org/10.15177/seefor.16-14
3. Csáki, P., Szinetár, M.M., Herceg, A., Kalicz, P., Gribovszki, Z.: Climate change impacts on the water balance-case studies in Hungarian watersheds. Időjárás **122**(1), 81–99 (2018). https://doi.org/10.28974/idojaras.2018.1.6
4. Csóka, G., Pödör, Z., Nagy, G., Hirka, A.: Canopy recovery of pedunculate oak, Turkey oak and beech trees after severe defoliation by gypsy moth (*Lymantria dispar*): Case study from Western Hungary. Lesn. Cas. For. J. **61**, 143–148 (2015). https://doi.org/10.1515/forj-2015-0022
5. Csóka, G., Hirka, A.: Alien and invasive forest insects in Hungary (a review). Biotic risks and climate changes in forest. Berichte Freiburger Forstliche Forschung **89**, 54–60 (2011)
6. Czimber, K., Gálos, B.: A new decision support system to analyse the impacts of climate change on the Hungarian forestry and agricultural sectors. Scand. J. For. Res. (2016). https://doi.org/10.1080/02827581.2016.1212088
7. Gálos, B., Führer, E., Czimber, K., Gulyás, K., Bidló, A., Hänsler, A., Jacob, D., Cs, Mátyás: Climatic threats determining future adaptive forest management—a case study of Zala County. Időjárás **119**(4), 425–441 (2015)
8. Hlásny, T., Mátyás, Cs, Seidl, R., Kulla, L., Mergaicová, K., Trombik, J., Dobor, L., Barcza, Z., Konopka, B.: Climate change increases the drought risk in Central European forests: what are the options for adaptation? Lesn. Cas. For. J. **60**, 5–18 (2014). https://doi.org/10.2478/forj-2014-0001
9. IPCC.: Summary for policymakers. In: Field, C.B., Barros, V., Stocker, T.F., Qin, D., Dokken, D.J., Ebi, K.L., Mastrandrea, M.D., Mach, K.J., Plattner, G.-K., Allen, S.K., Tignor, M., Midgley, P.M. (eds.) Managing the Risks of Extreme Events and Disasters to Advance Climate Change Adaptation. A Special Report of Working Groups I and II of the Intergovernmental Panel on Climate Change, pp. 1–19. Cambridge University Press, Cambridge, UK, and New York, NY, USA (2012)
10. IPCC.: Summary for policymakers. In: Stocker, T.F., Qin, D., Plattner, G.-K., Tignor, M., Allen, S.K., Boschung, J., Nauels, A., Xia, Y., Bex, V., Midgley, P.M. (eds.) Climate Change 2013: The Physical Science Basis. Contribution of Working Group I to the Fifth Assessment Report of the Intergovernmental Panel on Climate Change. Cambridge University Press, Cambridge, United Kingdom and New York, NY, USA (2013)
11. Jacob, D., et 38 coauthors EURO–CORDEX, 2013.: New high–resolution climate change projections for European impact research. Reg. Environ. Change **14**, 563–578 (2013)
12. Jacob, D., Kotova, L., Teichmann, C., Sobolowski, S.P., Vautard, R., Donnelly, C., Koutroulis, A.G., Grillakis, M.G., Tsanis, I.K., Damm, A., Sakalli, A., van Vliet, M.T.H.: Climate impacts in Europe under +1.5 °C global warming. Earth's Future **6**, 264–285 (2018). https://doi.org/10.1002/2017EF000710
13. Keenan, R.J.: Climate change impacts and adaptation in forest management: a review. Ann. For. Sci. **2**(72), 145–167 (2015). https://doi.org/10.1007/s13595-014-0446-5
14. Kjellström, E., Nikulin, G., Strandberg, G., Christensen, O.B., Jacob, J., Keuler, K., Lenderink, L., van Meijgaard, E., Schär, C., Somot, S., Sørland, S.L., Teichmann, C., Vautard, R.: European climate change at global mean temperature increases of 1.5 and 2 °C above pre-industrial conditions as simulated by the EURO-CORDEX regional climate models. Earth Syst. Dynam. **9**, 459–478 (2018)
15. Lakatos, F., Molnár, M.: Mass mortality of beech on Southwest Hungary. Acta Silv. Lign. Hung. **5**, 75–82 (2009)
16. Lindner, M., Maroschek, M., Netherer, S., Kremer, A., Barbati, A., Garcia-Gonzalo, J., et al.: Climate change impacts, adaptive capacity, and vulnerability of European forest ecosystems. For. Ecol. Manage. **259**, 698–709 (2010)
17. Mátyás, C., Berki, I., Czúcz, B., Gálos, B., Móricz, N., Rasztovits, E.: Future of beech in Southeast Europe from the perspective of evolutionary ecology. Acta Silv. Lign. Hung. **6**, 91–110 (2010)

18. Mátyás, C., Berki, I., Bidló, A., Csóka, G., Czimber, K., Führer, E., Gálos, B., Gribovszki, Z., Illés, G., Hirka, A., Somogyi, Z.: Sustainability of forest cover under climate change on the temperate-continental xeric limits. Forests **9**(8), 489 (2018). https://doi.org/10.3390/f9080489
19. Móricz, N., Rasztovits, E., Gálos, B., Berki, I., Eredics, A., Loibl, W.: Modeling the potential distribution of three climate zonal tree species for present and future climate in Hungary. Acta Silv. Lign. Hung. **9**, 85–96 (2013)
20. Móricz, N., Garamszegi, B., Rasztovits, E., Bidló, A., Horváth, A., Jagicza, A., Illés, G., Vekerdy, Z., Somogyi, Z., Gálos, B.: Recent drought-induced vitality decline of black pine (*Pinus nigra* Arn.) in South-West Hungary—is this drought-resistant species under threat by climate change? Forests **9**(7), 414 (2018). https://doi.org/10.3390/f9070414
21. Rasztovits, E., Berki, I., Mátyás, Cs, Czimber, K., Pötzelsberger, E., Móricz, N.: The incorporation of extreme drought events improves models for beech persistence at its distribution limit. Annals For. Sci. **71**, 201–210 (2014)
22. Somogyi, Z.: Projected effects of climate change on the carbon stocks of European beech (*Fagus sylvatica* L.) forests in Zala County. Hungary. Les. Cas. For. J. **62**, 3–14 (2016)
23. Vautard, R., et 25 coauthors 2013.: The simulation of European heat waves from an ensemble of regional climate models within the EURO–CORDEX project. Clim. Dynam. **41**, 2555–2575 (2013)
24. Vautard, R., Gobiet, A., Sobolowski, S., Kjellström, E., Stegehuis, A., Watkiss, P., Mendlik, T., Landgren, O., Nikulin, G., Teichmann, C., Jacob, D.: The European climate under a 2 °C global warming. Environ. Res. Lett. **9**, 034006 (2014)
25. WMO.: WMO Statement on the State of the Global Climate in 2017. WMO-No. 1212, 40 p (2018). ISBN 978-92-63-11212-5

Agriculture and Climate Change

Márton Jolánkai, Márta Birkás, Ákos Tarnawa and Katalin M. Kassai

1 Impacts Related to Climate Change

The World's climate has never been constant. There were profound past changes during the history of the planet, but there is increasing evidence that human activities are altering our climate at an unprecedented rate. When assessing climate change, natural variability must also be considered. Climate change is a fundamental threat to global food security, sustainable development and poverty eradication [1, 8]. In the following some major fields of climate change impacts in the field of agriculture are summarised briefly.

Most agricultural regions will be threatened by climate change, while some others may benefit. The impact on crop yields and productivity will vary considerably [1]. Added heat stress, shifting monsoons, and drier soils may reduce yields in the tropics and subtropics, whereas longer growing seasons may boost yields in northern tracts of the temperate zone. Projections of regional climate change and the resulting agricultural impacts, however, are still full of uncertainties.

Due to climate change agricultural zones are likely to shift towards the poles. Because average temperatures are expected to rise more near the north and south poles than near the equator, the shift in climate zones will be more pronounced at higher latitudes. In the mid-latitude regions (45°–60°), present temperature zones could shift by 150–550 km. Since each of today's latitudinal climate belts is optimal for particular crops, such shifts could strongly affect agricultural and livestock production [2, 5]. Shifting crops poleward could be limited by the crop site conditions like suitability of soils, radiation, length of days etc.

Soil moisture will be affected by changing precipitation patterns. Based on a global warming of 1–3.5 °C over the next 100 years, climate models project that both evaporation and precipitation will increase, as will the frequency of intense

M. Jolánkai (✉) · M. Birkás · Á. Tarnawa · K. M. Kassai
Szent István University, Gödöllő, Hungary
e-mail: Jolankai.Marton@mkk.szie.hu

M. Palocz-Andresen et al. (eds.), *International Climate Protection*,
https://doi.org/10.1007/978-3-030-03816-8_10

rainfalls. While some regions may become wetter, in others the net effect of an intensified hydrological cycle will be a loss of soil moisture. Some regions that are already drought-prone may suffer longer and more severe dry spells [2]. The models also project seasonal shifts in precipitation patterns: soil moisture will decline in some mid-latitude continental regions during the summer, while rain and snow will probably increase at high latitudes during the winter [7].

Higher temperatures will influence production patterns. Plant growth and health may benefit from fewer freezes and chills, but some crops may be damaged by higher temperatures, particularly if combined with water shortages. Certain weeds may expand their range into higher-latitude habitats. There is also some evidence that the poleward expansion of insects and plant diseases will add to the risk of crop loss [4, 5].

More carbon dioxide in the atmosphere could boost productivity. In principle, higher levels of CO_2 should stimulate photosynthesis in certain plants. This is particularly true for most C3 plants because increased carbon dioxide tends to suppress their photo-respiration, making them more water efficient. C3 plants make up the majority of species globally, especially in cooler and wetter habitats, and include most crop species, such as cereals and legumes. The response of C4 plants would not be as dramatic. C4 plants include such tropical crops as maize, sugar cane, sorghum and millet, which are important for the food security of many developing countries, as well as pasturage and forage grasses. Elevated CO_2 concentrations may increase crop yields. This effect could be enhanced or reduced, however, by accompanying changes in temperature, precipitation, pests, and the availability of nutrients [6, 9].

The productivity of grasslands and pastures would also be affected. Climatic factors profoundly influence grazing abilities and the yields of them. The main threat is the high variation in the amount and distribution of precipitation as well as the alterations in the species composition of grasses. The availability of fodder production may influence livestock. Animal husbandry in general and pasturage in particular are to be adapted to the crop site conditions [6].

Food safety and security risks are primarily local and national. Studies suggest that global agricultural production could be maintained relative to the expected baseline levels over the next century. However, regional effects would vary widely, and some countries may experience reduced output even if they take measures to adapt. Also, there is a unpredictable hazard of food safety due to possible effects of climate change, including changes in the occurrence of pests, weeds and diseases. A specific field of hazard may be related to the adaptable land use and soil tillage that may provide or fail to establish climate durability of crop sites [6, 10].

The most vulnerable people are the landless, poor, and isolated. Poor terms of trade, weak infrastructure, lack of access to technology and information, and armed conflict will make it more difficult for these people to cope with the agricultural consequences of climate change. Many of the world's poorest areas, dependent on isolated agricultural systems in semi-arid and arid regions, face the greatest risk [1, 6]. Climate change phenomena may contribute to global migration.

Reliable policies can help to improve food security. The negative effects of climate change can be limited by changes in crops and crop varieties, improved water-management and irrigation systems, adapted planting schedules and tillage practices, and better watershed management and land-use planning. In addition to addressing the physiological response of plants and animals, policies can seek to improve how production and distribution systems cope with fluctuations in yields [6].

Climate policy goals need to be regularly reassessed in the light of shifting social, economic, political and scientific contexts and this is best achieved through a reflexive policy appraisal process. At present, there is a lack of such reflexivity in climate policy appraisal processes in Europe. To enhance reflexivity in the appraisal process the following issues are suggested [3]: boundary organisations be created to conduct reflexive appraisals; new platforms be secured to enable inclusive and deliberative stakeholder processes; and "windows of opportunity for learning" caused by shocks or changes to wider society systems be recognised and grasped.

2 Adaptation Measures in the Field of Crop Production

Adaptation is not just attaining a physical outcome, but is a dynamic process that relies on institutional mechanisms to enable implementation of selected measures and to build local capacity. Involving stakeholders in adaptation and risk management processes is a key component of building adaptive capacity [3]. In the following, results in three fields of adaptation are highlighted, based on recent Hungarian climate change research.

3 Field Crop Species

Twelve field crop species were involved in an aridity assessment study. 50 years' data of twelve meteorological stations representing all regions of Hungary were used as a basis of evaluation. PAI indices of each station were processed with vulnerability indices of the given field crops. The results obtained suggest, that susceptibility of cereals proved to be the lowest, however maize and potato were highly affected by aridity x vulnerability interactions [10]. The strongest climatic influence could be detected in the case of alfalfa and sugar beet. Regional differences in aridity were detectable. Lowland regions were found to be more arid than those of mountainous areas and northern regions.

The average vulnerability indices (VI) are presented in Table 1. The twelve species studied cover some 92% of arable land in Hungary [10].

Table 1 Climatic vulnerability of twelve field crop species based on 50 years' averages in Hungary

Crop species	Spectral bands	VI indices	Susceptibility
Winter wheat	Triticum aestivum	5.6	low
Maize	Zea mays	7.3	high
Winter barley	Hordeum vulgare	5.8	low
Spring barley	Hordeum vulgare	5.7	low
Rye	Secale cereale	6.7	medium
Oats	Avena sativa	5.8	low
Peas	Pisum sativum	6.3	medium
Sunflower	Helianthus annuus	6.1	medium
Oilseed rape	Brassica napus	6.4	medium
Alfalfa	Medicago sativa	7.6	high
Sugar beet	Beta vulgaris	7.7	high
Potato	Solanum tuberosum	6.5	medium
Mean		6.5	

4 Carbon Sequestration by Plant Biomass

Atmospheric rise of CO_2 highlights crop production regarding both adaptation and mitigation. The negative effects of climate change can be limited by changes in crops and crop varieties, improved water-management and irrigation systems, adapted plant nutrition, protection and tillage practices, and better watershed management and land-use planning [1, 7]. The global potential of carbon sequestration through crop production, land use and soil management practices may offset one-fourth to one-third of the annual increase in atmospheric CO_2. Table 2 summarizes the carbon content of some essential plant compounds and substances. The annual carbon sink of plant biomass exceeds carbon source releases. Also renewable energy provides a chance for replacing fossil fuel uses.

The global potential of carbon sequestration through crop production, land use and soil management practices may offset one-fourth to one-third of the annual increase in atmospheric CO_2. Major conclusions of the study.

Improved management techniques and practices may utilise and reduce part of the annual CO_2 increase.

Energy uses based on fossil fuel combustion should be controlled Globally.

Atmospheric C budget can be balanced by photosynthetic dry matter production of natural vegetation and agricultural crops. The photosynthetic activities of plants determine soil organic matter (SOM). The organic matter manufactured by plants is originated from atmospheric CO_2. All plants, including field crops capture C and produce vegetative material, a biomass that comprises yield and by products. The latter can be influenced by agronomic applications. Soil organic matter is based on

Table 2 Carbon content of some amino-acids, carbohydrates and polypeptides [7]

	C wt	H wt	N wt	O wt	S wt	Mol wt	C %
glycine	24	5	14	32	0	75	32,000
alanine	36	7	14	32	0	89	40,449
serine	36	7	14	48	0	105	34,286
cysteine	36	7	14	32	32	121	29,752
phenylalanine	108	11	14	32	0	165	65,455
tyrosine	108	11	14	48	0	181	59,669
tryptophan	132	12	28	32	0	204	64,706
valine	60	11	14	32	0	117	51,282
threonine	48	9	14	48	0	119	40,336
methionin	60	11	14	32	32	149	40,268
leucine	72	13	14	32	0	131	54,962
isoleucine	72	13	14	32	0	131	54,962
proline	60	6	14	32	0	112	53,571
histidine	72	9	42	32	0	155	46,452
arginine	60	14	56	32	0	162	37,037
lysine	72	14	28	32	0	146	49,315
asparaticacid	48	7	14	64	0	133	36,090
asparagine	48	7	28	48	0	131	36,641
glutamicacid	60	10	14	64	0	148	40,541
glutamine	60	10	28	48	0	146	41,096
carbohydrate	12	2	0	16	0	30	40,000
glutatione	120	17	42	96	32	307	39,088

the sequestration of C derived from plant residues. In our studies carbon sequestration of some field crops ranged between 0.12 and 0.61 kg/m^2 SOM in relation with the crop species and agronomic treatments applied [9].

5 Land Use and Soil Management

Soil factors highly influence the life of plants [2, 7]. Soil structure is composed by its ingredients, i.e. organic and inorganic components, their amount, size, distribution and geometrical orientation. Soils as materials are made up of three phases; solid (organic and mineral), liquid (water and solutions) and gaseous materials. The geometric orientation of soil particles determine the soil texture. Soil is a medium providing water and nutrients to plants.

At the geographic latitude of the temperate zone plants require 250–400 g of water to build 1 g of dry matter. This water/dry matter ratio is named as transpiration

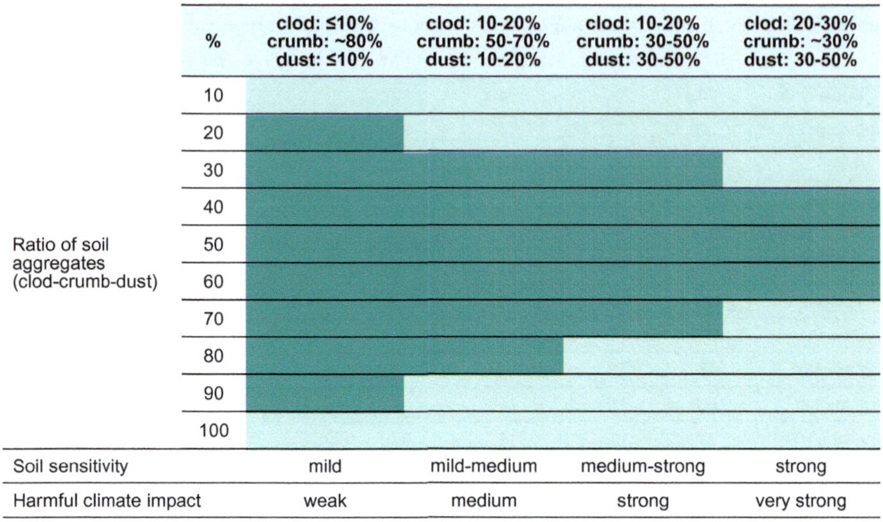

Fig. 1 Agronomic value of soil conditions regarding harmful climatic impacts (*Source* Jolánkai and Birkás [7])

coefficient. Since cultivated plants produce far more dry matter compared to natural vegetation, consequently their water demand is higher as well. From an aspect of soil management water storage and supply abilities of soils of a certain field should be determined Fig. 1.

Sensitivity of soils to climate change impacts is determined mainly by the cultural state of the soil [7]. The most durable soils have adequate ratio concerning the particle size of soil aggregates.

6 Conclusions

Undesirable effects of climate change may be limited by changes in the cropping structure and crop varieties, improved water-management, adapted plant nutrition, protection and tillage practices. Carbon sequestration through crop production, land use and soil management practices may utilise and so reduce the annual increase in atmospheric CO_2.

Climate change impacts on soil sensitivity are determined mainly by the cultural state of the soil. The most durable soils have adequate ratio concerning the particle size of soil aggregates. Actually climate resistance depends on the regulation of pore space capacity .

References

1. FAO.: Climate Change (2015). http://www.fao.org/climate-change
2. IPCC.: Climate Change 2001: The Scientific Basis, 881 p. IPCC, Cambridge University Press (2001)
3. Hulme, M., Neufeldt, H., Colyer, H., Ritchie, A. (eds.).: Adaptation and Mitigation Strategies: Supporting European Climate Policy. The Final Report from the ADAM Project. Tyndall Centre or Climate Change Research, University of East Anglia, Norwich, UK (2009)
4. Jolánkai, M., Szentpétery, Z., Nyárai, H.F., Kassai, K., Tarnawa, Á.: Impact of spatial distribution of water availability on the vulnerability of field crop species. Növénytermelés **64**(Suppl.), 171–174 (2015)
5. Jolánkai, M.: Agriculture, soil management and climate change. In: Faragó T., Láng, I., Csete, L. (eds.) Climate Change and Hungary: Mitigating the Hazard and Preparing for the Impacts—The VAHAVA Report, pp. 38–45. HAS, Budapest (2010)
6. Jolánkai, M., Láng, I., Csete, L.: Effects of global climate change on agriculture. In: Hidvégi, S., Gyuricza, C. (eds.) Proceedings of the 3rd Alps-Adria Scientific Workshop, pp. 20–24. Akaprint, Dubrovnik (2004)
7. Jolánkai, M., Birkás, M.: Global climate change impacts on crop production in Hungary. Agriculturae Conspectus Scientificus **72**(1), 17–20 (2007)
8. Jolánkai, M.: Climate change impacts on crop production. In: Palocz-Andresen, M., Németh, R., Szalay, D. (eds.) Támop-Humboldt College for Environment and Climate Protection, pp. 66–70. University Press, University of West Hungary, Sopron (2011)
9. Kassai, M.K., Tarnawa, Á., Nyárai, H.F., Pósa, B., Jolánkai, M.: The effect of crop species and N fertilization on soil organic matter. Columella **2**(2), 23–28 (2015)
10. Tarnawa, Á., Klupács, H., Sallai, A., Szalay, K., Kassai, M.K., Nyárai, H.F., Jolánkai, M.: Study on the impact of main climatic factors of crop production in a mathematical model. In: Celková, A. (ed.) Transport of Water, Chemicals and Energy in the Soil-Plant-Atmosphere System, pp. 566–571. Institute of Hydrology, Bratislava (2010)

Adaptation of Methods and Technologies in Agriculture Under Climate Change Conditions

Josef Eitzinger, Angel Utset and Miroslav Trnka

1 Introduction

Climate change affects crop growth processes and related agricultural production in direct and indirect pathways with respect to regional conditions [2]. For example, the increase in mean temperatures accelerates directly crop development, the change in seasonal precipitation amounts together with increasing evaporative demand can indirectly lead to more drought stress for crops. Given regional climate variability and potential climate changes strongly determine the severity and frequency of extreme weather events. Especially heat waves, droughts and heavy precipitation are expected to increase in future climates, but with distinct regional variations [5, 10, 24]. Therefore an early recognition of risks in the short (warning) and long term (advice) and the related implementation of adaptation strategies is crucial for all stakeholders in the whole food production chain. It is proved that anticipatory, precautionary adaptation is more effective and less costly than forced, last minute, emergency adaptation or retrofitting [4]. Regional climate change impact and adaptation studies demonstrate high spatial variability of potential impacts, depending on the specific agroecosystems, soil conditions, climate regions and farming systems pointing to the importance of regionally fitted adaptation measures.

Farming methods and technologies—either newly developed or already known or indigenous technologies developed over many generations—offer many oppor-

J. Eitzinger (✉)
Department of Water, Atmosphere and Environment, Institute of Meteorology, University of Natural Resources and Life Sciences, Vienna, Austria
e-mail: josef.eitzinger@boku.ac.at

A. Utset
ClimaRisk Ltd., Salford, UK

M. Trnka
Global Change Research Institute, Academy of Sciences of the Czech Republic & Mendel University of Agriculture and Forestry, Brno, Czech Republic

M. Palocz-Andresen et al. (eds.), *International Climate Protection*,
https://doi.org/10.1007/978-3-030-03816-8_11

tunities for adapting to existing climatic and weather conditions in crop production i.e. in closing existing yield gaps [6]. Because of ongoing climate change, the optimization of farming methods and technologies becomes even more important for ensuring sustainable productivity of various agricultural production systems at different farm input levels. The use of available farm technologies should secure sustainable production within given climatic and weather conditions through the proper management of natural resources or conditions fitted to specific farming systems, including water, soil (including nutrients), crops and microclimatic conditions [17]. In all agro-ecosystems farmers have developed specific strategies since centuries, mainly with the application of different farm technologies and related management options to survive in the given environment, but for various reasons not always with sustainability in mind. The development and improvement in farming methods and technologies has been responsible for most of the increase in productivity [15] and yields in agricultural production, especially in the developed countries. Although this trend is slowing down in high input farming systems due to minimized yield gap potentials, there is still a big yield gap potential especially in the less developed countries and agricultural regions. In that sense adaptation to climate change in agriculture has not only the aim to ensure existing yields but also increase yield levels in a sustainable sense in the less developed farming systems.

2 Optimizing Farm Technologies in Respect to Agricultural System

In order to analyze optimization strategies in various agricultural systems we propose to make a distinction between the most important and climate-sensitive agricultural resources to be managed, such as water, soil (including nutrients), crop (including management) and microclimatic conditions in relation to low, medium and high agricultural input systems. Of course, many farm technology optimization strategies can affect more than one of these resources at the same time. Low-input systems may be characterized as small farm structures and with low income in a less developed socioeconomic environment as is found in developing countries (almost no financial reserves for investment in farm technologies available). Medium input systems might be characterized as small farm structures with acceptable farm income in a good socioeconomic environment, as in small farms in Western Europe (limited financial reserves for investment in farm technologies available). High input systems might be characterized as farms with high income levels in any socioeconomic environment, where there is theoretically no limitation to investment in new farm technologies.

2.1 Farm Technologies and Water Resources

In many developed agricultural production regions with frequent drought occurrence new technologies for water management have been successful introduced and increased agricultural productivity. For instance, modern sprinkler and drip-irrigation systems, which can significantly reduce crop water use, have been introduced at great expense in some Mediterranean European regions such as Spain, leading to better productivity increase. Water availability could well be the most important agricultural constraint in Mediterranean agriculture in the future [1, 9] and in many other agricultural regions worldwide. Adaptation studies by Rosenzweig et al. [16] have shown that only few regions can expand irrigated land in a changed climate due to irrigation water shortage. Adapted crop cultivars, irrigation and drainage technology and water management are therefore recommended such as water saving "Deficit Irrigation" or drip irrigation. However, the introduction of deficit irrigation and similar techniques to improve irrigation efficiency need to be adapted for local conditions, taking into account climate variability [25, 26].

Irrigation investments include optimizing channel designs, water distribution systems and pumping devices. The engineering effort involved is usually significant, costing several million euros. Complete sprinkler coverage usually involves underground PVC or metal tubes all over the agricultural field. Automatic control devices also need solar cells, modems, computer systems and other related technology. Furthermore, irrigation advice services call for government investment in trained personnel, as well as laboratory infrastructures and supporting technological facilities such as agrometeorological weather stations (Fig. 1). Despite the large investment involved in these three potential measures, it can pay out in few years, particularly in view of the anticipated increase in water prices due to policy measures. The irrigation advice service could also become independent and self-funding in near future. A government-directed effort, providing loans and supporting funds is essential in the first stage. The total amount involved is however very high, which makes these potential measures affordable only by developed countries.

For many farmers in developing countries expensive new technologies are not affordable without external support and are therefore not applicable in low-input agricultural systems with weak infrastructure and poor socio-economic conditions. The adaptation and use of traditional methods should recommended in these cases. Indigenous techniques for agricultural water use in semi-arid regions are known, for example, already from the Incas. Similar systems of ancient underground pipes for the transfer of irrigation and drinking water through arid areas are in place in Iran. Ancient surface canal systems and surface tanks over large areas for the transfer, distribution, collection and storage of water from the monsoon periods can be found in India and Sri Lanka, which also still work effectively and are used for crop production in low-input farming systems. As another example Stigter et al. [22] reports, based on traditional knowledge, newly adapted effective crop water use methods in Africa using planting pits with improved soil water storage through the addition of manure.

Fig. 1 Agrometeorological weather station for use of crop management planning (Austria). *Photo* Eitzinger

Beside the revival of original or modified traditional or indigenous methods, new low-cost technologies are an increasingly promising option for low-input farming systems, especially for countries in transition such as India or China. Even simple low-cost technologies could significantly improve irrigation scheduling and crop water use compared with flood irrigation. Even methods and technologies, based on simple measurements (e.g. of precipitation) and algorithms to estimate water demand for irrigation, needs training and education measures, ideally organized as a bottom-up approach [23]. Moreover, a basic and stable infrastructure for local companies and technical support should exist or be built up, which is not the case in many regions of developing countries, especially in Africa. This could also act as an incentive for technological change to be driven more by environmental objectives and farmer innovations operated through the market as recommended by Norse and Tschirley [8], among others.

An important management option for low (and all level) input farming systems regarding water resources is the change to cropping methods with better water use efficiency as well as crops with better drought tolerance. This is especially important in regions where pressure on water reserves is increasing owing to human activities, climate change and variability. For example, the change from wetland rice to dry land rice or other crops can have enormous effects on agricultural water reserves, as demonstrated in northern China [28].

2.2 Farm Technologies and Soil Resources

Soil types developed over many centuries are determined among other factors by the climatic conditions. Soil conditions under prevailing crop management interact permanent with climate and climate variations. In this sense, farm technologies and soil and crop management have to be adapted to maintain important soil functions to secure sustainable agricultural production, which is basis for food security and the welfare of many countries. In many regions with extreme weather conditions, for example, soil functions can react very quickly to agricultural practices. Unfortunately, this can lead to rapid and irreversible degradation of soil functions and further to desertification, which has become a significant problem in many agroecosystems in the world.

For example, improper irrigation schemes and use of saline irrigation water can lead to increasing salinity of soils, making them unusable for agricultural production. Other examples are land use practices which endanger soils or soil functions such as overgrazing, leading to wind erosion and desertification processes. Crop production in warm semi-arid zones with frequently strong winds can easily lead to wind erosion triggered by soil degradation. In tropical regions the high soil temperatures combined with high precipitation leads to high decomposition and leaching rates and an inappropriate change in soil use for agricultural crop production can lead to fast soil degradation [18]. In climates with frequent extreme precipitation events, such as the Asian monsoon regions, soil water erosion, especially in hilly terrains, has already caused enormous soil degradation in the past. This is the case where the soil surface is not always fully covered and no terrace systems are in place. Under climate change and changing climate variability, these problems will become a more significant threat for the soils in many agroecosystems through more frequent droughts or extreme precipitation (for example, with too few organic matter residues or manure) [27].

The first important aim of soil cultivation is to control weeds and to optimize root growth conditions. This is still an important argument for ploughing in many agricultural areas and in ecological farming. However, because soil cultivation is an important cost factor, less intensive methods have been developed, such as reduced soil cultivation or minimum to no soil cultivation and tillage systems. These methods also reduce risk of soil water and wind erosion. It can also increase soil water-holding capacity and water infiltration. Finally, it has been demonstrated experimentally that increasing soil water-holding capacity by reducing soil cultivation in combination with mulch can lead to a significant positive yield effect in dry growing seasons.

Although climate change and climate variability directly affect soil erosion, changes in crops or crop rotation driven by climate change may influence indirectly soil erosion risk in vulnerable regions [14]. For example, O'Neil et al. [11] reports that increasing precipitation and decreasing soil cover from temperature-stressed maize are important factors for increasing soil erosion in the Midwest of the United States of America. In any case, soil erosion leads at the end to a decrease in soil fertility and hence to a reduction in crop productivity because of loss of organic matter, nutrients and lower water-holding capacity. This can be the start of self-forcing desertification processes and permanent loss of agricultural, fertile land.

In high-input agricultural systems often heavy machinery is used. In combination with too high soil water content this can contribute to soil compaction, decreasing water infiltration, increasing runoff and therefore water erosion. In humid regions of Europe, these problems are apparent for summer crops such as sugar beet and maize, where soils are not or only partly covered by a canopy for weeks in the early growing season. In case of late harvest heavy machinery has a devastating and often irreversible effect on soil structure during wet harvest periods in the autumn. This problem accelerates with increasing slopes of fields, as are frequently found in Europe. Perennial crops in various climatic regions such as vineyards, orchards, tea or coffee, which are often grown in hilly regions, are also subject to water erosion, especially during extreme precipitation events. Mulching technologies such as grass or straw mulch or other crop residues are therefore often applied and are sometimes mandatory. In some cases, even the more costly or manpower-intense terrace systems have been re-established in order to stop long-term soil erosion.

2.3 *Farm Technologies and the Microclimate of Crops*

Changes in climate variability and climate affect microclimatic conditions is many ways [19], however, the design and management of crops stands can strongly modify these impacts. For example, in semi-arid low-input systems there are known examples not only for improving water resources but also for optimizing the temperature and radiation regimes of crop stands [20, 21]. A classic example is oasis agroecosystems with complex crop mixing and patterns to permit efficient use of radiation within a small area, to increase air humidity for the shaded crops and to avoid extreme diurnal temperature variations.

Such Agroforestry systems are well known option of farm management to improve microclimatic conditions and not just to reduce wind and evapotranspiration. As crops respond especially to climatic extremes, any measure to reduce these extremes in most cases has had an accumulating positive effect on the yield level. For example, heat stress on crops can be reduced by shading, which has been reported as a significant yield factor for many crops. Different types of agroforestry systems related to specific climates and agroecosystems and their effects are described [19]. These systems were optimized for the specific characteristics of the relevant agroecosystem (climate, soils, crop production, farm input level, socio-economic framework). Agroforestry systems are used in many regions, especially in subtropical and tropical climates with extreme temperatures and/or weather variability.

Tree shading, on the other hand, can also prevent frost damage to crops and reduce nocturnal radiation cooling on the crop surfaces. Other frost production methods, such as covering plants with sheets or foil, are also used in small plots for low-input systems. For orchards or large fields methods such as frost irrigation, foil covering, or applying aerosols are costly and are therefore found mainly in medium- and high-input farming and for cash crops. The selection of crop stand location in relation to orography (avoiding cold air lakes) is very important for protection against radiation frost damage. These measures are often ignored, especially when frost occurs seldom, but the effect on perennial crops can be more devastating than hail damage, as whole plantation can be destroyed.

2.4 *Farm Technologies and Crop Resources*

Crop yield and crop production within a certain territory can be seen as an interaction of many factors. However, crops adapted to certain conditions are an important local resource for crop productivity with a significant influence on yield risk. Crop physiological processes normally respond nonlinearly to changes in their growing conditions (especially temperature), exhibit threshold responses and are often affected by combinations of stress factors. Higher temperature and precipitation variability increase the risk of lower yield, as many experimental and simulation studies have shown [13].

Farmers over centuries have selected the best cultivars for their use, creating locally well adapted crops, some of which are still in use in agricultural systems and are an important genetic resource for modern crop breeding. Arable farmers every year have the option to select crop type and cultivar but also modify crop management (i.e. sowing date according to the expected seasonal weather, for example). Seasonal precipitation pattern (onset of rain, duration of rainy season, distribution during crop growing period) is one of the most important information for farmers in monsoon-type climates using rain-fed cropping. Such climate conditions often occur under low-input systems in developing countries [19]. Efficient seasonal rain forecast and the related information transfer or warning to local farmers enables them to adapt their sowing dates and crop selection. Although already successfully applied, for example, in developed countries such as Australia [7], there is still a deficit when it comes to making such information useable for farmers in low-input agriculture [17].

Upcoming new technologies ("precision farming") permit the monitoring of crop conditions on a much smaller scale, such as field scale and less. This technology is under permanent development, using methods such as remote sensing, GPS and GIS. Because of still relatively high costs it is applicable at the farm scale only for high-input farming [12]. However, specific applications such as observing spatial variabilities of crop conditions in crop fields, such as nitrogen content of leaves for within-field fertilization optimization become more and more common. Observing drought status for irrigation scheduling, disease/pest occurrence for better pest management are other practical examples. Using the observed information the farmer can apply measures based on the actual site-related status, considering field-level variations. This can significantly decrease costs for fertilizers and chemicals and enhance crop yield and productivity.

In case of irrigation, applying water according to spatially changing soil conditions can increase significantly water use efficiency and avoid over-watering. The related equipment is still costly and not appropriate for low-input farming and small farms, but on a larger scale and on an institutional basis such technologies are applied already for early warning of, for example, drought and food risk [3]. Locally adapted crop management of low-input systems may use other options to precisely adapt management to spatially changing soil and crop conditions. In small farms, not technologically driven farms, such options tend in any case to be based on the experience or Know-How transfer or exchange to/of the local famers [23].

3 Conclusions

Optimization of farm technology and management plays an important role in reducing negative impacts of climate variability and extreme weather events on agricultural production. The relevant measures strongly influence the availability of the most important resources for agricultural production, namely water, soil (and nutrients), crops and microclimatic conditions. In low-, medium- and high-input farming systems various technologies are available ranging from traditional or indigenous methods to high-tech methods, low-cost technologies to high-cost technologies such as precision farming methods. Many authors have reported that high-input farming, especially in temperate regions, has the best prospect for adaptation to current or changing climate variability or for protection against extreme weather events. In these farming systems, usually located in developed countries with good infrastructures, technological developments and the availability of technologies in crop and animal production provide a rich toolkit enabling decision-makers to select measures from several options. Additionally in developed countries better infrastructure, extension services, warning and monitoring services for farmers already exist.

On the other hand, many low-input farming system farmers depend on traditional methods or low cost technologies and—in many cases not yet available—external inputs such as institutional forecasting and warning methods or investment in irrigation and other infrastructure. Both the re-establishment of locally adapted traditional (indigenous) farming technologies and warning/forecasting methods together with institutional support, by means of seasonal weather forecasts for example, (especially for precipitation) may help farmers in low-input agricultural systems to sustain or improve their productivity, food production and income. There is an urgent need, however, to consider sustainability of existing agroecosystems in developing adaptation options. This has to be a policy-driven basic strategy applying bottom-up participatory approaches in training farmers or stakeholders for the long term as water, land and soil as key resources for food production can be efficiently protected only by the direct land users.

References

1. European Environment Agency (EEA).: Vulnerability and Adaptation to Climate Change in Europe. EEA Technical Report No. 7/2005. EEA, Copenhagen, 84 pp (2005)
2. Eitzinger, J., Trnka, M., Semerádová, D., Thaler, S., Svobodová, E., Hlavinka, P., Siska, B., Takáč, J., Malatinská, L., Nováková, M., Dubrovský, M., Zalud, Z.: Regional climate change impacts on agricultural crop production in Central and Eastern Europe—hotspots, regional differences and common trends. J. Agric. Sci. **151**(6), 787–812 (2013)
3. Enenkel, M., Steiner, C., Mistelbauer, T., Dorigo, W., Wagner, W., See, L., Atzberger, C., Schneider, S., Rogenhofer, E.A.: Combined satellite-derived drought indicator to support humanitarian aid organizations. Remote Sens. Basel **8**(4) (2016)
4. Hayes, M., Wilhelmi, O., Knutson, C.: Reducing drought risk: bridging theory and practice. Nat. Haz. Rev. **5**(2), 106–113 (2004)
5. IPCC. Managing the Risks of Extreme Events and Disasters to Advance Climate Change Adaptation. A Special Report of Working Groups I and II of the Intergovernmental Panel on Climate Change [Field, C.B., Barros, V., Stocker, T.F., Qin, D., Dokken, D.J., Ebi, K.L., Mastrandrea, M.D., Mach, K.J., Plattner, G.-K., Allen, S.K., Tignor, M., Midgley P.M. (eds.)]. Cambridge University Press, Cambridge, UK, and New York, NY, USA, 582 pp (2012)
6. Lobell, D.B., Cassman, K.G., Field, C.B.: Crop yield gaps: their importance, magnitudes, and causes. Annu. Rev. Environ. Resour. **34**, 179–204 (2009)
7. Meinke, H., Stone, R.: Seasonal and inter-annual climate forecasting: the new tool for increasing preparedness to climate variability and change in agricultural planning and operations. Clim. Change **70**, 221–253 (2005)
8. Norse, D., Tschirley, J.B.: Links between science and policy making. Agric. Ecosystem Environ. **82**, 15–26 (2000)
9. Olesen, J.E., Bindi, M.: Consequences of climate change for European agricultural productivity, land use and policy. Eur. J. Agron. **16**, 239–262 (2002)
10. Orlandini, S., Nejedlik, P., Eitzinger, J., Alexandrov, V., Toulios, L., Calanca, P., Trnka, M., Olesen, J.E.: Impacts of climate change and variability on European agriculture: results of inventory analysis in COST 734 countries. Ann. N. Y. Acad. Sci. **1146**, 338–353 (2008)
11. O'Neal, M.R., Nearing, M.A., Vining, R.C., Southworth, J., Pfeifer, R.A.: Climate change impacts on soil erosion in Midwest United States with changes in crop management. CATENA **61**, 165–184 (2005)
12. Pedersen, S.M., Fountas, S., Blackmore, B.S., Gylling, M., Pedersen, J.L.: Adoption and perspectives of precision farming in Denmark. Acta Agric. Scand. Section B Soil Plant Sci 54(1), 2–8 (2004)
13. Porter, J.R., Semenov, M.A.: Crop responses to climatic variation. Philos. Trans. R. Soc. Lond. B Biol. Sci. **360**(1463), 2021–2035 (2005)
14. Rounsevell, M.D.A., Evans, S.P., Bullock, P.: Climate change and agricultural soils: impacts and adaptation. Clim. Change **43**(4), 683–709 (1999)
15. Rounsevell, M.D.A, Ewert, F., Reginster, I., Leemans, R., Carter, T.R.: Future scenarios of European agricultural land use: II. Projecting changes in cropland and grassland. Agric. Ecosyst. Environ. **107**(2–3), 177–135 (2005)
16. Rosenzweig, C., Strzepek, K.M., Major, D.C., Iglesias, A., Yates, D.N., McDluskey, A., Hillel, D.: Water resources for agriculture in a changing climate: international case studies. Glob. Environ. Change **14**(4), 345–360 (2004)
17. Salinger, M.J., Sivakumar, M.V.K., Motha, R.: Reducing vulnerability of agriculture and forestry to climate variability and change: workshop summary and recommendations. Clim. Change **70**, 341–362 (2005)
18. Sivakumar, M.V.K., Brunini, O., Das, H.P.: Impacts of present and future climate variability on agriculture and forestry in the arid and semi-arid tropics. Clim. Change **70**, 31–72 (2005)
19. Sivakumar, M.V.K., Motha, P. (ed.): Managing Weather and Climate Risks in Agriculture. Springer (2007). ISBN 978-3-540-72744-6

20. Stigter, C.J.: Microclimate Management and Manipulation in Traditional Farming. CagM Report No. 25, WMO/TD-No. 228, World Meteorological Organization, Geneva, p. 20, VI Appendices (1988)
21. Stigter, C.J.: Management and manipulation of microclimate. In: Griffiths, J.F. (ed.) Handbook of Agricultural Meteorology. Oxford University Press, Chapter 27, pp. 273–284 (1994)
22. Stigter, C.J., Dawei, Zheng, Onyewotu, L.O.Z., Xurong, Mei: Using traditional methods and indigenous technologies for coping with climate variability. Clim. Change **70**, 255–271 (2005)
23. Stigter, C.J. (ed.): Applied Agrometeorlogy. Springer (2010). ISBN 978-3-540-74698-3
24. Trnka, M., Rotter, R.P., Ruiz-Ramos, M., Kersebaum, K.Ch., Olesen, J.E., Zalud, Z., Semenov, M.A.: Adverse weather conditions for European wheat production will become more frequent with climate change. Nat. Clim. Change (2014). https://doi.org/10.1038/nclimate2242
25. Utset, A., Farre, I., Martínez-Cob, A., Cavero, J.: Comparing Penman-Monteith and Priestley-Taylor approaches as reference-evapotranspiration inputs for modelling maize water-use under Mediterranean conditions. Agric. Water Manag. **66**, 205–219 (2004)
26. Utset, A.: Introducing Tools for Agricultural Decision-Making Under Climate Change Conditions by Connecting Users and Tool-Providers (AGRIDEMA). Commission of the European Communities. Research Directorate General. EC Contract No. 003944 (GOCE) (2005)
27. Yeo, A.: Predicting the interaction between the effects of salinity and climate change on crop plants. Sci. Hortic. **78**, 159–174 (1999)
28. You, S.C.: Agricultural adaptation of climate change in China. J. Environ. Sci. (China) **13**, 192–197 (2001)

Climate Change Effects on Apple Phenology and Phenometry in an Apple Gene Bank Plantation of Hungary

László Lakatos

1 Adaptation Strategies in Hungarian Fruit Growing

The horticultural plant cultivation has to be considered of high value; the occurrence of extreme climatic influences may cause huge harms. The Hungarian fruit production is largely influenced by extreme climatic effects. We are in the climatic buffer zone of Europe. Due to this, in this region the weather is greatly changeable. This is a great advantage, because almost all the continental climate zone fruits can be grown in different regions of the country. While in the moderate climate zone, the effects of the expected global warming will not bring about such change, which would enable the safe and competitive growing of subtropical and tropical fruits.

Our potential solution against the extreme weather effect was the practical preparation on expected global climate change effects. In this matter our tasks are various. Among many the emphasis on product safety, improvement of activities which help to resolve the ecological, technological and biological issues and gathering information about climatic characteristics of the cultivating region are the most crucial. We should take into consideration all the favourable and disadvantageous weather conditions and requirements of all the main cultivating areas. According to our calculations climate change may reduce the volume of fruit production areas by 15%. This fact has less influence on the amount of fruit production, because the unfavourable country sides did not play determinant role in our fruit supply. Those fruits which will allocate Hungary's production in the future, can be produced with the greatest safety.

According to our prediction the distribution of Hungarian fruit cultivating areas will be the following in the near future: apple 40%, sour cherry 15%, plum 10%, sweet cherry 10%, peach 8%, apricot 7%, pear 5%, walnut 2%, other fruits altogether 3%. I used the PRECIS A1B and A2 scenarios daily weather data which is originated from

L. Lakatos (✉)
Eszterházy Károly University, Eger, Hungary
e-mail: lakatos.laszlo@uni-eszterhazy.hu

M. Palocz-Andresen et al. (eds.), *International Climate Protection*,
https://doi.org/10.1007/978-3-030-03816-8_12

Fig. 1 Micro-meteorological station in the apple orchard

the University of Eötvös Loránd at the Department of Meteorology. We calculated the future changes of apple blooming time. The horticultural plant cultivation is having to be considered of high value; the occurrence of extreme climatic influences may cause huge harms. We can reduce the cultivation risk if we use micro-meteorological station inside the orchard very close to the trees (Fig. 1).

Extreme weather phenomena occurred in the past as well, but the occurrence probability has been increasing during the last decades [5]. The influence of climate change on fruit development can be verified by only long phenological data series.

1.1 Phenological and Phenometrical Features of Apple Cultivars

Growth and development of plant is fundamentally coded in their inherited genetic constitution, but manifestation of the expression of the particular genes depends on the complex effects of the immediate environment. This environment is subject within wide limits to periodically repeated changes, especially through the repeating seasons of the successive years. The seasonal life cycle of plants is divided into distinct phases recognized easily by their appearance (phenophases), which are often closely related. Most of weather adversities in Hungarian fruit growing are due to the temperature minima during the winter and spring, although excessive heat may become deleterious as well. The effect of temperature on the development of buds has been studied by Szabó [11], and he stated that during the period of endodormancy, buds and aerial, lignified parts of fruit trees are practically undamaged by low temperatures of −20 °C in Hungary. As endodormancy finishes, any temperature above 0 °C stimulates the life processes of the trees and abolishes, gradually, the frost resistance experienced during endodormancy, [4]. The regression of frost resistance in the flower buds and flowers continues until the end of bloom and fruit set, which is the most frost susceptible period of the fruit trees [8]. The relation of phenophases and meteorological phenomena, especially temperature, has been mostly explored around the bloom time of fruit trees. The blooming process has been analysed meticulously by Nyéki [7] and divided into the following sub-periods

1. Start of bloom (1–5% of flowers opened on the tree),
2. Main bloom (the ratio of open flowers is 50% or more),
3. The day of full bloom (the ratio of open florets achieves a maximum),
4. End of bloom (when 95–100% of flowers shed their petals).

The length of endodormancy (a phenophase) of apple cultivars is determined by the demand for the occurrence of chilling hours, i.e. until that demand is fulfilled, the rising temperature does not trigger the process of flower development. Mild winters may cause reversion of the customary blooming order of cultivars in the following spring because endodormancy of some cultivars is still not completed. If in all cultivars the endodormancy expires regularly (which would be the case in normal or rather long winters), the blooming order is not disturbed. Excessively high temperatures during blooming shorten the length of the blooming period, since pollen can be quickly released but the drying out of the stigmatic fluid lowers the probability of the pollen grains being caught thus fertilization ensured. The chance of pollinating bees visiting the flowers declines at the same time. As a result, the chance of flowers being pollinated and ovules being fertilized is low if there are high temperatures during bloom.

The phenometry of fruit development has deep roots in Hungary as the first attempts are dated back to the 1950s, when Berényi and Justyák [1] published their study first focusing on orchards and hillside vineyards with grapes. The first study on the climate of vineyards and orchards appeared in 1961 by Bognár and Kozma [2]. In the 1960s, studies appeared analysing the interactions between phenological phasis-lengths and meteorological parameters, mainly focusing on apple and grape cultures as documented by Csöbönyei and Stollár [3]. By means of that technique, the security of fruit yields has been improved successfully as well as the performance of fruit varieties revealed. During the 1970s, couple of publications dealt with the relationship of growth and soluble solids with weather conditions. In the Studies about the prediction of the time of maturity by meteorological parameters for a certain variety of apple 'Jonathan' was the main stream of the research activity by Stollár [10]. In the 1980s, the meteorological characterisation of growing sites and their effects on different fruit varieties took place by Stollár [9].

1.1.1 Significance of Plant Phenology Studies

The importance of plant phenology and fruit phenology studies is that they can show us the effects weather have on the plant and fruit trees expressively. Of course, through ongoing measurement and analysis of weather conditions the fact of climate change and its scale can be detected, but we can not answer what the change will lead to, what it will change in the state of our environment. At phenological times the weather character appears in a complex way, and so as the effect of climate change in the long run. As a result, flowering dates or the times of ripening shift a few days earlier. Some cultivars and varieties of fruit are more sensitive to changes in environment, while others are less so. The occurrence of phenological phases have strong dependance on the species. There are early, middle and late flowering varieties that have different flowering dynamics. Generally we can not declare that the early flowering varieties can necessarily be characterized with a longer flowering period than late-flowering varieties. The weather during flowering period can significantly influence the duration of flowering. It is especially advantageous in phenological studies to have several varieties available, as in this case, so the results can be widely used and generalized. It is well known that the increase in temperature accelerates the development of the plant, and due to the drought we can also expect the earlier occurrence of certain phenological phases (of course, if in the meantime the plant is not destroyed). When higher than average temperatures and drought simultaneously occur it considerably speeds up the plant physiological processes. However, the cooler than optimal weather slows the times of occurrence of phenological phases. In case of cool and wet weather during the flowering dates may occur up to 10 days later, and the flowering time can be prolonged for days. The so-called cooling irrigation the flowering time can be delayed for up to two weeks. Looking at the future database produced by the climate models the occurrence of phenological phases dates, we can get feedback on the nature of the long-term processes, with which we can calculate risk for decades ahead.

1.1.2 Changes in Spring Frost Occurrences

At the occurrence of spring frost we can state that during the researched period the temperature was below −17.4 °C (Table 1). On the basis of a1b scenario we can say that between 2031 and 2060 the late spring frost risk will disappear from the apple cultivation. During the late future period both a1b and a2 scenario we can expect slight frost again during the blooming time of apple cultivation. Most severe frost was −6.2 °C during the blooming time between the period 1951–1980. The average of frost days were just 2 days in April between the period 1981–2010 which will disappear from 2031. If the beginning of blooming time of apple varieties start earlier in the future because of global warming, we have to take the March frost occurrences into consideration. In March the lowest temperature was −17.4 °C in the period of 1951–1980. We used the "frost days" as a climatic category, which is calculated from

Table 1 Absolute minima and frequency of frost days' occurrences (Újfehértó 1951–2100)

	1951—1980	1981—2010	2031—2060 a1b scenario	2071—2100 a1b scenario	2071—2100 a2 scenario
Spring frost days number	18	14	6	3	6
Autumn frost days number	16	15	5	2	6
Winter frost days number	68	64	45	34	44
March frost days number	15	12	6	3	6
April frost days number	3	2	0	0	0
May frost days number	0	0	0	0	0
Spring absolute min	-17,4	-15,4	-5,5	-1,6	-3,8
Autumn absolute min	-13,4	-17,0	-4,0	-2,3	-3,0
Winter absolute min	-27,6	-25,5	-11,4	-10,0	-7,9
March absolute min	-17,4	-15,4	-19,5	-7,8	-6,8
April absolute min	-6,2	-3,8	0,3	-0,5	-1,3

the daily minimum temperature. If the minimum temperature is less or equivalent with 0 °C we call that day a "frost day".

If we analyse the "frost days' number on the basis of two climate change scenarios (a1b, b2) we can see that we can expect 3 or 6 days frost occurrence between the periods of 2031–2060 and 2071–2100. So the frost risk of apple cultivation will remain in the future.

1.1.3 Start of Blooming

The temporal distribution of the start of blooming in the population of 586 apple cultivars which started blooming between 30th March and 26th June (between 90 and 147 days from 1st January) over the 33 years observation. The histogram is a normal distribution which means that earlier and later blooming cultivars are the same rate in the sample. Approximately half of the cultivars examined (46%) started blooming between 23rd and 30th April (between 114 and 121 days from 1st January). After 9th May and before 30th March blooming occurs very rarely. Considering the ripening dates, i.e. the elapsed until fruit maturity, it turned out that the start of blooming is not necessarily associated with the time of maturity (Fig. 2).

However a slight tendency of covariation is detected in the interval, namely between 98th and 105th day, i.e. April 7–14. The early ripening cultivars represent earlier blooming at a rate of 2–3% more than late blooming one.

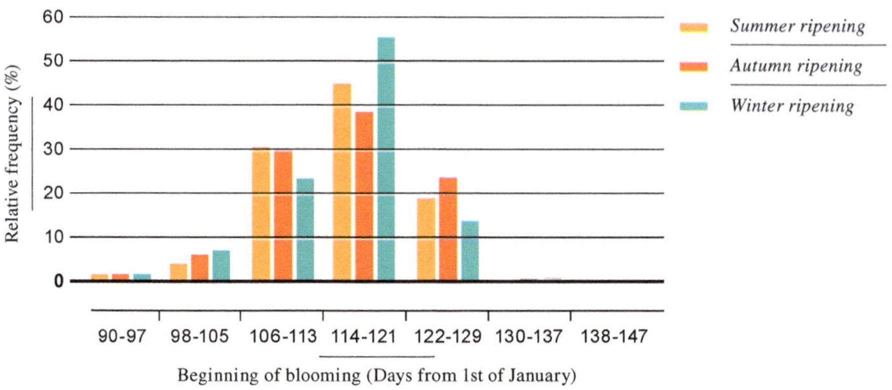

Fig. 2 Distribution of the beginning of blooming at different maturity groups apple cultivars (1984–2016)

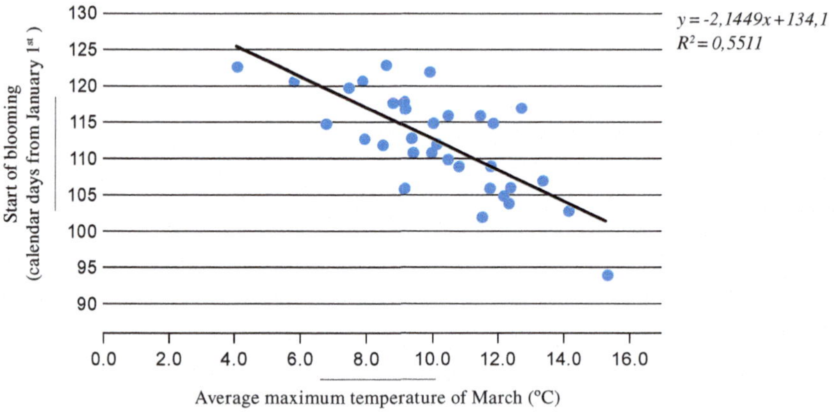

Fig. 3 Relationship between start of blooming of Idared apple cultivar and average maximum temperature of March (Újfehértó 1984–2016)

It is not surprising because at later flowering the temperature is generally higher than earlier. This is the closest correlation we have found between March maximum temperature and the start of blooming. you can see the linear regression equation in Fig. 3.

We can state that at higher spring temperature the beginning of blooming occurs in an earlier time. On the basis of predicted data base (a1b, a2 scenarios) we can state that on next 80 years the average maximum of March will be increasing significantly (Fig. 4). It means that at the end of this century the beginning of blooming time of main cultivated apple varieties will shift 9–10 day earlier than nowadays.

In a mild spring, blooming starts earlier. The large differences between daytime and night time temperatures stimulate the start of blooming. The large amplitude of temperature changes is associated with high daytime maxima during spring. In the

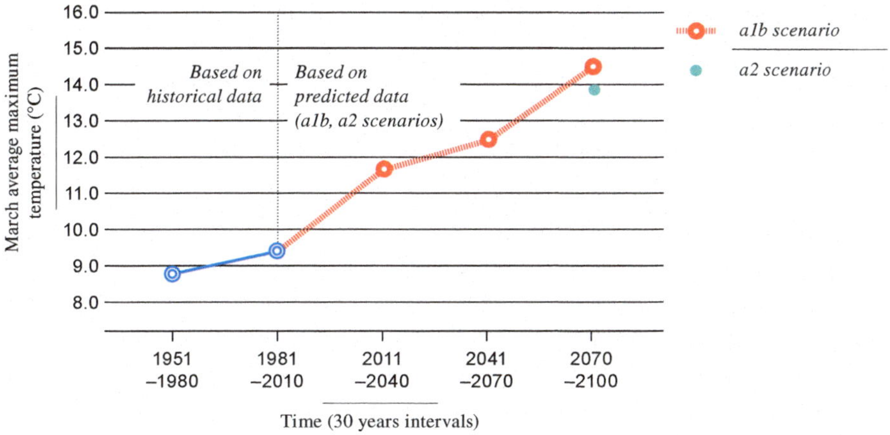

Fig. 4 Time series of historical and predicted average maximum temperature of March (Újfehértó 1951–2100)

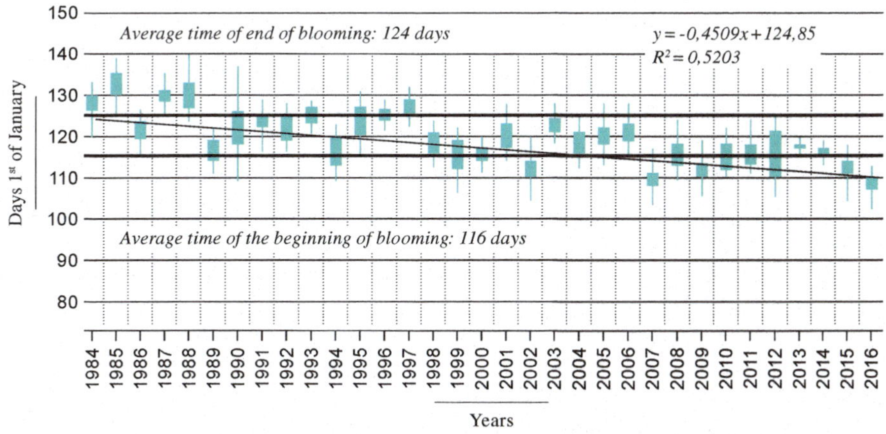

Fig. 5 Complex blooming time features of most popular Hungarian apple cultivars (Újfehértó 1984–2016)

other cultivar groups too, the two mentioned meteorological variables are decisive in influencing the start of bloom. In addition, cultivars of intermediate blooming date are significantly influenced by the minimum temperatures of spring. At the most popular Hungarian varieties as (Starking, Golden R., Jonagold, Idared) we found intensive decreasing tendency at all the blooming characteristics (as beginning of blooming, duration of blooming, end of blooming) of apple cultivars. The beginning of blooming shifted on earlier time (Fig. 5).

1.1.4 Significance of Plant Phenometry Studies

The phenometry, fruitphenometry studies are essential in quality cultivation. In the case of a lot of fruit today the most important tasks of the producers is not to increase the quantity but to achieve and maintain a certain level of the quality indicators. The state of measurable properties of fruit is of importance to fruit producers because of saleability. If you know under what weather conditions a fruit size reaches the optimum that is typical of the variety, or the fruit lid color in what day and night temperature difference will be favorable, then we have the opportunity to improve the quality of the fruit in case of unfavorable weather conditions, by water supplement irrigation or cooling from irrigation. If the microclimate modification options are not available, by examining the climatic conditions of the production site favorable vintages for fruit quality can be specified. Through statistical analysis past and future risk of quality production can be expressed numerically. Fruit phenometry studies can reduce risk of production for producers and improve variety selection in that area.

The size of the fruit is influenced significantly by meteorological effects during the growing period. The conditions around the growing site: exposition, soil, climatic and microclimatic moments are individually different also regarding to the phenophase of the same plant. Not only the meteorological elements could differ but also their appearance and distribution of their effect expressed in the parameters and expression of phenometry of fruit development.

The relation of developmental phases and meteorological elements should always be regarded as a complex reaction of many factors representing the whole physical environment of the climate. Thus effects of a unique factor are always combined with the other moments of the climate. It is not recommended to evaluate the effects of a single meteorological component regardless of the complex background, as most of phenomena are the results of interactions. For example, at low temperatures, the requirement for water is low (the equivalent of water deficiency is low). At high temperatures, all equivalents of precipitation increase substantially. The optimum is included as well because evaporation from the soil and from the leaves increases the water requirement, not only the transpiration.

Among the phenometrical traits, the length of fruit is one of the most important parameter. Mainly the sum of daily maximum temperatures determines the length of fruits. The increase in temperature lengthens the form of fruits.

The width of the fruits is influenced by the sum of daily temperature maxima. In summer-ripe varieties the correlation is significant. Higher sum of temperature maxima increases the width of fruits. In contrast the autumn- and winter-ripe apple varieties, showed no significant correlation between the width of fruits and the daily mean maxima.

The thickness of fruits is also influenced significantly by the daily mean maximum temperatures. Fruit thickness is dependent on the amount of precipitation of the period before maturity. In autumn-ripe varieties, the sum of precipitation increases the thickness of fruits. Interestingly, small doses have more remarkable effects than additional rainfalls according to the law of diminishing efficacy [6].

1.1.5 Conclusions

The results presented above prove that the dynamics of weather variables exert measurable effects on the development of fruits. There is a significant correlation between the phenology and phenometric values and meteorological parameters as temperature and precipitation. After analysing the main weather characteristics in many cultivating areas, we can determine the climatic risk of fruit growing of different species. These results help us to work out the defence strategies. Conventional growing technologies need to be updated. These methods serve the rise of ecological tolerance of fruit species, thus avoiding the extreme weather influences, and its damaging effect. As a future goal, we would pay more attention to the time interval between bloom and maturity, where several periods are to be distinguished as it is critical to improve the qualitative and quantitative characteristics of fruits.

References

1. Berényi, D., Justyák, J.: Fenológiai felvételezés hegyvidéki szőlőállományban. Időjárás **2**, 104–111 (1956)
2. Bognár, K.: Kozma F (1961) Együttes szőlő-gyümölcstermesztés mikrometeorológiai vizsgálatáról. Időjárás **6**, 366–369 (1961)
3. Csöbönyei, I.: Stollár A (1969) Az alma rügyfakadása és a rügyfakadás-virágzás fenofázis összefüggése a léghőmérséklettel. Kisérletügyi Közlemények Kertészet **1–3**, 19–23 (1969)
4. Holdefleiss, P.: Agrometeorologie, p. 234. Berlin (1930)
5. Lakatos, L.: Climatic model for dry matter production of winter-wheat in Hungary. Agric. Forest Meteorol. **83**(3/4), 231–246 (1997)
6. Lakatos, L., Szabó, T., Zhongfu, S., Wang, Y., Racskó, J., Szabó, Z., Soltész, M., Nyéki, J.: Relationship between several meteorological features of apple cultivars. Int. J. Hortic. Sci. **14**(1–2), 13–19 (2008). (Agroinform Publishing House, Budapest, Printed in Hungary, ISSSN 1585-0404)
7. Nyéki, J.: A gyümölcstermő növények virágzása, megporzása és termékenyülése. In: Gyuró, F. Gyümölcstermesztés. Mezőgazdasági Kiadó, pp. 61–90. Budapest (1990)
8. Soltész, M. (ed.): Integrált gyümölcstermesztés. Mezőgazda Kiadó, p. 843. Budapest (1997)
9. Stollár, A.: A gyümölcstermesztés agrometeorológiai vonatkozásai a Duna-Tisza közén. Légkör **4**, 8–10 (1984)
10. Stollár, A.: A meteorológia elemek hatása a jonathán alma érésére. OMSZ. Beszámolók az 1977-ben végzett tud. kut.-ról. 214–219 (1977)
11. Szabó, Z.: A kedvezőtlen meteorológiai hatások mérséklése. In: Soltész, M., Integrált gyümölcstermesztés. Mezőgazda Kiadó, pp. 353–359. Budapest (1997)

Application of UAVs in Precision Agriculture

Gábor Milics

1 UAVs in Agriculture

Application of UAVs in agriculture started in the early 21st century. In the 2000s satellite images provided information for the overview about a field with a relatively bad spatial resolution, low frequency in time and long time between observation and usage of the images. Aerial imaginary from airplanes or balloons as the next step provided better resolutions in space and time; however the cost of the data collection was too high for agricultural applications only. Within ten years, usage of UAVs became a daily routine for the most advanced farmers or rather the most advanced professional advisors. Accessibility of the aerial images nowadays are immediate, even for large areas, mosaicked images are available within a reasonable timeframe.

1.1 Types of UAVs

When talking about UAVs we have to separate at least three different types: Multi-Rotor, Fixed-Wing and Single-Rotor (Helicopter) types. There are several developments trying to integrate the benefits for each type, which are called hybrids; however due to their complexity they are not commercialized yet. Each type of vehicle has its pros and contras especially concerning take-off and landing requirements, area coverage, flight time, ease of use, accessibility, price, payload capability, etc.

G. Milics (✉)
Széchenyi István University, Győr, Mosonmagyaróvár, Hungary
e-mail: milics.gabor@sze.hu

M. Palocz-Andresen et al. (eds.), *International Climate Protection*,
https://doi.org/10.1007/978-3-030-03816-8_13

1.2 Sensor Availability for UAVs

The primary function of the UAVs in agriculture is to collect high quality images, therefore sensors are required to be mounted. The sensors have to be chosen carefully, depending on the weight, type of captured image, optical quality, resolution, or price. The most common sensors are available in the visible scale range (with relatively high resolution), however for some applications other ranges of the electromagnetic spectrum such as infrared of near infrared has to be detected as well.

1.2.1 Visible Light Sensors (Red-Green-Blue, RGB Sensors)

Visible light sensors capture the red, green and blue channels of the visible light, which provides the color image that is usual for the human eye. Some basic requirements are there for the RGB sensors. Unlike most small size action cameras UAV sensors have to have the minimum distortion, high quality optics (preferably with fixed lens) in order to perform the required images. Mosaicking the many collected image provide an overview picture about the investigated area.

1.2.2 Near Infrared Sensors (NIR Sensors)

The Near Infrared Sensors collect information in the range where human eyes are less sensitive. In agriculture this electromagnetic range (from 720 to 1,000 nm) is highly interesting as this is the range for collecting information about vegetation. Various indices are performed using the NIR and RED bands, sometimes even using the red edge (around 680–730 nm) which is a narrow range of NIR in order to collect information about vegetation stress, chlorophyll content, etc.

1.2.3 Infrared Sensors (IR Sensors)

Infrared sensors provide information of the electromagnetic spectrum which is invisible for the human eye; however infrared range can be very useful to obtain further information about vegetation. In case we choose to have a radiometric camera, we are able to measure accurate temperature. This is valuable information. In the practice sometimes a relative temperature difference helps in mapping, therefore non-radiometric cameras are used. Images are visualized with false colors. Thermal imaging is more widespread in industrial applications, however water management and animal management also collects information by this means.

1.2.4 Multispectral Sensors

Multispectral sensors have multiple channels and collecting data in different range or specific wavelength of the electromagnetic spectrum. The most common cameras include the visible RGB plus the NIR channels, some in a wider spectrum, some only in specific wavelengths. The difference in the multispectral cameras compared to the earlier mentioned ones is that data collection happens in different channels, unlike for instance RGB cameras where true color pictures have to be split into three different channels order to receive the separate R, G and B bands.

1.2.5 Hyperspectral Sensors

Hyperspectral sensors are collecting information in many separate wavelengths that are providing the spectral signature of an investigated object. These sensors can cover the electromagnetic spectrum from 350 to 2,500 nm, up to over 350 different wavelengths, therefore these multiply images require further knowledge about analysis. Hyperspectral sensors in UAVs are very expensive, and mostly are used for scientific research.

1.3 Application of UAVs in Agricultural Practice

In agriculture UAV technology and applicability is still in research stage. Some users—especially up-to-date farmers—are already applying the technology in everyday practice, but majority of the farmers do not see the possibilities the UAV based remote sensing provides. The potentials are there in a very wide range, here are some examples how UAV technology can be utilized in /precision/ agriculture practice.

1.3.1 Farm Overview, Checking the State of the Field

One interesting possible application out of many of UAVs is farm or field overviews or mid-season vegetation health monitoring. In a particular field in order to check the status of the plants with a relatively cheap UAV, a true color image can provide sufficient information.

1.3.2 Mapping Inland Water

Inland water causes loss in yield. In a true color image, inland water clearly shows the boundary of the area where the soil is saturated, therefore survival for vegetation (seeds) is impossible. Inland water appears only in particular time of the year, however the areas limited by this phenomena remains the same or very similar in long term.

Fig. 1 Observation of missing vegetation due to inland water at early stage in maize

Therefore farm management has to think about some kind of solution for meliorate the area or an alternative utilization (Fig. 1).

1.3.3 Mapping Vegetation Status and Wildlife Damage

Vegetation status mapping requires near infrared imaging. In some cases within field differences can be visualized by true color images such as greenness, but reliable information on vegetation status is provided by the near infrared (NIR) channel. In most cases NIR alone does not show the stress of the plants, therefore combined NIR and red images are used to visualize the differences. These combinations vary depending on the available bandwidth or available channels. The calculations provide the so called vegetation indexes. There are several indexes that show different vegetation status (such as greenness, nitrogen level, moisture content in the plant, stress, etc.). In-field measurements are also available for instance for variable rate head fertilizing, however in this case the sensors are measuring NIR and red reflectance and calculating the Normalized Differential Vegetation Index for the visible range of the field. Applying UAVs, in a relative short time period, NDVI index is available for the full field, therefore fertilizing can be planned, and valuable data can be saved at the same time.

Wildlife damage can also be detected by UAVs due to the resolution the sensors provide. There are typical patterns that are detectable in an image which are "drawn"

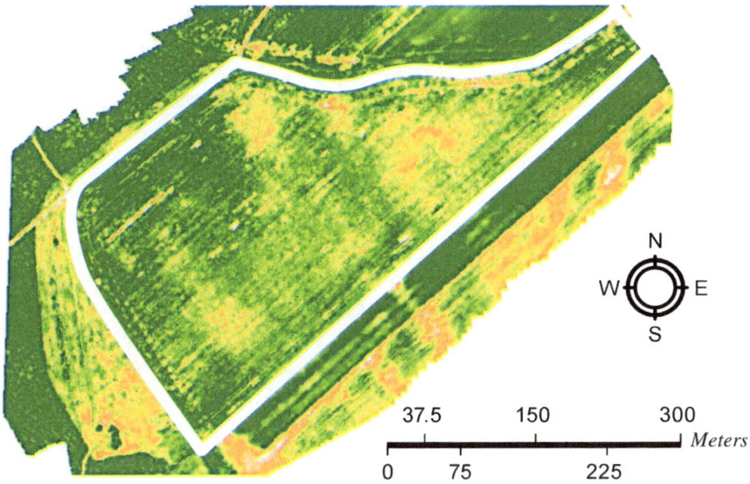

Fig. 2 NDVI index showing different ripening status and missing vegetation in maize

by the animals. Analysis of the images when looking for wildlife damage requires further knowledge about the habits of the animals, or expert field investigations have to be carried out in order to avoid misinterpretation of the missing vegetation (Fig. 2).

1.3.4 Weed Mapping

Weed mapping and applying herbicides are also possible by UAVs. With satellite or near field imaging systems weed mapping was not possible due to the insufficient resolution. UAV sensors create fine resolution pictures, where individual plants can be detected, therefore weeds can be observed. UAV based sprayers are available in the market, so weed management is also possible by applying this technology.

Striving Towards Abiotic Stresses: Role of the Plant CDPK Superfamily Members

Abu Imran Baba, Gábor Rigó, Norbert Andrási, Olaf Tietz, Klaus Palme, László Szabados and Ágnes Cséplő

1 Impact of Auxin in Root Growth and Abiotic Stress Responses with Special Emphasis on Gravitropism

Plant root development is highly responsive to changing environmental conditions. Root growth and differentiation in plants are controlled by a number of regulatory genes including transcription factors, protein kinases and transporters controlling local hormone content [1]. Root development is regulated by various plant hormones, out of which auxin is considered to be the master regulator [2, 3]. Auxin is involved in virtually every aspect of plant growth and development such as embryogenesis, organogenesis, tissue patterning; it controls various stages of root development, root elongation, architecture and tropisms [4]. Besides auxin, several other phytohormones such as cytokinin, brassinosteroids, abscisic acid (ABA) and gibberellin have also been implicated in root development. These hormones modulate root development by interacting with auxin signals, altering auxin biosynthesis and transport [5].

Extreme environmental conditions such as drought, high soil salinity or nutrient deprivation and toxicity cause alterations in phenotypic and molecular responses in roots. Despite extensive research on regulatory pathways controlling responses to environmental stimuli, little is known about the way how these stimuli affect root development. For example, salinity affects root growth in all developmental zones: cell division is suppressed and cell expansion is attenuated, lateral root formation

A. I. Baba · G. Rigó · N. Andrási · L. Szabados · Á. Cséplő (✉)
Biological Research Centre, Hungarian Academy of Sciences, Szeged, Hungary
e-mail: cseplo.agnes@brc.mta.hu

G. Rigó
Department of Plant Biology, University of Szeged, Szeged, Hungary

O. Tietz · K. Palme
Faculty of Biologie II, Albert-Ludwigs Universität, Freiburg, Germany

M. Palocz-Andresen et al. (eds.), *International Climate Protection*,
https://doi.org/10.1007/978-3-030-03816-8_14

is reduced, root hair outgrowth is suppressed and agravitropic growth emerges [6, 7]. All these responses appear to be mediated by changes in auxin distribution and are influenced by ABA and ethylene [6, 8]. Root architecture was recently shown to affect drought avoidance in crops [9]. Moreover, adaptation to salinity in *Arabidopsis* may be mediated in part by an auxin/redox interaction [10, 11].

Alterations in environmental signals such as gravity, light, water and touch stimulate changes in the direction of roots and shoots. The gravity vector change is perceived by specific starch-containing statocytes in root columella and stem endodermis cells. The molecular mechanisms of gravity response and their connection with auxin signaling are still largely unknown [12, 13]. However, gravitropic responses, directing downward and upward bending of horizontally placed roots and shoots, respectively, are known to be controlled by the asymmetric distribution of the plant hormone auxin [13, 14]. In response to an altered gravity stimulus, auxin is transported from the upper to the lower sections of bending organs, stimulating differential cell elongation [15]. Cellular transport of auxin is controlled by the AUX/LAX influx [16] and PIN-FORMED (PIN) efflux carriers [17–19], and the PGP/ABCB (P-glycoprotein/ATP binding cassette protein subfamily B) transporters [15]. The agravitropism is due to repression of transcription and localization of the PIN2 auxin efflux transport protein, resulting in reduced shootward auxin reflux in the lateral root cap and in epidermal cells [6]. Protein phosphorylation is instrumental in early signaling events [15, 20, 21]. In addition to auxin, other signals like Ca^{2+}, 1,4,5-trisphosphate (InsP3), nitric oxide (NO) and reactive oxygen species (ROS) are also involved in root gravitropism [13]. The movement of the gravity-sensing statocytes along the gravity vector stimulates various plasma membrane-localized mechanoreceptors, further activating several signaling pathways including secondary messengers such as Ca^{2+} ions, the Ca^{2+} and calmodulin (CaM) sensing mechanism and the consequent phosphorylation cascade [15]. In plants, calcium-dependent protein kinases (CDPKs) are the main players in Ca^{2+} signaling, and CPDKs and the Ca^{2+} calmodulin-dependent kinase-related kinases (CRKs) have regulatory functions in many diverse processes like plant growth and development as well as abiotic and biotic stress responses [8, 22, 23]. Various abiotic stresses such as heat, cold, drought, salt/osmotic stress and mechanical stimuli involving touch, wind, ozone, hypoxia and oxidative stress rapidly lead to elevation of the cellular Ca^{2+} level. Biotic stresses from various pathogens and insect feeding can also cause changes in cellular Ca^{2+} levels [8].

2 Distribution and Function of the CDPK Superfamily

In plants, the following four families of serine/threonine kinases are able to decode Ca^{2+} signals as a response to different environmental effects: the Ca-dependent protein kinases (CDPKs), the CDPK-related protein kinases (CRKs), the Ca^{2+} and calmodulin-activated kinases (CCaMKs) and the sucrose non-fermenting1-related kinases3 (SnRK3s) [23, 24]. The size of kinase families broadly varies in different plant species. For example, the *Arabidopsis thaliana* genome analysis revealed

34 CDPKs, 8 CRKs and 38 SnRKs [25, 26]. Amongst these, the largest and best characterized kinase family is the CDPK family [23]. The CDPKs are the most important receptors in Ca^{2+} signalling [8, 22]. They play a prominent role in the regulation of plant growth and development and are widely involved in abiotic/biotic stress responses, in pollen tube growth and in cell death [23, 27]. The CDPK family is important not only in higher plants but also in unicellular protists and green algae [28]. CDPKs contain a variable N-terminal domain and also possess a specific N-myristoylation site important for their subcellular localization [23]. Some *Arabidopsis* SnRK members are activated by osmotic and salinity stresses [29].

The CDPK-related kinase CRK family is closely related to CDPKs and consists of eight members in *Arabidopsis* [26]. Unlike CDPKs and SnRKs, limited information is available about the functional role of plant CRKs in vivo in *Arabidopsis* plants. For a decade, only predictions of their membrane localization have been published [30, 31]. The substrate specificities of *Arabidopsis* CRKs are not well known till now. However, exceptions exist, e.g. AtCRK1, which binds CaM in a Ca^{2+}-dependent manner but phosphorylates itself and its substrates in a Ca^{2+}-independent manner [32], and positively regulates tolerance to heat and salt stresses [33, 34]. AtCRK3 is another protein kinase which is known to interact with a cytosolic glutamine synthetase AtGLN1;1 involved in leaf senescence [35]. AtCRK5 was first shown to have a direct role in the regulation of root gravitropic response and polar auxin transport (PAT) by phosphorylating the PIN2 auxin transport protein [20]. Recently, it was reported that most AtCRK genes influence root growth, root and hypocotyl gravitropism, while AtCRK1—additionally its above described functions—is implicated in responses to continuous light and cellular redox homeostasis [36]. The tomato LeCRK1 was described to participate in fruit ripening [37]. Recently, a more detailed description of the whole tomato CRK protein kinase family consisting of six members was reported, and the first tomato CRK gene (SlCRK6) carrying disease resistance against plant pathogen was published [22]. N-myristoylation sites predicting membrane localization have also been described for all of the CRK family members in *Arabidopsis* and tomato [20, 22, 23].

2.1 *Regulatory Role of an* **Arabidopsis thaliana** *CRK Protein Kinase in Root Growth, Gravitropic Response and Nutrient Uptake*

Roots are essential for growth and development of plants performing various functions, e.g. in addition to their storage, aeration and transfer functions, they perform anchorage, support the plant body and absorb water, oxygen and nutrients from the soil. Ameliorating the root architecture and the subsequent root response may improve aerial plant development under abiotic stresses. This may convert into enhanced yield [38]. Various plant hormones including auxin influence root development. The main function of auxin is to help plants to grow. Auxin located in the apical

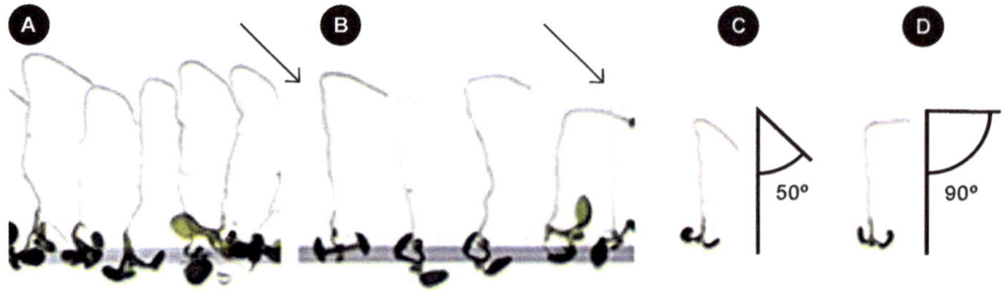

Fig. 1 Comparison of the root gravitropic responses of wild type Col-0 (**a**) and the knock-out At*crk5-1* mutant (**b**) in *Arabidopsis thaliana* (note delayed gravitropic response of the At*crk5-1* mutant). Root bending is indicated at 24 h after rotation of the 7 days old vertically grown seedlings by 135°. The gravity direction is shown by black arrows. Measurement of root angles were carried out according to (**c**) for wild type and (**d**) for the mutant seedlings using ImageJ

meristems of plant roots or shoots stimulates plant cells to elongate. Upon gravitropic response of roots, auxin elongates the cells of the upper section of the root turning down into the soil, resulting in plants with a strong underground support system and absorbing nutrients from the ground. The molecular mechanism of the graviresponse of an *Arabidopsis* CRK protein kinase is in detail described in Rigó et al. [20] supporting the fundamental role of protein kinases in abiotic stress responses [8, 22, 23]. The CRK5 protein kinase is a member of the *Arabidopsis thaliana* Ca^{2+}/calmodulin-dependent kinase-related kinase family. Detailed functional analysis of this CRK5 protein kinase revealed that inactivation of the protein kinase increases the number of knock-out At*crk5-1* mutant lateral roots and inhibits normal graviresponse of roots (Fig. 1).

Impaired gravitropic response is a consequence of altered auxin distribution in the At*crk5-1* mutant's root tips as compared to the wild type. Immunolocalization pattern of the auxin efflux protein PIN2—which is a key facilitator of basipetal auxin transport upon *Arabidopsis* root graviresponse—exhibited a considerable alteration in the At*crk5*-1 mutant comparing to the wild type. The AtCRK5 protein kinase is able to phosphorylate the PIN2 auxin efflux protein in vitro [20]. Phosphorylation of PIN2 in roots of the At*crk5-1* mutant is impaired, which further hinders and ultimately delays auxin transport. Thus, a direct functional correlation was found amongst asymmetrical auxin distribution, impaired PIN2 phosphorylation and gravitropic response delay in the roots of the At*crk5*-1 mutant. Intracellular localization of the green fluorescence protein (GFP) tagged CRK5-GFP fusion protein demonstrated mostly U-shaped plasma membrane distribution of this protein (Fig. 2), a localization pattern that is indicative of proteins involved in nutrient uptake, as it was observed e.g. for boron transporters in *Arabidopsis* [39, 40]. This specific localization pattern may predict a novel role for AtCRK5 protein kinase in microelement uptake as well as in water transport regulation.

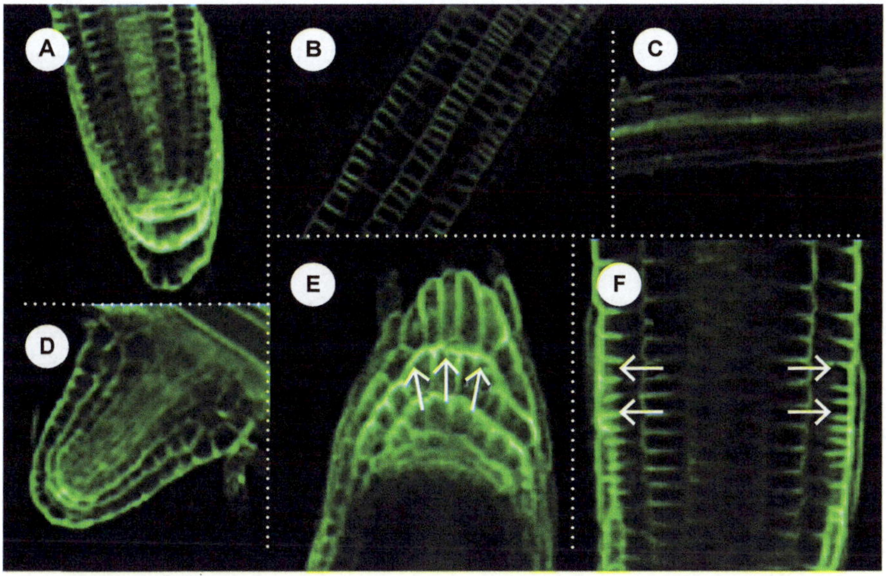

Fig. 2 Plasma membrane localization of AtCRK5-GFP in various parts of *Arabidopsis thaliana* in 7 days old seedlings visualized by laser scanning microscopy (LSM). **a** Root cap, **b** root meristematic zone, **c** root stele region, **d** young lateral root. Bars = 50 μm in (**a**), 25 μm in (**b**), 100 μm in (**c**) and 25 μm in (**d**). U-shaped plasma membrane localization pattern of AtCRK5-GFP in the main root cap (**e**) and lateral root cap cells (**f**) indicated by arrows. Bars = 25 μm in (**e**) and (**f**)

3 Conclusions

Continuous climate fluctuations as a realistic tendency have severe effects on plant ecosystems, causing a reduction in plant productivity [8]. The latter is influenced by many environmental factors like abiotic and biotic stresses. Therefore, it is very important to increase the abiotic stress adaptability of plants. Main players of decoding the environmentally triggered Ca^{2+} signals are calcium-dependent protein kinases (CDPKs) in plants. The members of this superfamily have been identified as participants of many diverse regulatory actions involving all major developmental processes in plants as well as coping with stresses [8, 23]. Since approximately 3% of the proteome is involved in Ca^{2+} signaling in the model plant *Arabidopsis thaliana*, it seems that plants prefer Ca^{2+} over other signals in regulating their stress responses [8]. Stress-induced responses of plants always lead to transcriptional reprogramming of genes involving the action of several transcription factors (TFs) with multiple functions. However, functions of only a few Ca^{2+}- and Ca^{2+}/CaM binding TFs have been clarified so far [8], e.g. AtWRKY8—which is able to bind CaM in a Ca^{2+}-dependent manner—is shown to be a positive regulator of salt tolerance and disease resistance as well [23]. Full understanding of CDPK superfamily participation in stress responses necessitates further functional studies with other Ca^{2+} and Ca^{2+}/CaM regulated CDPKs not only in model species [23] but also in other crop plants [22]. It is also important to puzzle out the role of the Ca^{2+}-regulated transcriptional networks in single and combined abiotic stresses. Identification of more Ca^{2+}/CaM-regulated TFs which regulate a set of several abiotic responses genes could be especially useful for the production of plants carrying multiple stress tolerance [8]. We hope that in-depth understanding of the functional role of the model plant *Arabidopsis thaliana* CDPK superfamily members in abiotic stress responses will aid in developing novel crop varieties with enhanced tolerance to several environmental stresses.

Acknowledgements This work was supported by the National Research, Development and Innovation Fund of the Hungarian Government via Hungarian-German TÉT_12_DE-1-2013-0015 (A.I.B., Á.Cs., G.R., N.A., L.Sz., O.T., K.P), by the Tempus Public Foundation, Hungary and the Biological Doctoral School, University of Szeged, Hungary (A. I. B.), by Hungarian Ministry for National Economy GINOP-2.3.2-15-2016-00001 (A.I.B., Á.Cs., G.R., N.A., L.Sz.) and PD OTKA Grant No. 115502 and No. PD128055 (G.R.).

References

1. Petricka, J.J., et al.: Control of *Arabidopsis* root development. Ann. Rev. Plant Biol. **63**, 563–590 (2012)
2. Saini, S., et al.: Auxin: a master regulator in plant root development. Plant Cell Rep. **32**, 741–757 (2013)
3. Ljung, K.: Auxin metabolism and homeostasis during plant development. Development **140**, 943–950 (2013)
4. Sauer, M., Robert, S.: Auxin: simply complicated. J. Exp. Bot. **64**, 2565–2577 (2013)
5. De Smet, S., et al.: Gene networks involved in hormonal control of root development in *Arabidopsis thaliana*: a framework for studying its disturbance by metal stress. Int. J. Mol. Sci. **16**, 19195–19224 (2015)
6. Sun, F., et al.: Salt modulates gravity signaling pathway to regulate growth direction of primary roots in *Arabidopsis*. Plant Physiol. **146**, 178–188 (2008)
7. Wang, Y., et al.: Auxin redistribution modulates plastic development of root system architecture under salt stress in *Arabidopsis thaliana*. J. Plant Physiol. **166**, 1637–1645 (2009)
8. Reddy, A.S.N., et al.: Coping with stresses: roles of calcium- and calcium/calmodulin-regulated gene expression. Plant Cell **23**(6), 2010–2032 (2011)
9. Uga, Y., et al.: Control of root system architecture by DEEPER ROOTING 1 increases rice yield under drought conditions. Nat. Genet. **45**(9), 1097–1102 (2013)
10. Miller, G., et al.: Reactive oxygen species homeostasis and signalling during drought and salinity stresses. Plant Cell Environ. **33**(4), 453–467 (2010)
11. Iglesias, M.J., et al.: Auxin signaling participates in the adaptive response against oxidative stress and salinity by interacting with redox metabolism in *Arabidopsis*. Plant Mol. Biol. **74**, 215–222 (2010)
12. Hashiguchi, Y., et al.: Mechanism of higher plant gravity sensing. Am. J. Bot. **100**, 91–100 (2013)
13. Sato, E.M., et al.: New insights into root gravitropic signalling. J. Exp. Bot. **66**(8), 2155–2165 (2015)
14. Blancaflor, E.B.: Regulation of plant gravity sensing and signaling by the actin cytoskeleton. Am. J. Bot. **100**, 143–152 (2013)
15. Baldwin, K.L., et al.: Gravity sensing and signal transduction in vascular plant primary roots. Am. J. Bot. **100**, 126–142 (2013)

16. Swarup, R., Péret, B.: AUX/LAX family of auxin influx carriers—an overview. Front. Plant Sci. **3**(Article 225) (2012)
17. Chen, M.K., et al.: ERECTA family genes regulate auxin transport in the shoot apical meristem and forming leaf primordia. Plant Physiol. **162**, 1978–1991 (2013)
18. Lusching, C., Vert, G.: The dynamics of plant plasma membrane proteins: PINs and beyond. Development **141**, 2924–2938 (2014)
19. Armengot, L., et al.: Regulation of polar auxin transport by protein and lipid kinases. J. Exp. Bot. **67**(14), 4015–4037 (2016). https://doi.org/10.1093/jxb/erw216
20. Rigó, G., et al.: Inactivation of plasma membrane-localized CDPK-RELATED KINASE5 decelerates PIN2 exocytosis and root gravitropic response in *Arabidopsis*. Plant Cell **25**, 1592–1608 (2013)
21. Nemoto, K., et al.: Members of the plant CRK superfamily are capable of trans,- and autophophorylation of tyrosin residues. J. Biol. Chem. **290**(27), 16665–16677 (2015)
22. Wang, J.P., et al.: Calcium dependent protein kinase (CDPK) and CDPK related kinase (CRK) gene families in tomato: genome wide identification and functional analyses in disease resistance. Mol. Genet. Genomics **291**(2), 661–676 (2016)
23. Simeunovic, A., et al.: Know where your clients are: subcellular localization and targets of calcium-dependent protein kinases. J. Exp. Bot. **67**(13), 3855–3872 (2016). https://doi.org/10.1093/jxb/erw157
24. Harper, J.F., et al.: Decoding $Ca^{(2+)}$ signals through plant protein kinases. Ann. Rev. Plant Biol. **55**, 263–288 (2004)
25. The *Arabidopsis* Genome Initiative. Analysis of the genome sequence of the flowering plant *Arabidopsis thaliana*. Nature **408**, 796–815 (2000)
26. Hrabak, E.M., et al.: The *Arabidopsis* CDPK-SnRK superfamily of protein kinases. Plant Physiol. **132**, 666–680 (2003)
27. Boudsocq, M., Sheen, J.: CDPKs in immune and stress signaling. Trends Plant Sci. **18**, 30–40 (2013)
28. Hamel, L.P., et al.: Ancient signals: comparative genomics of green plant CDPKs. Trends Plant Sci. **19**, 79–89 (2014)
29. Umezawa, T., et al.: SnRK2C, a SNF1-related protein kinase 2, improve drought tolerance by controlling stress-responsive gene expression in *Arabidopsis thaliana*. Proc. Natl. Acad. Sci. USA **101**, 17306–17311 (2004)
30. Podell, S., Gribskov, M.: Predicting N-terminal myristoylation sites in plant proteins. BMC Genom. **5**, 37–52 (2004)
31. Rigó, G., et al.: Suspension protoplasts as useful experimental tool to study localization of GFP-tagged proteins in *Arabidopsis thaliana*. Acta Biol. Szeged. **52**, 59–61 (2008)
32. Wang, Y., et al.: Characterization of a calmodulin-regulated Ca^{2+}-dependent-protein-kinase-related protein kinase, AtCRK1, from *Arabidopsis*. Biochem. J. **383**, 73–81 (2004)
33. Liu, H.T., et al.: The calmodulin-binding protein kinase 3 is a part of heat-shock signal transduction in *Arabidopsis thaliana*. Plant J. **55**, 760–773 (2008)
34. Tao, X.C., Lu, Y.T.: Loss of AtCRK1 gene function in *Arabidopsis thaliana* decreases tolerance to salt. J. Plant Biol. **56**, 306–314 (2013)
35. Li, R.J., et al.: *Arabidopsis* cytosolic glutamine synthetase AtGLN1; 1 is a potential substrate of AtCRK3 involved in leaf senescence. Biochem. Biophys. Res. Commun. **342**, 119–126 (2006)
36. Baba, A.I., et al.: Functional analysis of the *Arabidopsis thaliana* CDPK-related kinase family: AtCRK1 regulates responses to continuous light. Int. J. Mol. Sci. **19**, 1282–1303 (2018). https://doi.org/10.3390/ijms19051282
37. Leclercq, J., et al.: Molecular and biochemical characterization of LeCRK1, a ripening associated tomato CDPK-related kinase. J. Exp. Bot. **56**, 25–35 (2005)
38. Salopek-Sondi, B., et al.: Improvement of root architecture under abiotic stress through control of auxin homeostasis in *Arabidopsis* and *Brassica* crops. J. Endocytobiosis Cell Res. **26**, 100–111 (2015)
39. Miwa, K., et al.: Plants tolerant of high boron levels. Science **318**, 1417 (2007)
40. Takanoa, J., et al.: Polar localization and degradation of *Arabidopsis* boron transporters through distinct trafficking pathways. Proc. Natl. Acad. Sci. U.S.A. **107**, 5220–5225 (2010)

Adaptation of Methods and Technologies in Agriculture and Forestry, in Water Resources Economy, and Changes in Biosphere

Role of Agro-forestry in Adaptation to Climate Change

Andrea Vityi and Marie Gosme

1 Introduction

Agriculture provides mankind with a range of products and services: food, fuel, fibre, landscapes of cultural and natural values, sites for recreation, and habitats for wildlife. With appropriate management it can also provide ecological services such as regulation of nutrient cycle, water flow and quality, or compensate extreme climate effects such as droughts and floodings. Many people take these services for granted, but agriculture is highly impacted by climate change and adaptations are necessary to make agriculture more productive (to meet the increasing needs of an increasing world population), sustainable (in particular reducing pollution and land degradation) and resilient, in particular towards climate change. In order to reach these objectives, farmers, researchers and agricultural extensionists are working together to design and test innovative cropping systems. Among them, agroforestry seems to give promising results.

2 What Is Agroforestry?

Agroforestry (AF) systems include both traditional and modern land-use systems where trees are managed together with crops and/or animal production systems in agricultural settings, [1] to optimise the benefits from their ecological and economic

A. Vityi (✉)
Institute of Forest and Environmental Techniques, University of Sopron, Sopron, Hungary
e-mail: vityi.andrea@uni-sopron.hu

M. Gosme
INRA, Montpellier, France

M. Palocz-Andresen et al. (eds.), *International Climate Protection*,
https://doi.org/10.1007/978-3-030-03816-8_15

interactions. As a multi-purpose mixed and integrated system, agroforestry encompasses a wide variety of practices with different combinations of woody, herbaceous and animal components, adapted to local conditions. Thus, AF practices vary both spatially and temporally [2] and include a wide range of systems (Table 1). These examples show that agroforestry systems from long traditions and comprises traditional practices as well as new technologies. In recent years agroforestry has been evolving, both as a land use practice and as a science.

Table 1 Examples of AF practices in Europe [46]

Agroforestry practice	Brief description	Example
Silvopasture	Combining trees with forage and animal production. It comprises forest or woodland grazing, open forest trees, and grazed orchards	Montado with pigs and cattle grazing in Portugal [3]  Intensive grazing in a forest in USA to reduce the height and density of the understory [4]  Grazed orchards in Northern Ireland, UK [5]. This practice includes fruit orchards, olive groves and vineyards which are grazed 

(continued)

Table 1 (continued)

Agroforestry practice	Brief description	Example
Silvoarable	It comprises hedge-rows and wind-break systems (line belts providing shelter, shade, and natural boundaries to a crop and/or livestock production system) and alley cropping systems (widely spaced trees inter-cropped with annual or perennial crops)	Windbreak system in North Dakota, USA [6]  Bocage in Normandy, France [7]  Wheat and walnut alley cropping system in Restinclières, France (photo by C. Dupraz) 
Multipurpose trees	Trees providing fruit (chestnut, fruits etc.), fuelwood, timber, nutrient for crops and fodder (e.g. leaves from *Morus alba*, *Fraxinus excelsior*, *Betula alba* or *acorn*) among other services	Traditional farming systems on the foothills of the Himalayas with multipurpose trees [8] 

(continued)

Table 1 (continued)

Agroforestry practice	Brief description	Example
Improved Fallow	Areas where during the fallow phase of shifting cultivation fast growing, preferably leguminous woody species are planted in order to improve soil fertility and to yield economic products	Improved fallow with Pigeon pea (*Cajanus cajan*) in Zambia [9] 
Riparian buffer strips	Strips of trees and shrubs separating croplands/pastures from water sources such as streams, lakes, wetlands, and ponds to protect water quality	Riparian buffer strips in Story County, Iowa [10] 
Forest farming	Forested areas used for production or harvest of natural standing specialty crops for medicinal, ornamental or culinary uses	Medical plants grown in a forest farm [11] 

3 Why Use Agroforestry?

Productivity:

The benefits from agroforestry come from the functional diversity of the plants and animals that are associated, leading to complementarities/facilitations and providing a large range of ecosystem services. The productivity of agroforestry systems is generally higher than the productivity of the same surface of land where agriculture and forest are separated. The productivity of a mixed system is measured by the area necessary to produce separately as much of each product as one hectare of the mixed

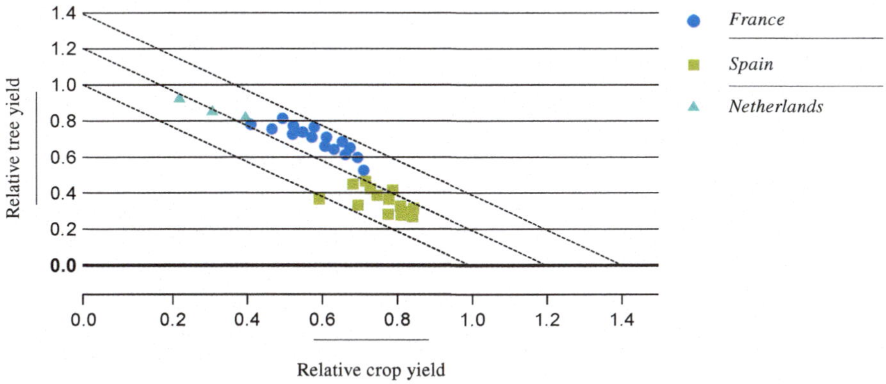

Fig. 1 Relative crop and tree yield obtained by simulating different agroforestry systems in three European pedo-climatic zones [41]

system. This is calculated as the sum of the ratios of each component's yield over the yield as a monoculture and the result is called the Land Equivalent Ration (LER) [12]. Combining field measurements and computer simulation, researchers in the European SAFE (Silvoarable Agroforestry For Europe) project (2001–2005) showed that LERs ranging from 1 to 1.4 could be expected from temperate agroforestry (Fig. 1), but literature sources and the results of a recent comparative study [13] indicate that the relative productivity of silvoarable systems compared to sole crop systems depends on the region or part of Europe.

One of the main explanations for the higher productivity of agroforestry systems compared to forests and crop monocultures is the better use of resources thanks to the complementarities between trees and crops (e.g. for light between winter crops and deciduous trees (Fig. 2), for water and nutrients between shallow-rooting crops and deep-rooting trees etc ...).

Other important effects are facilitations due to modifications by one species of the environment of the other species. Such modifications include modifications of the microclimate. In dry areas, trees improve pasture microclimate: with proper proportion of canopy cover, forage production period is extended in summer and the pasture's nutritional quality for animal breeding is improved [14–17]. By providing shade for animals and thus reducing heat stress, trees also improve animal performance [18].

In windy areas, trees are often used as windbreaks to improve crop yield. For example, Canadian researchers concluded that the average yield increase due to wind protection was 15% in winter wheat systems, 25% in soybeans and 12% in maize [19]. Similarly, wheat grain yields were increased in windbreak and alley-cropping sites during drought, compared to the control [20]. Supporting scientific evidences from Europe have been reported about similar increases in crop yields on lands protected by shelterbelts [21–23, etc.].

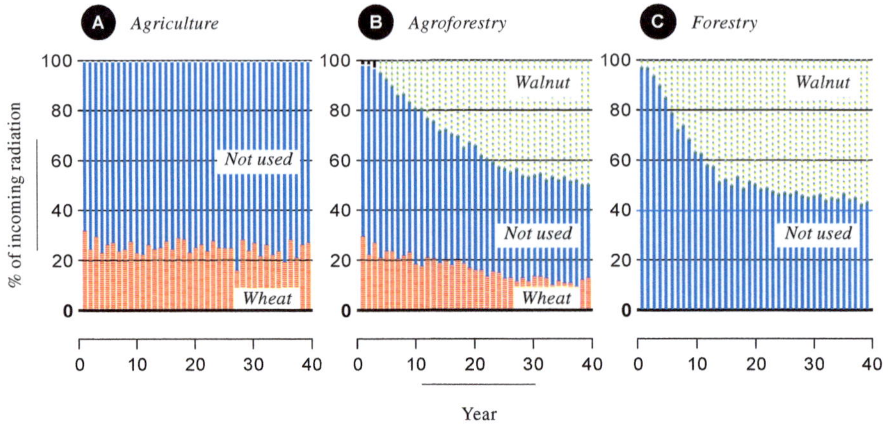

Fig. 2 Modelled proportion of solar radiation intercepted by a wheat monoculture (**a**), a wheat-walnut agroforestry system (**b**), and a walnut forestry system (**c**) over 40 years [42]

Furthermore, the diversity within agroforestry systems improves their resilience, making them more stable than large scale single crop agricultural lands.

Environmental benefits:

The integration of trees, agricultural crops, and/or animals into an agroforestry system has the potential to enhance soil fertility and reduce erosion [24, 25], maintain watershed hydrology and improve water quality [24, 25], enhance aboveground and below-ground biodiversity, which in turn, increases nutrient cycling [24, 25], promotes biological control of pest and favours bees and other pollinators [26], increase aesthetics [24], sequester carbon and serve as CH_4 sinks [24, 27, 28, 29] protect natural landscape elements and man-made infrastructure and prevent emergencies and losses caused by adverse weather conditions—e.g. snow drifts and sand storms [30].

Economical benefits:

Higher diversity and multifunctionality of a production system also means a wider variety of products thus diversifying incomes compared to monoculture systems. Not only yields but also quality of woody, crop or animal-based products can potentially improve because of the improved ecosystem services. Under appropriate market conditions and the existence of relevant value chains, agroforestry results in a more stable farm economy together with a higher security of production. Trees in landscapes also provide scenery that may augment land prices and/or improve products marketability thanks to a positive image.

On the whole, the beneficial impacts of agroforestry could help attaining food security and securing land tenure both in developing part of the world and in the developed countries, although many aspects of agroforestry still need to be optimised before these systems can develop to their full potential. Among other things, research is now focusing on identifying the best tree/crop combinations, crop varieties, width of the alley, and management of tree stands to improve both production and provision of ecosystem services by agroforestry systems.

4 The Role of Agroforestry in Climate Change

Vegetation exchanges carbon dioxide between the atmosphere and the terrestrial biosphere through photosynthesis and plant/soil respiration. This natural exchange has been occurring for hundreds of millions of years. Human activity is changing the rate of natural carbon exchange through land use, land-use change, and forestry activities [18]. Agriculture is responsible for 10% of total EU GHG emissions [31] while forests and agricultural lands currently cover more than three-quarters of the EU territory and naturally hold large stocks of carbon, preventing its escape into the atmosphere [46]. Therefore the importance of the agricultural sector on climate change is huge. Agriculture impacts climate change both negatively through greenhouse gas (GHG) emissions into the atmosphere (e.g. through mineralisation of carbon due to land use change, nitrous oxide emissions from N fertilization, methane emissions from livestock), and positively, through the removal of carbon from the atmosphere thanks to photosynthesis. Agricultural practices determine the balance between the negative and positive effects, and therefore the overall impact of agriculture: whereas draining of peat land, felling of forest or ploughing up grassland generates GHG emissions, afforestation, conversion of arable land into grassland or agroforestry system can result in protection of carbon stocks or even carbon sequestration.

There is a growing body of literature that indicates that agroforestry systems have the potential to store more carbon than conventional agricultural systems, both as plant biomass (if wood is used for construction rather than fuel), and as soil carbon [23, 27, 32, 33]. According to Aertens et al. [34], agroforestry was the agricultural practice showing the highest carbon sequestration potential in Europe among the 4 tested measures (agroforestry, introduction of hedges along agricultural plots, introduction of cover crops in the rotation and practices of low or no tillage). One cubic meter of wood biomass represents approximately 0.5 Mg of dry matter (depending on the species specific wood density) containing 45% carbon so wood used for construction or furniture stores 0.225 Mg cm^{-3}. The amount of carbon sequestration in the soil is still uncertain and varies greatly in the literature. For example, Sharrow and Ismail [35] estimated a sequestration potential of 0.4 Mg CO_2e ha^{-1} $year^{-1}$ while Eagle et al. [36] estimated a value of 4.97 Mg CO_2e ha^{-1} $year^{-1}$. Besides methodological difficulties in estimating agroforestry sequestration potential [37], this variability may also result from the fact that sequestration potential depends on many factors such as tree species, tree density or pedoclimatic conditions. For example, a simulation study [38] showed a sequestration potential ranging between 5.55 Mg CO_2e ha^{-1} $year^{-1}$ for slow-growing (50 years) trees at 50 trees ha^{-1} and 14.8 Mg CO_2e ha^{-1} $year^{-1}$ for fast-growing (15 years) trees at 100 trees ha^{-1}.

To meet the target of a 40% GHG reduction within the 2030 GHG mitigation framework, new polices should be established. Therefore European institutions[1] are nowadays discussing the possibilities for integrating LULUCF into the mitigation and adaptation strategies [39].

Certainly, the net effect of LULUCF[2] activities on the atmosphere depends not only on changes in on-site carbon stocks but also on [27]:

The lifetime of carbon in agricultural and forest products and how they replace other products that require more or less energy to produce and use Concomitant changes in the net fluxes of other GHGs (especially CH_4 and N_2O) Changes in GHG emissions resulting from changes in the fossil fuel energy needed to maintain the new land-use practices Changes in non-GHG-related radiant forcing (such as changes in albedo).

In addition to carbon-dioxide, nitrous oxides and methane also have a considerable impact on global warming, therefore they are also worth considering. Agroforestry has potential in affecting these components—for instance in silvopastoral areas or by reduced application of fertilizers—thus examination of nitrogen and carbon cycle in AF systems is one of the hot research topics of today.

The Special Report "Land Use, Land-Use Change and Forestry" of the Intergovernmental Panel on Climate Change (IPCC) discusses the global carbon cycle and how different land use and forestry activities currently affect standing carbon stocks and emissions of greenhouse gases. It also looks forward and examines future carbon uptake and emissions. According to the report, agroforestry systems—when the technology is appropriate—can be superior to other land uses at the global, regional, watershed, and farm scales because they optimize trade-offs between increased food production, poverty alleviation, and environmental conservation [27]. Therefore agroforestry practices must play an important role in strategies for mitigation and adaptation to climate change.

A public consultation on accountability of GHG in the EU from 2020 concluded that water, agriculture and forestry sectors were reported to be the top three priority sectors for adaptation [39]. Indeed, for the major crops (wheat, rice, and maize) in tropical and temperate regions, climate change without adaptation will negatively impact production for local temperature increases of 2 °C or more above late-20th-century levels [40]. The use of traditional and/or innovative land management practices through agroforestry has indirectly positive effects on carbon storage and improves micro- and meso climate. By maintaining and protecting national resource services, AF systems can contribute to land afforestation and reduce the pressure on natural forests. Reaching the goals of enlarging tree covered areas may be possible through agroforestry programmes carried out in cooperation with rural populations [41]. Due to their diversified structure and economy, AF systems are a bit more adaptable and therefore resilient to risk, compared to monoculture systems. In this way, combination of trees with crop or trees with forage and livestock contributes to the social-economic sustainability of rural areas [42], but also can positively influence

[1]The European Commission, the European Parliament, and the European Council.

[2]Land use, Land Use Change and Forestry (sector).

food production in developing countries and reduce risks of production in developed countries while meeting the carbon storage goals.

The potential of agroforestry to contribute to sustainable development has been recognized in international policy meetings, including the United Nations Framework Convention on Climate Change (UNFCCC) and the Convention on Biological Diversity (CBD), justifying increased investment in its development. Yet agroforestry continues to face challenges such as unfavourable policy incentives, inadequate knowledge dissemination, legal constraints and poor coordination among the multiple sectors to which it contributes [43]. Therefore several international organisations[3] and European funded projects[4] aim at promoting agroforestry by bridging the gaps in knowledge and overcome the existing barriers to agroforestry developments.

5 Summary

Nowadays, adaptation to climate change is necessary to make agriculture more productive, sustainable and resilient. Agroforestry, which encompasses a wide variety of practices with different combinations of woody, herbaceous and animal components adapted to local conditions, is among the most promising practices. The advantages of agroforestry discussed above—e.g. the potentially higher productivity both in biological and economic terms—derives from the functional diversity of the system, leading to a large range of ecosystem services and a better use of resources thanks to the complementarities between tree, crop and animal components. Due to the diversified structure and economy, AF systems are more adaptable and therefore resilient to risk, which can positively influence food production while meeting the carbon storage goals.

References

1. FAO: Advancing Agroforestry on the Policy Agenda: A Guide for Decision-Makers, by G. Buttoud, in collaboration with Ajayi, O., Detlefsen, G., Place, F., Torquebiau, E. Agroforestry Working Paper no. 1, Rome, 37 (2013)
2. Mosquera-Losada, M.R., Santiago-Freijanes, J.J., Rois, M., Moreno, G., Pisanelli, A., Lamersdorf, N., den Herder, M., Burgess, P., Fernández-Lorenzo, J.L., González-Hernandez, P., Rigueiro-Rodríguez, A.: Cap and agroforestry practices in Europe. In: 3rd European Agroforestry Conference—Book of Abstract. European Agroforestry Federation, Montpellier, pp. 428–432 (2016b)
3. https://www.agforward.eu/index.php/en/montado-in-portugal.html
4. http://smallfarms.cornell.edu/2016/01/28/learn-better-grazing-for-land-animal-health-and-profit/
5. https://www.agforward.eu/index.php/en/grazed-orchards-in-northern-ireland-uk.html

[3] ICRAF, EURAF, CIRAD.

[4] SAFE, AGFORWARD, AGROFE, and AGROF-MM.

6. https://en.wikipedia.org/wiki/Windbreak
7. https://www.mesopinions.com/petition/nature-environnement/sauvons-bocage-normand/10271
8. https://leisaindia.org/agroforestry-for-ecological-and-economic-benefits/
9. https://saminzambia.wordpress.com
10. https://en.wikipedia.org/wiki/Riparian_buffer
11. https://articles.extension.org/pages/64919/forest-farming-and-non-timber-forest-products-defined
12. Mead, R., Willey, R.W.: The concept of a "Land Equivalent Ratio" and advantages in yields from intercropping. Exp. Agric. **16**, 217 (1980). https://doi.org/10.1017/S0014479700010978
13. Ivezić, V., van der Werf, W.: Relative crop yields of European silvoarable agroforestry systems. In: 3rd European Agroforestry Conference—Book of Abstract. European Agroforestry Federation, Montpellier, pp. 292–294 (2016)
14. Campos, P., Huntsinger, L., Oviedo. J.L., Starrs, P.F., Diaz, M., Standiford, R.B., Montero, G.: Mediterranean Oak Woodland Working Landscapes: Dehesas of Spain and Ranchlands of California. Landscape Series. Springer, Berlin, pp. 273–309 (2013)
15. Mosquera-Losada, M.R., McAdam, J., Rigueiro-Rodríguez, A.: Silvopastoralism and sustainable land management. In: Proceedings of an International Congress on Silvopastoralism and Sustainable Management held in Lugo, Spain, in April 2004
16. McAdam J.H., Burgess P.J., Graves A.R., Rigueiro-Rodríguez A., Mosquera-Losada M.R.: Classifications and functions of agroforestry systems in Europe. In: Rigueiro-Rodróguez, A., McAdam, J., Mosquera-Losada, M.R. (eds.) Agroforestry in Europe. Adv. Agrofor. **6**. Springer, Dordrecht (2009)
17. Obrador-Olán, J.J., García-Lópe z, E., Moreno, G.: Consequences of dehesa land use on nutritional status of vegetation in central-western spain Schnabel. In Gonçalves, S.A. (eds.) Sustainability of Agrosilvopastoral Systems—DehesaMontadosCatena VerlagReiskirchen, pp. 327–340 Germany (2004)
18. Nair, P.K.R., Rao, M.R., Buck, L.E.: New Vistas in Agroforestry: A Compendium for 1st World Congress of Agroforestry, 2004. Advances in Agroforestry, vol. 1 (2004)
19. Kort, J.: Benefits of windbreaks to field and forage crops. Agric. Ecosyst. Environ. **22**(23), 165–190 (1988)
20. Rivest, D., Lorente, M., Olivier, A., Messier, C.: Soil biochemical properties and microbial resilience in agroforestry systems: effects on wheat growth under controlled drought and flooding conditions. Sci. Total Environ. **463–464**(C), 51–60 (2013)
21. Kanzler, M., Böhm, C., Mirck, J.: Microclimate effects of short rotation tree-strips in Germany. In: 3rd European Agroforestry Conference—Book of Abstract. European Agroforestry Federation, Montpellier, pp. 321–324 (2016)
22. Mirck, J., Kanzler, M., Boehm, C., Freese, D.: Sugar beet yields and soil moisture measurements in an alley cropping system. In: 3rd European Agroforestry Conference—Book of Abstract. European Agroforestry Federation, Montpellier, pp. 282–285 (2016)
23. Vityi, A., Frank, N.: Shelterbelt as a best practice of improving agricultural production In: 3rd European Agroforestry Conference—Book of Abstract. European Agroforestry Federation, Montpellier, pp. 211–212 (2016)
24. Jose, S.: Agroforestry for ecosystem services and environmental benefits: an overview. Agrofor. Syst. **2009**(76), 1–10 (2009)
25. Rigueiro-Rodríguez, A., McAdam, J., Mosquera-Losada, M.R.: Agroforestry in Europe. Current Status and Future Prospects. Advances in Agroforestry, vol. 6. Springer, Berlin (2009)
26. Clément, H., Canet, A., Asfaux, D., Balaguer, F.: Without trees no bees: agroforestry for a productive and bee-smart agriculture. In: 3rd European Agroforestry Conference, Montpellier, 23–25 May 2016
27. IPCC: Land Use, Land-Use Change and Forestry. Summary for policymakers. IPCC Special Report, 2000, Intergovernmental Panel on Climate Change (2000)
28. Kumar, B.M., Nair, P.K.R.: Sequestration potential of agroforestry systems: opportunities and challenges. Springer Science (Advances in Agroforestry, n°8), Dordrecht,

326 p. http://library.uniteddiversity.coop/Permaculture/Agroforestry/Carbon_Sequestration_Potential_of_Agroforestry_Systems-Opportunities_and_Challenges.pdf (2011)
29. Moreno, G., et al.: Agroforestry systems of high natural and cultural value in Europe: structure, management, goods and services. In: 3rd European Agroforestry Conference—Book of Abstract. European Agroforestry Federation, Montpellier, pp. 277–280 (2016)
30. Frank, N., Takács, V.: Hó- és szélfogó erdősávokminősítése szeélsebesség-csökkentő hatásuk alapján/Windbreaks and shelter-belts examination by their effect on decreasing the windspeed. Erdészettudományi Közlemények **2**(1), 151–162 (2012)
31. Greenhouse gas emission statistics. Eurostat Statistics, 2015. http://ec.europa.eu/eurostat/statistics-explained/index.php/Greenhouse_gas_emission_statistics. Downloaded on 29 June 2016
32. Cardinael, R., Chevallier, T., Barthès, B.G., et al.: Impact of alley cropping agroforestry on stocks, forms and spatial distribution of soil organic carbon—a case study in a Mediterranean context. Geoderma **259–260**, 288–299 (2015). https://doi.org/10.1016/j.geoderma.2015.06.015
33. Peichl, M., Thevathasan, N.V., Gordon, A.M., et al.: Carbon sequestration potentials in temperate tree-based intercropping systems, Southern Ontario, Canada. Agrofor. Syst. **66**, 243–257 (2006). https://doi.org/10.1007/s10457-005-0361-8
34. Aertsens, J., De Nocker, L., Gobin, A.: Valuing the carbon sequestration potential for European agriculture. Land Use Policy **31**, 584–594 (2013). https://doi.org/10.1016/j.landusepol.2012.09.003
35. Sharrow, S.H., Ismail, S.: Carbon and nitrogen storage in agroforests, tree plantations, and pastures in Western Oregon, USA. Agrofor. Syst. **60**, 123–130 (2004). https://doi.org/10.1023/B:AGFO.0000013267.87896.41
36. Eagle, A.J., Olander, L.P., Henry, L.R., Haugen-Kozyra, K., Millar, N., Robertson, G.P.: Greenhouse Gas Mitigation Potential of Agricultural Land Management in the United States: A Synthesis of the Literature, 3rd edn. In: Eagle, A.J., Olander, L.P., Henry, L.R., Haugen-Kozyra, K., Millar, N., Robertson, G.P. 2012 Greenh. Gas Mitig. Potential Agric. Land Manag. U. S. Synth. Lit. Rep. NI R 10-04 Third Ed. Durh. NC Nicholas Inst. Environ. Policy Solut. Duke Univ. https://nicholasinstitute.duke.edu/ecosystem/land/TAGGDLitRev. Accessed 25 July 2016 (2013)
37. Nair, P.K.R.: Carbon sequestration studies in agroforestry systems: a reality-check. Agrofor. Syst. **86**, 243–253 (2012). https://doi.org/10.1007/s10457-011-9434-z
38. Hamon, X., Dupraz, C., Liagre, F.: L'agroforesterie, outil de séquestration du carbone en agriculture. AGROOF, INRA, AFAF (2009)
39. Mosquera-Losada, M.R., Santiago-Freijanes, J.J., Lawson, G., Balaguer, F., Vaets, N., Burgess, P., Rogueiro-Rodríguez, A.: Agroforestry as a tool to mitigate and adapt to climate under LULUCF accounting. Presentation available at https://euraf.isa.utl.pt/files/pub/docs/climatechange_1_mosquera_lawson.pdf. Downloaded on 29 June 2016 (2016a)
40. Porter, J.R., Xie, L., Challinor, A.J., et al.: Food security and food production systems. In: Climate Change 2014: Impacts, Adaptation, and Vulnerability. Part A: Global and Sectoral Aspects. Contribution of Working Group II to the Fifth Assessment Report of the Intergovernmental Panel on Climate Change. Cambridge University Press, pp. 485–533 (2014)
41. Szedlak, T.: Plantation and climate change. Trop. For. Update **3**(5), 9 (1993). ISSN 1018-5690
42. Vityi, A., Marosvölgyi, B.: Role of agroforestry in the development of the Hungarian rural areas. Rural resilience and vulnerability: XXVth Congress of the European Society for Rural Sociology. In: eProceedings: The Rural as Locus of Solidarity and Conflict in Times of Crisis Universita degli Studi, Firenze, 2013, pp. 271–272 (2013)
43. Buttoud, G.: Advancing Agroforestry on the Policy Agenda: A Guide for Decision-makers. Food and Agriculture Organization of the United Nations Rome (2013)
44. Dupraz, C., Burgess, P.J., Gavaland, A., Graves, A.R., Herzog, F., Incoll, L.D., Jackson, N., Keesman, K., Lawson, G., Lecomte, I., Mantzanas, K., Mayus, M., Palma, J., Papanastasis, V., Paris, P., Pilbeam, D.J., Reisner, Y., van Noordwijk, M., Vincent, G., van der Werf, W.: SAFE (Silvoarable Agroforestry for Europe) Synthesis Report. SAFE Project, August 2001–January 2005

45. Dupraz, C., Liagre, F.: Agroforesterie: des arbres et des cultures. France Agricole Editions (2008)
46. Mosquera-Losada, M.R., Santiago-Freijanes, J., Pisanelli, A., Rois, M., Smith, J., Herder, M., Moreno, G., Malignier, N., Mirazo, J.R., Lamersdorf, N., Ferreiro-Domínguez, N., Balaguer, F., Pantera, A., Rigueiro-Rodríguez, A., Gonzalez, P., Lorenzo, J.L., Romero, R., Chalmin, A., García de Jalón, Silvestre Burgess, P.: How can policy support the uptake of agroforestry in Europe?. Deliverable 8.24 for EU FP7 Research Project: AGFORWARD 613520. (7 September 2017) 21 pp.

Hydrological Impacts of Climate Change on Forests

Zoltán Gribovszki, Péter Csáki and Márton Szinetár

1 Introduction

Climate change can generally be characterized by a global rise of average temperatures (global warming) and the subsequent strong impact this has on the hydrological cycle. Temperatures, which have already increased by 0.6 °C on average during the last century, are predicted to increase even more [8].

Global warming scenarios forecast a temperature increase of 3–5 °C by the end of the 21st century, which will intensify the driving forces that influence the hydrological cycle. These effects will probably modify rainfall patterns and evapotranspiration processes at multiple scales [21, 25]. Extreme events like thunderstorms and droughts are expected to occur more often [8] and it is widely accepted that such impacts will considerably affect forest ecosystems, especially in the forest/grassland transition zone (dry edge of the closed forest belt) [16].

2 Forest Water Balance

A portion of the precipitation in a forested area is intercepted by the canopy (canopy interception), from where the precipitation returns to the atmosphere through evaporation. Water that flows down along the stem (streamflow) or falls and drips through the canopy (throughfall) reaches the ground where another portion of it is intercepted by the forest floor through litter interception. Water that is not intercepted infiltrates

Z. Gribovszki (✉) · P. Csáki
Faculty of Forestry, Institute of Geomatics and Civil Engineering, University of Sopron, Sopron, Hungary
e-mail: gribovszki.zoltan@uni-sopron.hu

M. Szinetár
University of Natural Resources and Life Sciences, Vienna, Austria

M. Palocz-Andresen et al. (eds.), *International Climate Protection*,
https://doi.org/10.1007/978-3-030-03816-8_16

into the ground from where it either flows downhill as subsurface runoff or is consumed by plants for transpiration and plant tissue formation. This remaining water either increases the soil moisture content and groundwater storage or, in the presence of an impermeable layer near the surface, flows downwards as subsurface flow.

Surface runoff occurs if the amount of precipitation reaching the ground is greater than its infiltration capacity, or if precipitation intensity is greater than the soil-infiltration rate (Fig. 1).

We can analyze the water balance of a forest by comparing the water amount that flows in and out of a system and considering the change in storage.

Based on the conservation of mass, the water balance equation describes the water regime of a system by stating that the difference of the input and output components has to be equal to the change in storage.

$$P + p + Rin_{s,g} - I - T_{soil} - T_{gw} - Rout_{s,g} = dS$$

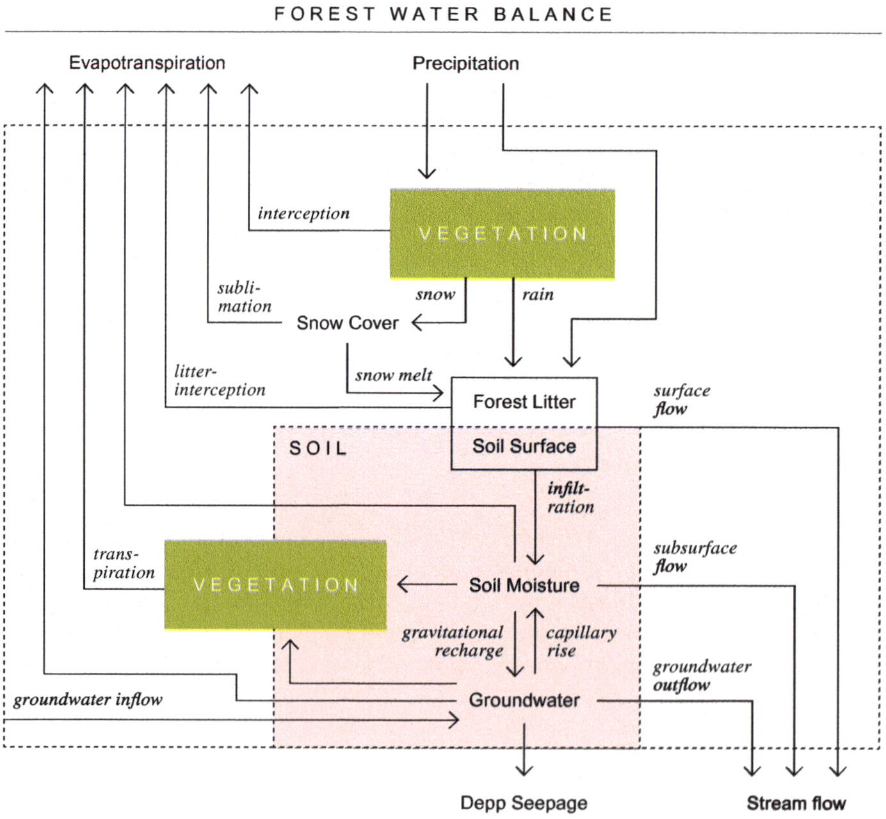

Fig. 1 Water balance of a forest [13]

where: P: fluid and solid macro-precipitation; p: fluid and solid micro-precipitation; Rins, g: surface and subsurface inflow; I: interception (canopy and litter); T_{soil}: transpiration from vadose zone; T_{gw}: transpiration from groundwater; $Rout_{s,g}$: surface and subsurface runoff; dS: change in storage.

In the following, we will describe the main components of the forest water cycle and their relation to the climate change.

3 Precipitation

Macro-precipitation is the main source of water for forested areas. Climate change greatly influences the amount and spatiotemporal resolution of precipitation, which has a direct effect on the primary production of specific forest types. The presence of forests in the xeric limit area could also be affected. The possible influence of forests on macro precipitation amounts is highly debated.

4 Interception

Interception refers to the process whereby the canopy retains a portion of precipitation and does not allow it to reach the soil. Rather than reach the soil, this intercepted precipitation evaporates back into the atmosphere from the canopy. According to Delfs [1], interception equals the amount of precipitation that is intercepted and evaporated back to the atmosphere by the canopy or any other aboveground vegetation.

Interception loss varies between 10 and 40% depending on the type of forested ecosystem [2]. Neglecting interception when modelling the water balance of a forested ecosystem could result in a major miscalculation [20]. Interception is often only considered in terms of canopy interception, but a more comprehensive definition includes the sums of both canopy and litter interception.

According to Járó [9], canopy interception rates in Hungary amount to 20–45% of annual precipitation. Generally, interception rates are higher in coniferous forests than in deciduous forests. Interception of forests with multi-layered canopies is higher than in single-layered ones. At similar canopy closures, the interception loss is higher for aged forests due to their bigger surface for evapotranspiration.

Litter interception does not necessarily have to be accounted for as interception loss since forest litter considerably reduces the evaporation of the underlying ground layer. In the United States, the annual litter interception was estimated to be 1–5% of the annual precipitation [6] whereas in a sessile oak forest in west Hungary this value varies between 5 and 7% according to Zagyvainé Kiss et al. [24]. One of the highest litter interception rates (34%) was estimated by Gerrits et al. [3] in a common European beech forest. The storage capacity of forest litter—a decisive factor of interception—is proportional to the mass of fallen leaves in a given area [19, 23].

4.1 Climate Change and Interception

On the one hand, reduction of interception loss is expected due to reduction of number of rainfall events. This means that annual precipitation sum will likely be concentrated into smaller amount of heavy rainfall events resulting in less interception and more available water for runoff. On the other hand, increasing temperature—in the presence of sufficient amount of plant available water—will enhance transpiration and, consequently, the leaf area index as well. This will lead to positive feedbacks including higher canopy storage capacity and, eventually, higher interception loss.

5 Transpiration

The active transport of liquid water into the air via transpiration is an essential plant life function. Transpiration is an evaporation-like process controlled by similar factors, but transpiration occurs on a different and, usually, larger surface where surface resistance/tension plays a more determinative role [14]. Relative to other vegetation types, the roughness and leaf area index of a forest determines its transpiration demand, which is increased by both factors. The root system of forested vegetation reaches deeper ground layers with higher water storage capacities making them less vulnerable to extended dry periods.

Járó [10] determined the water use of forests in Hungary as a context of biological production and found that species growing in more humid climates (beech, spruce) have less water demand than those living in less humid environments (common oak, native poplars). Based on this statement, forests living in the xeric environment of the Hungarian Great Plain generally have a higher transpiration water demand. However, some species located here (scotch pine, black pine, black locust) have very low water demands in dry climate conditions.

5.1 Climate Change and Transpiration

As long as the available amount of water does not limit it, transpiration rates will increase as temperatures increase. Soil moisture content supplies the water demand of forests in mountain and hilly areas; therefore, water storage capacity of soil is a crucial factor in overcoming the water stress of dry periods and securing the long-term viability of a forest. The required volume of the moisture storage space depends on the rooting depth and physical properties of the soil.

6 Groundwater Utilization/Usage

In the Hungarian Great Plain, forest and groundwater interaction was the most examined topic; in this context, the results are the following.

Forest evapotranspiration (both transpiration and interception) is generally higher than the evapotranspiration of neighboring grasslands because of the enhanced LAI and root depth of the woody vegetation. This difference is especially true in the semi-arid climate of the Hungarian Great Plain where precipitation is generally inadequate to support woody vegetation; trees there survive drought periods by utilizing groundwater resources. Throughout the year the groundwater level under forests (provided trees are able to reach it) can be detected at levels deeper than under grassland and under agricultural land, with bigger differences detectable during the growing season [7].

Major [15] has reported that under a middle-aged pine forest in the Danube-Tisza Sand plateau region the groundwater was on average 0.8–1.1 m deeper than in the surrounding non-forested areas. This actively growing black pine forest has a mean annual ET rate of 712 mm a-1. On average, this forest uses 130 mm more water than it receives from precipitation annually.

Gribovszki et al. [4] compared the groundwater balance of two neighboring plots (a common oak forest and a pasture) in a sandy soil environment. The water table under the oak forest was 0.44 m lower, and the groundwater uptake of the oak was more than twofold in the very dry summer of 2012. The larger groundwater use of the forest was not parallel with salt uptake; therefore, salts accumulated in both the soil and groundwater (Fig. 2). The measured differences in salt content, however, were small when compared to similar research results for clayey soils [18].

A good example for comparing the water balance of different land uses is an analysis in the northeastern part of the Hungarian Great Plain (Nyírség). Móricz et al. [17] describe that a lowland common oak forest has approximately 30% more evapotranspiration (758 mm/year) than a neighboring fallow (623 mm/year). The difference in the groundwater use of the two different vegetation types is even more significant (threefold: oak: 243 mm, fallow: 85 mm). Groundwater consumption was approximately 40% less in the wet growing season than in the drier growing season, despite the groundwater level being deeper during the dry period. Thus, both vegetation covers relied considerably on available groundwater resources during the dry season.

Remote sensing provides expanding possibilities to map and analyze the spatial heterogeneity of landscape unit hydrology. From a hydrological point of view, forests are very important landscape elements. Szilágyi et al. [22] analyzed the evapotranspiration (determined by linear transformation of the MODIS daytime land surface temperature) in the Danube-Tisza Sand Plateau region of the Hungarian Great Plain. According to land cover types, the largest ET, about 505 mm/year, was found over deciduous forests (meanwhile the regional annual precipitation was 550 mm/year). In some locations (mostly forest cover areas), ET is estimated to be larger than precipitation. Often the dense and deep root systems of forests can tap the shallow

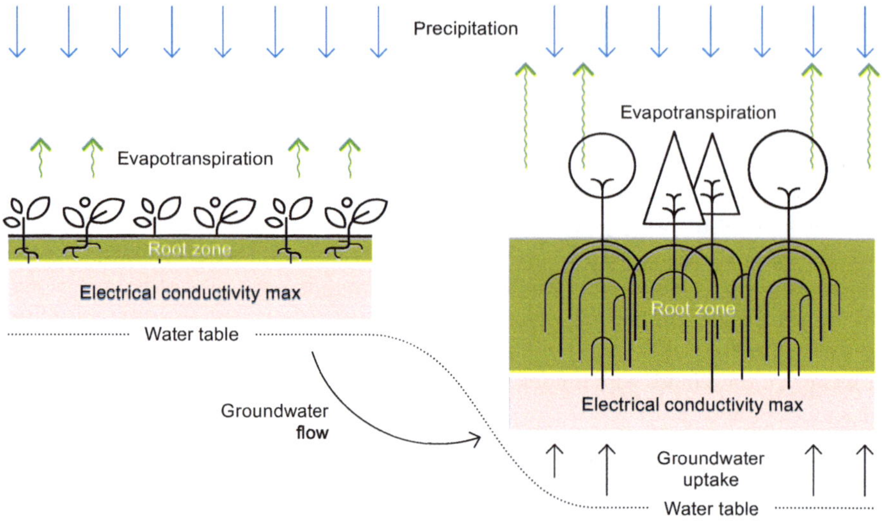

Fig. 2 Forest vegetation impact on water balance in a shallow groundwater environment

groundwater level (if it exists), thus leading to a high ET rate frequently exceeding the precipitation rate of the area. In the groundwater discharge areas, the average annual ET for the forests is 620 mm/year, which is about 70–80 mm more than the mean annual precipitation rate of the region. This negative water balance can be maintained if forests create a local depression in the water table to induce groundwater flow toward them.

6.1 Climate Change and Groundwater Dependent Forests

Driven by rising temperatures, increasing transpiration demand in the future will likely induce an enhanced groundwater uptake by plant communities. Eventually this could lead to the lowering of the groundwater table and significant salt accumulation. If this occurs, the existence of groundwater-dependent forest communities in these areas is questionable since the root structures of younger forests will not be able to reach the additional water source.

7 Surface Runoff and Flood Peak Mitigation Effect

Forests, as terrestrial habitats with the highest water consumption and water storage capacities, minimize surface runoff with their effective infiltration capacities. Due to their high water demand, they reduce long-term runoff, and their significant storage capacity helps reduce the extremes of runoff by mitigating and shifting flood peaks.

Hydrological studies focusing on streamflow difference between forested and non-forested catchments in a changing climate are scarce in Hungary. Based on the little data that exists, it can be stated that 7–10% of annual precipitation (606 mm) can be measured as streamflow (so evapotranspiration is 90–93%) in two neighboring small mixed forested catchments (100% forest cover) of western Hungary in the dry year of 2001 [5]. The average annual evapotranspiration in this forest covered region was 615 mm (85%) on a 9 year average (when yearly average precipitation was 726 mm in the same period between 2000 and 2008) [12].

Kalicz et al. [11] evaluated the runoff data sets of three different catchments (forested, country side and urban) around Sopron, Hungary. He concluded that specific discharges of floods in urban areas induced by major rainfall events exceed those in a forested drainage basin by as much as two magnitudes.

7.1 Climate Change and Runoff

Extreme weather conditions such as large storms and longer dry periods will likely be more frequent and more intense due to human-induced climate change. Forests could mitigate the adverse effects of surface erosion caused by heavy rainfall-related surface runoff. Furthermore, the water storage capacity of the soil layer under a forest—which is usually higher than it is for other surface cover types—becomes less saturated, or even empty during extended dry periods. This watershed storage effectively retains water from even a big rainfall event; in addition to this, water can percolate down to deeper aquifers quicker in the typical forest soil profile.

Acknowledgements This research has been supported by "Agroclimate 2 (VKSZ_12-1-2013-00-34)" project. The research of Zoltán Gribovszki was supported by the European Union and the State of Hungary, co-financed by the European Social Fund in the framework of TÁMOP 4.2.4. A/2-11-1-2012-0001 'National Excellence Program'.

References

1. Delfs, I.: Die Niederschlagzurückhaltung im Walde/Interzeption/. Mitteilungen des Arbeitskreises "Wald und Wasser", Nr.2.Koblenz: p. 54 (1955)

2. Dingman, S.L.: Physical Hydrology, 2nd edn. Prentice Hall, New Jersey (2002)
3. Gerrits, A.M.J., Savenije, H.H.G., Hoffmann, L., Pfister, L.: Measuring forest floor interception in a beech forest in Luxembourg. Hydrol. Earth Syst. Sci. Dis. **3**(4), 2323–2341 (2006)
4. Gribovszki, Z., Kalicz, P., Balog, K., Szabó, A., Tóth, T.: Comparison of groundwater uptake and salt dynamics of an oak forest and of a pasture on the Hungarian Great Plain. Acta Silvatica et Lignaria Hungarica **10**(1), 103–114 (2014)
5. Gribovszki, Z., Kalicz, P., Kucsara, M.: Streamflow characteristics of two forested catchments in Sopron Hills. Acta Silvatica et Lignaria Hungarica **2**, 81–92 (2006)
6. Helvey, J.D., Patric, J.H.: Canopy and litter interception of rainfall by hardwoods of eastern United States. Water Resour. Res. **1**, 193–206 (1965)
7. Ijjász, E.: A fatenyészet és az altalajvíz, különös tekintettel a nagyalföldi viszonyokra. Erdészeti Kísérletek **42**, 1–107 (1939). (Connection between the forest production and the ground water in the Great Hungarian Plain)
8. IPCC: Climate Change 2014: Synthesis Report. Contribution of Working Groups I, II and III to the Fifth Assessment Report of the Intergovernmental Panel on Climate Change, p. 151. [Core Writing Team, Pachauri R.K., Meyer L.A. (eds.)]. IPCC, Geneva, Switzerland (2014)
9. Járó, Z.: Intercepció a gödöllöi kultúerdei ökoszisztémában. Erdészeti kutatások **73**, 7–17 (1980). (Interception of the experimental forest ecosystems in Godollo)
10. Járó, Z.: A hazai erdők vízfogyasztása. Agrártudományi közlemények **40**, 353–356 (1981). (Water use of the Hungarian forests)
11. Kalicz, P., Erős, M., Gribovszki, Z., Markó, G., Primusz, P.: A soproni Rák-patak egy városi szakaszának hidrológiai és hidrodinamikai vizsgálata. In: Albert, L., Bidló, A., Jancsó, T., Gribovszki, Z. (eds). Városok öko-környezetének komplex vizsgálata a nyugat-dunántúli régióban, Nyugat-magyarországi Egyetem Kiadó, p. 261. Sopron (2012). ISBN 978-963-334-084-4. (Hydrological and hydrodynamical evaluation of an urban section of the Rak-brook in Sopron)
12. Kovács, Á.D.: Tó-és területi párolgás becslésének pontosítása és magyarországi alkalmazásai. (Specifying lake and areal evapotranspiration rates in Hungary) Ph.D. Thesis, p. 101. TUB, Budapest (2011)
13. Kucsara, M.: Csapadék és lefolyás erdészeti kisvízgyűjtőn. Ph.D. Thesis, Sopron (1996) (Rainfall and runoff of small forested catchments)
14. Lee, R.: Forest hydrology. Columbia University Press, New York (1980)
15. Major, P.: Síkvidéki erdők hatása a vízháztartásra. Hidrológiai Közlöny **82**(6), 319–324 (2002). (Effect of the lowland forests on the water balance)
16. Mátyás, C., Sun, G.: Forests in a water limited world under climate change. Environmental Research Letters **9**(8), 085001 (2014)
17. Móricz, N., Mátyás, C., Berki, I., Rasztovits, E., Vekerdy, Z., Gribovszki, Z.: Comparative water balance study of forest and fallow plots. iForest-Biogeosci. For. **5**(4), 188 (2012)
18. Nosetto, M.D., Jobbágy, E.G., Tóth, T., Di Bella, C.M.: The effects of tree establishment on water and salt dynamics in naturally salt-affected grasslands. Oecologia **152**, 695–705 (2007)
19. Putuhena, W., Cordery, I.: Estimation of interception capacity of the forest floor. J. Hydrol. **180**, 283–299 (1996)
20. Savenije, H.G.: The importance of interception and why we should delete the term evapotranspiration from our vocabulary. Hydrol. Process. **18**(8), 1507–1511 (2004)
21. Szilágyi, J., Józsa, J.: Klímaváltozás és a víz körforgása. Magyar tudomány **6**, 698–703 (2008). (Climate change and water-cycle)
22. Szilágyi, J., Kovács, Á., Józsa, J.: Remote sensing based groundwater recharge estimates in the Danube-Tisza Sand Plateau region of Hungary. J. Hydrol. Hydromech. **60**(1), 64–72 (2012)
23. Zagyvainé Kiss, K.A., Kalicz, P.: Gribovszki Z (2013) Az erdei avar tömege és víztartó képessége közötti összefüggés. Erdészettudományi közlemények **3**(1), 79–88 (2013). (Dry weight-dependence of water capacity of the forest litter)

24. Zagyvainé Kiss, K.A., Kalicz, P.: Forest litter interception model for a sessile oak forest. Acta Silvatica et Lignaria Hungarica **1**, 91–101 (2014)
25. Zhou, G., Sun, G., Wang, X., Zhou, C., McNulty, S.G., Vose, J.M., Amatya, D.M.: Estimating forest ecosystem evapotranspiration at multiple temporal scales with a dimension analysis approach. JAWRA **44**(1), 208–221 (2008)

China's Actions on Adaption to Climate Change

Yijing Li

1 Problems Arising from Climate Change in China

China is one of the developing countries suffering severest adverse effects from climate change. According to "The 3rd National Assessment Report on Climate Change" in 2015, China's temperature on land areas has increased averagely by 0.9–1.5 °C during the century (1909–2011), which was taken as one of the self-explanatory phenomenon caused by climate change. Upon such circumstances, extreme climate events, e.g. droughts, floods, hot weave, and typhoons, become to be more frequent and concurrent in contemporary China. For example, the catastrophic freezing rain and sleet in January 2008 in the southern China, the severe drought disaster in southwest China in 2010, and the extreme mudflow disasters in recent years, have caused destructive damages and losses to citizens' lives and properties, as well as the national development.

Along with the global warming trend, the sea levels in coastal China had increased with the pace at 2.9 mm annually from 1980 to 2012, which was higher than the global average increasing speed. During the period, the glacier area has shrunk by 10.1%, and the area of permafrost has shrunk by 18.6%. Meanwhile, average annual precipitation has not seen significant changing trends in the past century. However, the spatio-temporal distribution patterns changed significantly in that, the precipitation in northern China decreased significantly ever since 1950s with aggravated droughts; while the precipitation in the south and the west increased constantly in previous 30 years [1], inducing to the frequent floods in some areas of southern China. The changing pattern of precipitation also incurred extreme weather situations, such as extremely hot in majority of China and extremely cold in certain places, and further caused much more severe losses. In general, effects caused by climate change in China have both pros and cons , but problems are much more obvious than the advan-

Y. Li (✉)
The Centre for Urban Science and Progress London (CUSP London), King's College London, Peking University UK Campus, London, UK
e-mail: yijing.li@kcl.ac.uk

M. Palocz-Andresen et al. (eds.), *International Climate Protection*,
https://doi.org/10.1007/978-3-030-03816-8_17

tages in following 8 main areas: agriculture, grassland livestock husbandry, water resources, ecosystem and biodiversity, forestry, coastal resources, human health and others.

Agriculture. Although the increase of light and heat resources due to global warming, were taken as helpful to the growing of some plants, it is discussed that the impacts from climate change were mainly disadvantageous to the development of agriculture in China. For example, climate change during 1980–2008 has induced the decreases of wheat and corn by 1.27% and 1.73% respectively; did adverse impacts to water resources, sea environment and urban ecology, etc.; the growing period has been prolonged, filling of grains became to be insufficient, and the normal growth and development process has been altered [2]; the production environment for agriculture became to be deteriorated in that, the organic matter and nitrogen in soil run off more quickly, soil and water loss became to be severer and severer, the soil degraded into more salinization and desertification, while the diseases and insect pests moved towards the northern China [3]; the planting patterns for agriculture have changed according to the changed climate, while the disasters relevant to agriculture, such as floods and droughts, frosts, and diseases increased dramatically.

Grassland livestock husbandry. The planting regions for corn, which is a main fodder crop, have been move towards the north, with increased varieties of late maturing species; diseases and insect pests have aggregated; the precipitation's distribution saw severe disparity; soil moisture content in some places reduced seriously; grassland's grazing capacity decreased due to the climate change, such as global warming and drying, hence aggregated the burden; the frequencies of droughts has increased, with expanded inter annual and lasting periods; it was also claimed of shortage of forage grass reserves after snow disasters.

Water resources. In recent 50 years, the actual measured runoffs of great rivers decreased dramatically over years; majority of regional water resources have seen obvious reduces in these 20 years, and sequent dry years appeared in northern China [4, 5]; the disparity in water resources between north and south is becoming wider and wider in that, the dry area in northern China expanded annually, while floods and in southern China aggregated especially in regions along Middle and Lower Yangtze River, and those southeast regions; the balance between supply and demand for water resources are becoming more and more intense [6] (Fig. 1).

Ecosystem and biodiversity. Ever since 1980s, the eastern temperature zone has been moved towards the north, the growing periods for plants have prolonged, and the phenological period has been brought forward. Timberlines have shift up, the border line of forest community in northern China has moved towards the north, the net primary productivity has increased; the spatial disparity became more and more obvious, the grassland degraded seriously, and the structure of grassland ecosystem community has changed [4, 5]. In the arid and semiarid regions, due to the decreases of natural lakes and wetlands, the drop of water level, the aggregated natural disasters from extreme climate events, and the intensified stony desertification of karst areas, animals distribution patterns, the phonological and migration routines have changed, together with their shrunk habitats and altered overwintering habits [7, 8].

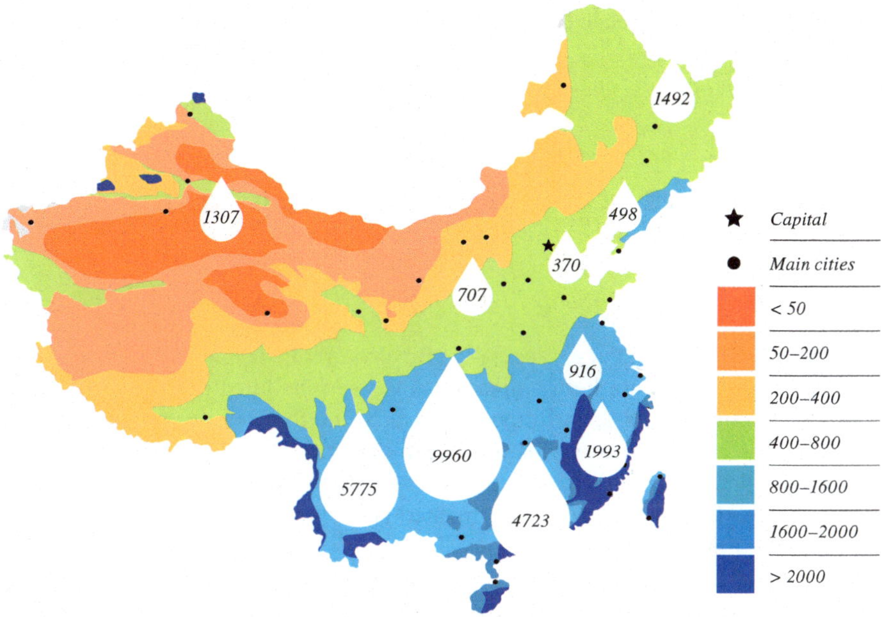

Fig. 1 Water resources distribution in China

Forestry. The growing period for forests has been prolonged, with increased productivity [9, 10]. Forest fires became more frequent and influential in key forestry areas [11, 12]. Extreme events have induced large volume of broken trees and deaths, as well as increased risks of getting on fire and getting ill by pests [13].

Coastal resources. The annual arising rate for sea levels in coastal China was 2.6 mm on average in previous 30 years; the ecosystem health for coastal wetlands, coral reeves, and mangroves have deteriorated in the meantime; typhoons and storm surges occurred in advance and more frequent, with broader regions been affected and intensified impacts; the temperature in coastal regions have increased obviously in recent 20 years, together with aggregated eutrophication, broader and more influential red tides; the salt tides have already influenced water supply safety in urban areas and the nutritious structure in wetlands.

Human Health. Intensified and more frequent hot waves have induced higher fatality of certain diseases; injuries and deaths caused by extreme climate events and secondary disasters have increased dramatically; high violent radiation from ozone holes incurred higher morbidity of skin cancer, cataract, and whiteout, vector borne diseases became more frequent in the north.

Others. The rise of annual temperature and thinning of frozen earth made great influences on projects in cold regions, in lowering the solidarity of bases and the designing standards; the change of precipitation has affected major water storage and transfer projects' functioning, as well as the energy supply patterns and progressing of projects; urban planning layout has been affected by climate change, while citizens' daily life has been influenced by the heat island effect; industrial structures also changed hereafter, especially the tourism industry and seasonal visiting; extreme events also made impacts on transportation, electricity infrastructure and city safety.

2 Effects Caused by Climate Change in Main Areas

2.1 Agriculture

Food security. Climate change, especially global warming, had induced severe water resources shortages in northern China, posing a serious threaten to food security. The extreme climatic disasters further aggravated the threaten, for example, losses of food yields caused by droughts took up more than 60% of all the losses from natural disasters. From empirical studies, climate change might induce the reduction of crop yields, drop of the quality, aggravated pests' impacts, increase of the costs, and decrease of the profits.

Meteorological disasters. It is taken that extreme climatic disasters are one of the most important reasons for crops' reduction of output. It is estimated that meteorological disasters had induced 34 million hm^2 area affected annually, causing more than 100 billion Yuan economic loss [14].

Pest and diseases damages. Pest and diseases damages to agriculture took up 20–25% of the gross production, and global warming seemed to be helpful to disease pests' overwintering, advancing of the first appearance period, the migration period and the burst period of some pest diseases. The adaption period and detrimental period of some pests have been prolonged hereafter, with more generations of reproduction.

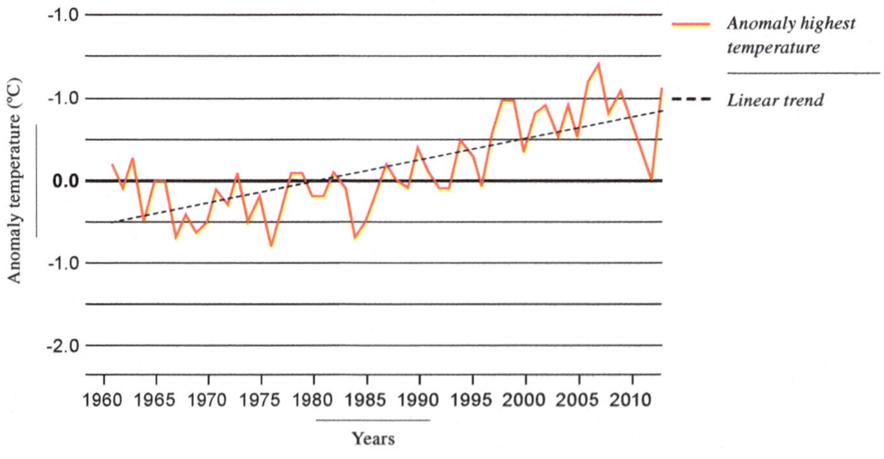

Fig. 2 Annual anomaly highest temperature in China (1960–2013)

Climate change also quickened the evaporation and decomposition of pesticides, together with the decrease of biodiversity and beneficial species, to increase the governing difficulty and costs (Fig. 2).

Severe degradation of grasslands. In recent 50 years, grasslands in north China have seen obvious temperature rises, high fluctuation of cumulative temperature, and decreases of annual precipitation [15]. In addition, climate change has affected animal husbandry through direct influences on animals, and indirect influences on grasslands' biomasses and plants structure.

2.2 *Water Resources*

Water safety now facing with extra pressures and challenges. Since 1980s, China's major rivers have seen observed decreases of runoffs. For example, the Haihe basin's volume of runoffs reduced by more than 40%, and the middle and lower reaches of Huang River saw extraordinary shortage of water resources [4, 5]. Severe droughts in some regions not only challenged the development of agriculture, but also did harmful influences on citizens' daily water requirements and industrial development. Decadal and inter-decadal available water changed dramatically. Global warming could probably initiate abnormal extremely meteorological events, and place heavy burdens on the coordination of water resources and relevant infrastructures. More difficult to counter the challenges of disasters, and new threats to water ecology and environment safety. Since 1990s, natural disasters, such as floods, hot weaves, and debris flows became more and more, with severer threats to citizens' lives and properties. It is also noticed that some rivers and lakes shrunk, with lower functions and aggregated pollutions [16].

2.3 *Ecosystem and Biodiversity*

The effects on ecosystem from climate change overlaid with those from human activities. Climate change has changed the structure of species and the succession of various ecosystems; while human activities in certain areas have exacerbated phenomena such as, desertification and migrants driven by climate change. Abnormal and extreme weathers and climate events further aggregated the instability of natural ecosystem. Some species have quitted from the original ecosystem because of maladjustment to newly ecosystem, and new species that are more suitable joined to change the structure, contents and distribution. Some vulnerable ecosystem degraded more quickly during this process, in facing with more ecological risks, with lower adaptability to regional climate change. Obvious drop of biodiversity, with some species' distinction and intrusion of some pests. The distribution extents for some species have seen obvious shrinks, with altered advantageous species and biodiversity. Some exotic species might threat local ecosystem and biodiversity, even the human health and property losses.

2.4 *Forestry*

Forest structure and species' distribution pattern have change. Under the influence of climate change, the distribution of forests in China has changed dramatically in previous decades. For example, some bushes in Yunnan have unhealed to the alpine meadow due to global warming, and the altitude of forest line increased by 8.5 m per decade [17]. The movement and change of forest ecosystem was comprised by the movements of species and their habitats, as well as some distinctions and disappears.

Forest productivity changed dramatically. Ever since the increase of carbon dioxide density, forest productivity in China has increased by 12–35% by regions. The forest ecosystem has turned from carbon source to carbon sink in recent 30 years, and the forest productivity has increased dramatically especially in the northeast [9].

Forest fires have increased obviously. Global warming and the change of precipitation have induced higher risks of thunder strikes, extreme fire conditions and seriousness. In 21 century, the acidification trend in Daxinganling became more and more obvious, high temperature worked to reduce the moisture content in soil surface and combustible materials in deep [11].

Forest pests' harms have increased. Since 1990s, forest diseases and pests occurred over 0.12 billion acres annually, equaling to 80% of the annual reforestation areas. At the same time, the occurrence of pests became more frequent and more broadly spread.

2.5 *Coastal Resources*

Rises of sea levels posed heavier threats to coastal areas, and expanded the area that might be inundated. It is estimated that if the sea level rises by 100 cm, 1,500 km^2 regions lower than 2 m in Yangtze River Delta would be inundated [18]; if the sea level rises by 70 cm, 1,500 km^2 regions lower than 0.4 m in Pearl River Delta would all be inundated [19].

Wind tides and great waves increased with heavier impacts. Under the joint influences of climate change and sea level rises, typhoons and cold tides were more frequent in coastal areas these years, bringing with heavier destructive losses to both villages and projects. The salt tides also posed threats to inland drinkable waters and farmlands with longer duration and larger extent.

Exacerbated erosions in coastal zones and degraded coastal ecosystem. Climate change and human activities worked together to reduce the sand supply from rivers and the rise of sea level, further inducing to aggregated erosion in coastal zones. At the meantime, global warming propelled the expansion of red forests towards salt wet areas with high altitude.

2.6 Human Health

Extreme weathers, such as extraordinary hot weathers and hot waves, and catastrophes posed serious threats to citizens' safety. Climate change also played as a trigger for some types of contiguous diseases [20], and increased the spread of transfer medias for such diseases. Furthermore, the nutrition contents and structures in agricultural products, as well as the dieting habits changed, both worked as influential factors to human health.

3 Actions on Adaption to Climate Change in Main Areas

3.1 Agriculture

In consideration of adverse effects from climate change, especially the extreme weathers; on agriculture production and food security, in order to realize the balance between farmland, intensify the agricultural infrastructure and farmland basic construction. China has taken solutions such as adjusting the structure of agriculture and planting system, increase the multiple cropping indexes appropriately; improving the prevention and treatment for animal and plant diseases and insect pests, control pests invasion and spread; improving the breeding of resistance variety, optimize the distribution of variety; and promoting research and development of water-saving technologies vigorously, improving the agricultural irrigation ratio and water utilization efficiency.

3.2 Grassland Livestock Husbandry

Actions concerning about grassland adaptions to climate change include: build cultivated pasture and Forage storage; insist on moderate enclosure and rotational grazing upon grassland-livestock balance, to increase the prairie coverage; adjust the structures of droves and cultivation, select and breed proper variety; improve cultivation facilities and environmental conditions; construct climate change adaption technology system for breeding industry, and for various plants and animals, in order to enhance the capability in resisting climate disasters; and capability building for agriculture, villages and farmers' adaption to climate change.

3.3 Water Resources

In facing with the uneven spatio-temporal distribution of water resources, aggregated frequency of floods and droughts, and declined water quality, we should enhance the protection of catchment areas; construction of irrigation and water conservancy infrastructure; monitoring, regulation and storage management of water resources in important drainage basins; enhance the capability in resisting extreme hydrological disasters; promote and develop water-saving technology, utilize the untraditional water resources, execute the strictest water resources management institutions, and improve the ability in adapting to future water deficits; solve the problems of drinkable water safety in the village, and the severe water shortage in some regions; enhance the capability in adapting to quickened melt of glaciers in the northwest; and improve the protection and renovation of water ecology, especially the lakes and rivers, in order to restrain the deterioration of water environmental gradually.

3.4 Ecosystem and Biodiversity

Actions in the field of ecosystem and biodiversity's adaptions to climate change are mainly: to promote natural ecosystem protection projects in the areas with high risks of climate change; to build up network in natural conservation area and corridors for species migration; to adjust the layout of functional areas within the natural conservation areas; to build monitoring and disasters defense system for natural conservation areas, develop adaptive risk management techniques; to enhance the in-place and migrating species' protection, as well as the protection of habitats for key species; to develop hereditary protection technology for specific species; to control the harm of pests, and intensify species' natural adaptability; to enhance the protection and management of wetland ecosystem, and the recovery and reestablishment of degraded ecosystem; and to improve natural ecosystem's resistance capability for climate change risks through self-adaption, and the ecosystem's stability.

3.5 Forestry

In view of the obvious impacts on forestry growth and forest ecosystem from climate change, forestry ecological projects have been emphasized on: adjusting the layout of forest categories, the timing of afforestation and its density; planting the proper types of trees on sites with suited soil quality, build up mixed forests combining with grass and shrubs, and develop near-natural forest and mixed management techniques; adjusting intermediate cuttings and circulation periods; developing classified forest management technology; transforming low-yield forests scientifically, improve the afforestation survival rate, and the capability for adapting to climate change of dominant tree species; and enhancing the prevention and control of forest fires and pest and diseases damages.

3.6 Coastal Resources

Actions taken in coastal regions are mainly strengthening the research on monitoring, forecast, simulation and impacts' evaluation of the sea level rise, hence to build up a comprehensive decision making system; enhancing capability in monitoring, forecasting and risk evaluation of storm surge, waves, ride tides, and sea ices, etc., and provide real time policy making supporting services; developing coastal ecosystem monitoring network, protect the biodiversity, and propel the construction of protection areas, as well as the projects of ecosystem renovations; revising standards for offshore fortification projects of adapting to climate change, and for the construction of coastal safeguards; and enhancing the protection of coastal areas and the comprehensive management of drinkable water resources, responding to the offshore erosion, upstream salt tide and the intrusion of sea.

3.7 Human Health

Actions regarding with human health are: to strengthen the public hygiene infrastructure construction in weak climate change areas; to enhance capability in monitoring, forecasting, prevention and control of diseases relevant to climate change, especially those infectious and the unexpected; to improve the supervision and control of medias for diseases transmission, and their impacts on human health upon climate change; to intensify the forecast and prevention mechanisms for extremely harmful weather and climate events, reduce the deaths and injuries during catastrophes, as well as mental harms; and to get the living environment improved, and revise the labor forces protection standards.

4 Conclusion

As the impacts from climate change became more and more prominent, especially those extreme disasters, China has been actively making its due contribution towards the goal to fight against climate change, and concrete efforts have been witnessed in recent years. In adherence with the principles of common but differentiated responsibilities, equity and respective capabilities in light of national circumstances, China has put environmental governance at the same important position as those of economic development and poverty reduction, and got ready to actively engage in international talks on climate change so as to yield successful results. Just as Chinese Premier Li Keqiang said, it is indeed a challenge for developing countries to address climate change, "It will bring pains and pressure in the process, yet, in the end, it will bring the continued and healthy economic development in return."

References

1. Ren, G., Chu, Z., Zhou, Y.: Progress of research on temperature change in China. Climate Environ. Res. **10**(4), 701–716 (2005)
2. Liu, Y., Lin, E.: Impacts of climate change on agriculture in different regions in China. Res. Progr. Climate Change **3**(4), 229–233 (2007)
3. Cai, Y.: China agriculture vulnerability and adaption strategies against global climate change. Acta Geogr. Sin. **51**(3), 202–212 (1996)
4. Zhang, J., Wang, G.: The impacts of climate change on water resources and hydrology. Science Press, Beijing (2007)
5. Zhang, J., Wang, G.: Climate change's impacts on hydrology and water resources. Science Press, Beijing (2007)
6. Hu, S.: Propel the construction of water-saving society. Speech on the symposium of national water saving society experience. China Water Energy Electr. **11**, 9–15 (2007)
7. Qi, R.: Animal phenology responses to climate change in Qinghai Province. Qinghai Meteorol. **1**, 28–31 (2006)
8. Li S., Tang W., Tang X.: Impacts of climate change on barn swallows phenology change. Anhui Agr. Sci. **37**(18), 8531–8532, 8640 (2009)
9. Fang, J., Piao, S., Field, C.B., et al.: Increasing net primary production in China from 1982 to 1999. Front. Ecol. Environ. **1**(6), 293–297 (2003)
10. Cao, L., Dou, Y., Zhang, D.: Effect of climate change on ecological environment of Heihe field. Arid Meteorol. **21**(4), 45–49 (2003)
11. Tian, X., Wang, M., Shu, L.: Trends and prevention strategies for forest fires in China in globalization. Forest fire prev. **3**, 32–34 (2003)
12. Tian, X., Shu, L., Ali, P.: Literature review on forest fires (III)—Impacts of ENSO on forest fires. World Forest Res. **16**(5), 22–25 (2003)
13. Zhao, F., Shu, L., Di, X.: Temporal patterns change of forest fires' frequency in Daxinganling in Inner Mongolia of China. Forestry Sci. **45**(6), 166–172 (2009)
14. Liu, L., Sha, Y., Bai, Y.: Spatial distribution of China major agricultural meteorological disasters and disaster relief strategies. Natural Disaster J. **12**(2), 92–97 (2003)
15. Yang, J., Pan, X.: Agricultural climate and its changing features in the agro-pastoral ecotone in the north foot of Yinshan mountain. Inner Mongolia Meterol. **4**, 3–5 (2008)
16. Zhang, J., Wang, G., Liu, J., et al.: Literature review on the impacts of climate change on water resources. Renmin Chang River **40**(8), 39–41 (2009)
17. Moseley, R.K.: Historical landscape change in northwestern Yunnan, China. Mt. Res. Dev. **26**, 214–219 (2006)
18. Ren mei'e.: Sea levels increasing trend in these 30 years in Huang River, Yangtze River and the Pearl River Delta, and the forecasting of increasing in 2030. Acta Geogr. Sinica **48**(5): 385–393 (1993)
19. Li, P., Fang, G., Huang, G.: Potential impacts and strategies from sea levels rise on economic construction in Pearl River Delta regions. Acta Geogr. Sin. **48**(6), 527–534 (1993)
20. Zhou, X., Zhang, Y., Sun, H.: Impacts of climate change on Quercus mongolica community dynamics in the north of Daxinganling. J. Ecol. **22**(7), 1035–1040 (2002)

Part III
Sustainable Development and Mobility

Introduction

Anne-Marie Coles

Social mobility has a direct impact on the so-called 'bottom triple line' of sustainable development, requiring protection not only for the environment, but also social and economic activities that rely on access to affordable and reliable transport systems [1]. One of the core issues when considering how to intervene in such a sector is to understand how it can contribute to sustainable development. Although this term has a range of interpretations, there appears to be an international consensus that it should promote social justice now and for future generations. This definition was first mooted in the 1987 report of the United Nations, World Commission on Environment and Development, 'Our Common Future' [2]. The theme has subsequently been adopted, in 1992 by the United Nations Conference on Environment and Development (UNCED) in Rio de Janeiro, with publication of the Agenda 21 action plan. Also, more recently, in 2015, in the adoption of a renewed 'Agenda for Sustainable Development' comprising a set of development goals to be implemented by 2030 [3]. Sustainable mobility is incorporated within these goals as it is included in aims to promote human well-being, affecting aspects such as access to work, the requirement for sustainable infrastructure as well as for sustainable urban regeneration. However, as transport currently accounts for 20% of energy use worldwide, it remains a challenge to the development of sustainable practices [4].

The reconfiguration of global, interrelated systems of mobility is an ambitious target. Transport systems typically comprise a network of heterogeneous actors that cut across stakeholder interests in both innovation of new artefacts and in the process of travel. There are a range of forms, in terms of air, water and road travel and these have differential geographical implications, with impacts at local, regional, national and international levels. In terms of understanding the complexity of mobility it is necessary to consider who or what is being transported, and how much distance is being covered. Also, this is an issue which impacts unevenly in urban and rural areas, with the latter possibly more dependent on the use of fossil fuels

A.-M. Coles (✉)
Greenwich University, London, UK
e-mail: A.Coles@greenwich.ac.uk

M. Palocz-Andresen et al. (eds.), *International Climate Protection*,
https://doi.org/10.1007/978-3-030-03816-8_18

for transportation due to a lack of viable alternatives. In terms of the promotion of environmental sustainability, short term decision making by the public can have a significant environmental impact. For example, campaigns to educate the public into walking or cycling for shorter distances carry both objectives of promoting a healthy lifestyle as well as reducing the impact of the local use of private automobiles. In addition, the choice of using public transport both locally and nationally can deliver environmental benefits. However, it is also clear that permanent change is dependent on sustainable innovation in the production of new vehicles and the development, maintenance and renewal of transport systems. There are examples where new types of public transport infrastructures have been trialed, such as the European fuel cell bus projects [5]. More recently, in Milton Keynes in the UK, an electric bus project has demonstrated an innovative system of battery charging based on induction power transfer [6]. However, despite these schemes sustainable public transport systems are still underdeveloped.

In terms of innovation for sustainable mobility there has also been much emphasis on new technologies for the privately owned automobile. Developments have mainly focused both on use of renewable materials in manufacture and on the potential for innovation in vehicle fuels. In terms of the former, research is being carried out into novel materials such as bio-plastics and plant fibre, as well as focusing on the need to reduce the volume of material use and recycling where appropriate [7]. Development of vehicle engines that avoid the use of fossil fuels has, it is fair to say, taken longer than expected. This is potentially due to the fact that there are a number of nascent alternative technologies, but not yet one that clearly has advantages of cost and performance over the incumbent technology. Fuel cells for vehicles, for example were pioneered by the small Canadian company Ballard, which was established in 1974, but are still not yet fully commercially viable [8]. Electric vehicles have just passed the 1% sales mark in Europe and remain entrenched in a market niche [9]. Both these fuel types face the major problem of slow infrastructural development for refueling (see, for example, Clarke [10]). In addition, the use of biofuels is still at an early stage worldwide. The sustainable transformation of private transport is still a long term goal, however, issues related to achieving sustainable development in other types of transport is, arguably, more difficult. A report by the Institute of Transport Studies at the University of Leeds, UK in 2010 noted that the amount of freight traffic in Europe is increasing. Moving to a more sustainable position will involve not only considerable investment in research and development but will also require incentives for the operators to adopt suitable solutions. A recent report for the European Parliament notes that a similar situation bedevils air flight [11].

Global energy consumption has stagnated in the last two years, achieving the goals of COP 21 (Paris 2015) is far behind schedule. On 6 April 2009, the European Council (EC) adopted the EU Energy and Climate Package (2009/28/EC), which provides 20% share for renewables in total energy consumption for the EU as a whole. On 30 November 2016, the European Commission presented a Clean Energy for All Europeans Package. To meet these requirements, it is necessary to develop and use renewable energy sources and improve methods of energy production. However, it is not enough to introduce new devices to the market, the education and co-operation of the population is an indispensable part of the design of green cities.

This chapter aims to contribute to emerging research in the area of sustainable mobility, renewable energies, green cities and new technology of environment protection. The following sections will focus on a range of issues relating to current environmental problems from existing transportation practices, energy use and urban development. In addition, moves towards potential solutions will be outlined. Dóra Szalay's section will review the current state of biofuels, by type and application and will evaluate the strengths and weakness of this fuel. Following this contribution, Venkatesan Kanagaraj and Martin Treiber report on research that has developed a model to produce instananeous information on individual vehicle fuel consumption and emissions, while Gowri Asaithambi and his co-authors made a comparison between the life cycle CO_2 emissions of internal combustion engine vehicles and electric vehicles. The next two contributions focus on policies and practical actions that can be taken to reduce greenhouse gas emissions and ecological damage more broadly. László Rácz, specifically considers potential for change in the industrial sector, while Eggo Bracker considers the environmental impact of shipping and maritime transport.

Béla Bakó analyses the importance of functionalist urban development to improve the quality of life of the people living in cities. The following sections define some of the new opportunities offered by advanced renewable energy production. Otto Horváth and Lajos Fodor describe the most relevant types of photocatalytic systems, which are driven by sunlight in the ultraviolet-visible range, producing environmentally friendly fuels such as hydrogen. The next section focuses on the energy concept by the University of Lüneburg, which achieved the first climate-neutral energy balance for heat, electricity, cars and business trips in 2014. László Bánhidi reports on study, which takes into account a thermal comfort in the dimensioning of closed spaces. The last part of this section presents the situation with use of geothermal energy in Hungary and the Pannonian basin. This section presents a historical review, and comments on the necessity and processes of thermal water re-injection.

In conclusion, it can be stated that, although transportation and renewable energy as a sector poses immense challenges to the emergence of a more sustainable paradigm, progress can be made through the co-operative interaction between individual action, infrastructure renewal and appropriate innovation strategies.

References

1. Elkington, J.: Cannibals with Forks: The Triple Bottom Line of Twenty-First Century Business. Capstone, Oxford (1997)
2. Brundtland Commission: Report of the World Commission on Environment and Development, United Nations (1987). https://en.wikipedia.org/wiki/Brundtland_Commission
3. United Nations General Assembly: Transforming Our World: the 2030 Agenda for Sustainable Development, United Nations (2015). https://en.wikipedia.org/wiki/United_Nations
4. EIA: International Energy Outlook 2016, U.S. Energy Information Administration, Office of Communications, EI-40 U.S, Forrestal Building Washington, DC 20585 (2016)
5. Chrysler, D.: The HyFLEET:CUTE—project and experiences out of the CUTE, ECTOS and STEP-projects, Brussels. http://gofuelcellbus.com/uploads/CUTE_Results_2005_reduced.pdf. Accessed 7 Mar 17
6. Kontoua, A., Miles, J.: Electric buses: lessons to be learnt from the Milton Keynes demonstration project. Procedia Eng. **118**, 1137–1144 (2015)
7. Gilbert, N.: Renewable materials for automotive manufacture—the next big thing in green motoring? AZO Mater. (2013)
8. Peters, S.R., Coles, A.-M.: Strategic innovation in sustainable technology: the case of fuel cells for vehicles. Int. J. Environ. Sustain. Dev. **5**(4), 338–354 (2006)
9. Fergusson, M.: Electric Vehicles in Europe 2016—Approaching Adolescence. European Federation for Transport and Environment, Brussels (2016)
10. Clarke, I.: Range anxiety and electric vehicles: 'battery swapping' as an example of technological and business model experimentation. Paper presented at Science Policy Research Unit 50th anniversary Conference, Sussex University, September (2016)
11. Aviation Steering Group: European Aviation Environmental Report 2016, EASA (2016)
12. Institute for Transport Studies: The Future of Sustainable Freight Transport and Logistics University of Leeds report for European Parliament Directorate-General for Internal Policies Transport and Tourism Workshop on the Future of Transport (2010)

Development of Biomass and Biofuel Usage

Dóra Szalay

1 Introduction—The Beginning

The operation of most mechanical devices requires energy that can be drawn from various sources. According to modern conceptions, transport can be defined as carrying people or goods from one place to another via a vehicle, aircraft, or ship (Oxford English Dictionary). The earliest form of transport was, of course, walking, the energy source of which is food. The unconscious use of biofuels began at the dawn of humanity's development around 3–3.5 million years ago. The first biofuels were grasses and succulent plants with meat being incorporated into the mix approximately 2.6 million years ago [1, 2]. The sole use of human power gave way to the harnessing of animal power following the domestication of animals in 15,000 BC. This development not only increased efficiency, but also gave humans more control over nature. As with humans, animals consume vegetation and/or meat for energy. The first known use of vessels dates back to about 10,000 years ago [3]. Today, human-powered transport remains a popular cost saving option in developed countries where it is most often linked to leisure, physical exercise and environmentalism; however, in underdeveloped regions or areas that are difficult to access with vehicles, walking and other forms of human-propelled transport often remains the only viable forms of transportation [4].

The first simple "machines" humanity invented were the lift and the roller. These were followed by the wheel around 3,500 BC. These developments initiated the mechanization of transportation equipment. Primitive modes of transport include the sled, the tumbril, the chariot, and the boat. By about 3,000 BC, ancient Egyptians learned how to assemble wooden planks into a hull [5]. These simple innovations were eventually improved and made suitable for transporting things and people. Boats are among the oldest vehicles in the world and ancient people built and used

D. Szalay (✉)
Institute of Forest- and Environmental Techniques, University of Sopron, Sopron, Hungary
e-mail: szalay.dora@uni-sopron.hu

M. Palocz-Andresen et al. (eds.), *International Climate Protection*,
https://doi.org/10.1007/978-3-030-03816-8_19

Table 1 Evolution of steam powered vehicles

Date	Inventor	Process	Fuel
1712	Thomas Newcomen	**first steam-engine**	coal
1769	James Watt	modern steam-engine	coal, wood
1769	Nicolas Cugnot	**first steam-car**	wood
1802	William Symington	steam boat	
1807	Robert Fulton	steam ship	coal, oil, wood
1825	George Stephenson	steam train	
1852	Henri Giffard	steam airship	lighting gas
1933	William & George Besler	steam airplane	oil

them mostly for fishing and transportation. The first boats had no real navigation controls, but eventually people discovered how to steer and control the movement of water craft. Boats were first propelled by rowing, but the first sails appeared in the Pacific and the Indian Ocean about 6,000 years ago [6]. The sail not only marks the start of wind energy utilization in transport, but is also an early example of the harnessing of a renewable energy source. From this point of view, wind cannot be considered a new alternative energy source because its application came much earlier than the utilization of fossil fuel did.

2 Barriers to the Early Alternative Fuel Use

Involving other energy sources for machine propulsion did not occur until the seventeenth century and the advent of the steam engine. Initially steam engines ran on coal and wood, but later gasoline, kerosene, oil, and gas were also used for fuel; see Table 1. Despite the invention of the steam engine, the technology did not become widespread until the Industrial Revolution.

The first steam car was built specifically for military purposes in 1769 and had the advantage of being able to transport people. Though wood was powered the first steam engines [7], later designs could use fuel such as petrol, kerosene, fuel oil, and coal. Steam traction engines were widespread and in the early 1900s, making up about 40% of the American vehicles in circulation at that time [8]; however, with the appearance of the internal combustion engine, steam engines became somewhat obsolete and their numbers decreased rapidly. Their ultimate end came during the 1973 oil crisis and the subsequent age of emission limits and modern, low fuel-consumption automobiles [9].

At the beginning of the nineteenth century, a new chapter in transport history began with the introduction of steamships. Steamships could travel long distances and deliver goods much faster than sailing vessels could. By the beginning of the

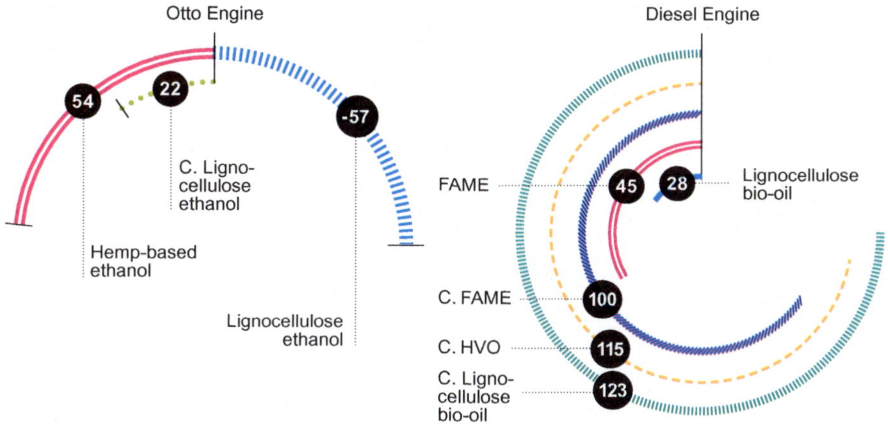

Fig. 1 Development of biofuel production (C–Commercial size)

twentieth century sailboats were all but displaced from commercial shipping. Within the same century, diesel-powered ships eventually displaced steamships.

3 Evolution of Renewable Fuel Use

Biofuels belong to alternative fuels group. In 1804, Isaac de Rivaz developed the world's first internal combustion engine, which used hydrogen and oxygen as fuel [10]. In 1839, Sir William Grove invented the fuel cell (Grove Gas Battery) [11], and in 1959, Harry Karl Ihring built the first propane powered fuel cell tractor [12].

Ethanol was known as far back as the 1100s when it was extracted from the winemaking process. The first Otto four-stroke engine in 1876 used ethanol as fuel. Though the engine for the first Ford Model T in 1908 was designed to run on gasoline, it could also run on other combustible fuels like ethanol and petroleum. The use of hemp-based ethanol became widespread in the 1930s with 54% of pre-war German fuel production being derived from non-petroleum sources; 8% of this volume being ethanol derived from renewable sources [13]. Nevertheless, big oil companies lobbied for and eventually succeeded in getting Ford's alcohol powered engine banned. In 1925, Henry Ford remarked: "The fuel of the future is going to come from fruit like that sumac out by the road, or from apples, weeds, sawdust—almost anything."

The development of the biodiesel fuel has taken a much longer time interval (Fig. 1). Originally, diesel engine fuel was produced from peanut oil, but Rudolf Diesel attempted to use coal dust and kerosene in the process as well. In the end, only fossil-based diesel fuel became widespread and it eventually took over the market. Concerning the matter, Rudolf Diesel made the following statement in 1912: "The use of vegetable oils for engine fuels may seem insignificant today, but such oils may become, in the course of time, as important as petroleum and the coal tar products of the present time." In 1901, Thomas Hugh Parker built the first wood-fuel powered car. With the mass production of Imbert's wood gas generators in 1931, about 9,000 wood gas vehicles were in use by the end of the 1930s, with almost all of them exclusively in Europe.

Wood gas generator technology became commonplace in many European countries during the Second World War as a consequence of fossil fuel rationing. Altogether, more than one million wood gas generators were used during World War Two. After the war, gasoline was once again readily available, which made the technology obsolete almost immediately. The development of electric cars began in 1828 when Hungarian engineer Ányos Jedlik invented the world's first electric car model. Davenport's first electric car ran on a short section of railway track. In 1859, Gaston Planté's solenoid battery made the development of this sector possible. Electric vehicles had a number of advantages over their early-1900s competitors—they lacked the vibration, smell, and noise associated with gasoline cars—but their high cost, low top-speed, and short range compared to later internal combustion engines quickly led to a worldwide decline in their use.

4 Expected Trends

Rigorous restrictions currently determine development trends. The European Union (EU) limits food crops for energy production. On 6 April 2009, the European Council (EC) adopted the EU Energy and Climate Package (2009/28/EC), which provides a 20% renewable share in total energy consumption for the EU as a whole. A further objective is a 10% share of renewable energy in the total energy use of transport by 2020, which is binding for all EU Member States (RED).

The 2015/1513 Directive (iLUC) represents a major tightening of previous regulations by limiting the share of biofuels from crops grown on agricultural land that can be counted towards the renewable energy targets of 2020 to 7%. RED II states that by 2030, at least 14% of transportation fuel must come from renewable sources. Crop-based biofuels are capped at 2020 levels—with an extra 1%—but cannot exceed 7% of the final consumption of road and rail transport for each EU member state. In addition, the share of advanced biofuels must be at least 0.5% in 2022, 1% in 2025, and at least 3.5% in 2030.

In 2016, 11% of the world's energy consumption was covered by biomass while biofuel use in the road transport sector was 4% [14, 15]. An important task will be to incorporate non-food by-products from agriculture, forests, and households in biofuel production. Due to recent technological developments, a wider range of base materials for biofuel production is currently available; see Fig. 2.

The conversion of applied new lignocellulose base materials into biofuels requires a much more complex technology than first-generation biofuel production did. The production of advanced fuels is not cost-effective due to several technical barriers that need to be overcome before the potential of advanced fuels can be realized. Five second-generation lignocellulose biofuel plants currently operate commercially in

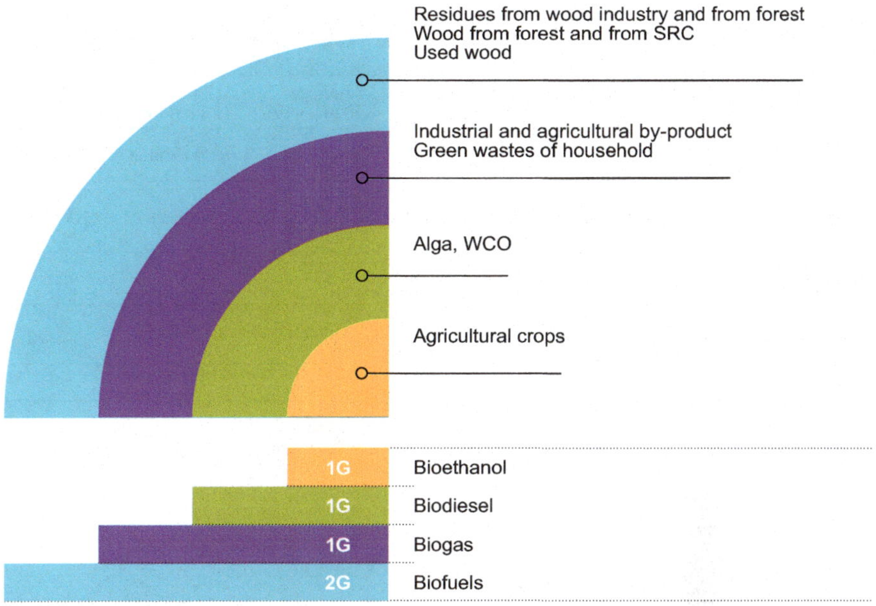

Fig. 2 Expanding the choice of base materials for advanced biofuels (SRC–Short Rotation Coppice, WCO–Waste Cooking Oil)

the EU, with most producing mainly heat and electricity. Their conversion rate and production capacity is lower than those found in first generation biofuel production; see the example noted in Table 2. Furthermore, the investment costs associated with lignocellulose biofuel plants is almost 10 times higher per unit of biofuel than it is for conventional biofuel production plants. However, second generation biofuels may be more advantageous regarding land use demand. In addition, air quality regulations can affect trends for the use of raw materials for passenger cars. More than half of all European passenger cars run on petrol, with about 41% running on diesel fuel in 2015. Diesel was a popular fuel among consumers, especially in Western Europe, but the 2009 economic crisis generally slowed down new car purchases of diesel-powered vehicles and an actively decreasing trend became visible after 2011; see Fig. 3.

Some reasons for the decrease in diesel fuel-powered car purchases is the diesel crash of 2015 and the ever-tightening regulations imposed by big cities in Western European countries. Several major cities are also planning to ban diesel cars from their streets within the next ten years, adding another impetus for decline. There are basically two types of planned restrictions:

- Restrictions according to vehicle age.
- Restrictions according vehicle classification.

Both regulations are primarily concerned with diesel-powered cars, and they have been or will be applied as intermittent or permanent traffic restraint measures in

Table 2 Example of production data of conventiona and lignocellulosic biofuel plants

	1st generation		2nd generation		
Local	Hungrana, Hungary	Rossi biofuel Hungary	Abengoa* Kansas, USA	Empyro Netherland	St1 Finland
Base material	corn	oilseed, UCO, animal fat	corn stalk	wood residue	sawdust
Technology	fermentation	transesterification	enzymatic hydrolisis	pyrolisys	enz. hydrolysis ferm
Base material need [10^3 t/yr]	1000	450	350	37	n.a.
Product [t/yr]	355 ethanol	150 biodiesel	78 ethanol	24 biooil**	40 ethanol
Yield	1:0,35	1:0,33	1:0,22	1:0,65	n.a.
Investment [m EUR]	120	30	200	19	40

** The plant is closed but not due to a technological problem*
*** Requires further treatment for using as biofuel*

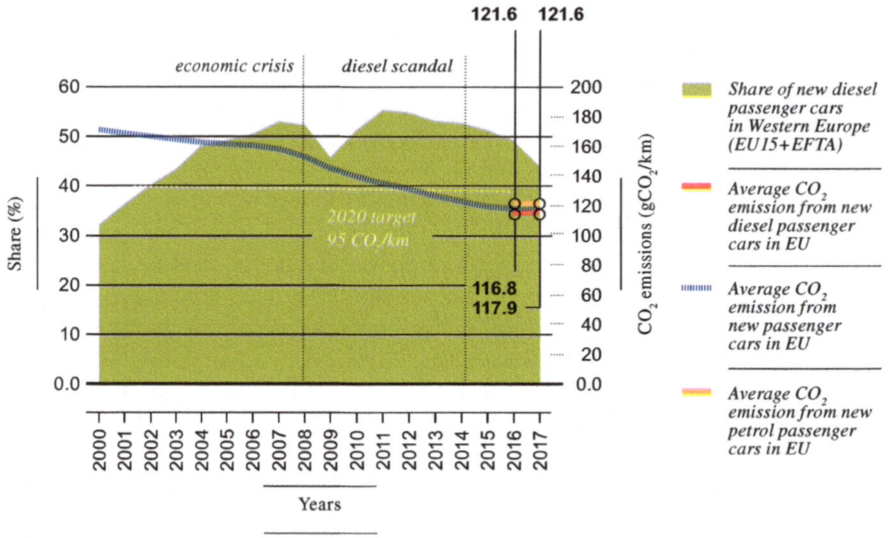

Fig. 3 Share of diesel in new passenger cars in the western countries and the average CO_2 emissions of a new passenger cars in the EU [16–18] (EFTA–European Free Trade Association)

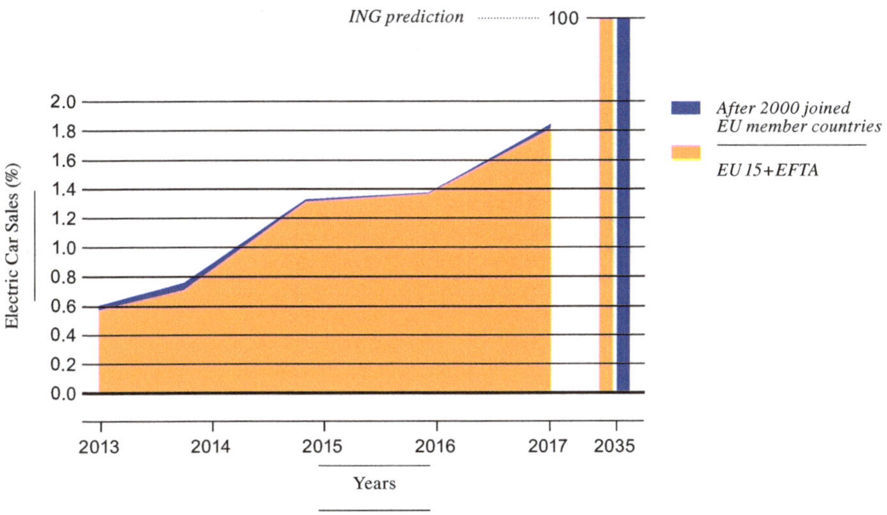

Fig. 4 New electric passenger car registration in Europe in proportion to all new passenger car registration [19–23] (EFTA–European Free Trade Association)

almost 200 cities across Europe. In addition, several major automotive companies have abandoned producing some diesel models or have no initiatives to introduce the production of new diesel models (Fiat, Porsche, and Toyota). At the same time, there has been growing interest in the electric vehicles launched in the 2010s. Global electric car stock surpassed two million vehicles in 2016 after crossing the one-million threshold in 2015 [15]. RED II encourages the spread and use of electric vehicles; the contribution of renewable electricity supplied to road vehicles shall be considered to be four times its energy content. By 2040, the electric vehicle sector, including hybrids, is expected to become the sector with the highest development; see Fig. 4.

5 Summary

In ancient times, people relied on their own bodies or on animals for transport. In the eighteenth century, technical developments opened the door for the construction of the first transportation means. Steam engines used coal, the first significant type of fossil fuel. The sailboat, which used the renewable energy of wind, became an innovation around 5,500 BC, and remained a primary technology for over 6,000 years until it was supplanted by steamboats in the 1800s. On the whole, nearly all early automobiles used alternative fuels such as hydrogen, ethanol, vegetable oil or wood gas, as energy sources. While the commercial production of vehicles needed only a few decades to evolve, the commercial production of biofuels took one or two centuries and the process is still ongoing today. However, biofuel demand will grow more intensively than it did in the past due the expected increases in automobile numbers. Despite the 7% limit, the amount of biofuel production from food crops continues to rise. After 2020, REDII (Renewable Energy Directive II) can promote greater use of agricultural and forest by-products. However, the use of these by-products is ecologically questionable due to concerns regarding organic matter removal. Furthermore, investment in this field currently carries a great deal of risk due to the lack of safe technology. Nevertheless, predictions for future applications are economically net positive. Further developments and new inventions will surely improve the cost situation of future biofuels.

References

1. Pobiner, B.: Evidence for meat-eating by early humans. Nat. Educ. Knowl. **4**(6), 1 (2013)
2. Wynn, J.G., Sponheimer, M., Kimbel, W.H., Alemseged, Z., Reed, K., Bedaso, Z.K., Wilson, J.N.: Diet of Australopithecus afarensis from the Pliocene Hadar Formation, Ethiopia. PNAS June 25, vol. 110, no. 26 (2013)
3. Niekus, M.J.L.T., de Roever, J.P., Smit, J.L.: Een vroeg-mesolithische nederzetting met tranchetbijlen bij Lageland (Gr.). Paleo-Aktueel **8**, 28–32 (1997)
4. Wikipedia: Transport (2016). https://en.wikipedia.org/wiki/Transport
5. Cheryl, W.: World's oldest planked boats. Archaeology **54**(3) (2001). ACEA European Automobile Manufacturers Association (2018): Share of Diesel in New Passenger Cars. http://www.acea.be/statistics/article/Share-of-diesel-in-new-passenger-cars
6. DAWN: Sailboat History Timeline. https://www.dawn.com/news/617729 (2011)
7. Foster, K.: 1771 Cugnot Fardier a Vapeur: the mother of all motorcars. Autoweek (2001). http://autoweek.com/article/car-news/1771-cugnot-fardier-vapeur-mother-all-motorcars
8. Purdy, K.W., Foster C.G.: History of the Automobile. https://www.britannica.com/technology/automobile/History-of-the-automobile
9. MM Mu˝velt Mérnök: Padlógőzzel. A Williams Steamer gőzautó (2014). http://gyartastrend.hu/autoipar/cikk/padlogozzel_a_williams_steamer_gozauto
10. HCN Hydrogen Cars Now: Hydrogen Fuel Cars 1807–1986 (2018). http://www.hydrogencarAQ3179snow.com/index.php/1807-1986/
11. NMAH National Museum of American History: Fuel Cell Origins (2004). http://americanhis165tory.si.edu/fuelcells/origins/orig1.htm
12. BPA Big Picture Agriculture: 1959 Fuel Cell Tractor (2015). http://bigpictureagriculture.blogspot.hu/2015/04/1959-fuel-cell-tractor.html
13. Egloff, G.: Motor fuel economy of Europe. American Petroleum Institute, Washington, D.C. (1939)
14. Enerdata: Global Energy Statistical Yearbook (2017). https://yearbook.enerdata.net/oil-prod169ucts/world-oil-domestic-consumption-statistics.html
15. IEA: Tracking Progress: Transport biofuels (2017). https://www.iea.org/etp/tracking2017/transportbiofuels/
16. ACEA European Automobile Manufacturers Association: Share of Diesel in New Passenger Cars (2018). http://www.acea.be/statistics/article/Share-of-diesel-in-new-passenger-cars
17. EEA: No Improvements on Average CO_2 Emissions from New Cars in 2017. https://www.eea.europa.eu/highlights/no-improvements-on-average-co2 (2018)
18. EEA: Average Carbon Dioxide Emissions from New Passenger Cars. https://www.eea.europa.eu/data-and-maps/daviz/average-emissions-for-new-cars-4#tab-chart_1 (2018)

19. Erich, M., Witteveen J.: Breakthrough of electric vehicle threatens European car industry. ING. https://www.ing.nl/media/ING_EBZ_breakthrough-of-electric-vehicle-threatens-European-car-industry_tcm162-128687.pdf (2017)
20. ACEA: New Electric Vehicle Registrations in the European Union. https://www.acea.be/uploads/news_documents/ACEA_Electric_Vehicle_registrations_Q4_2014-2013.pdf
21. ACEA: New Passenger Car Registrations by Alternative Fuel Type in the European Union. https://www.acea.be/uploads/press_releases_files/AFV_registrations_Q4_2015_FINAL.PDF
22. Electric Car Sales Per European Union and EFTA Country in 2017 (Q1–Q4). https://www.best-selling-cars.com/europe/2017-full-year-europe-electric-hybrid-vehicle-sales-per-eu-efta-country/
23. ACEA: Annual Car Registrations in the EU and EFTA. https://www.best-selling-cars.com/europe/2018-full-year-europe-car-sales-per-eu-and-efta-country/

Fuel-Consumption and CO_2 Emissions Modells for Traffic

Venkatesan Kanagaraj and Martin Treiber

1 Introduction

Generally, models for fuel consumption and for emissions (CO_2, CO, NO_x, particulate matter, and others) have the same structure, so they can be discussed together. These models establish a relationship between the exogenous variables (traffic demand, properties of traffic flow, vehicle composition and infrastructure) to following one of two sets of endogenous variables: Local emission factors describe fuel consumption or emissions in kg per meter (or liters per meter). Instantaneous emission factors describe fuel consumption or emissions in terms of kg per second per vehicle (or in liters per second per vehicle). Generally, two different types of models are available for fuel consumption and emissions, namely macroscopic and microscopic models. Depending on the aggregation level and level of detail, there are several model categories (Fig. 1).

1.1 Macroscopic Models

1.1.1 Area-Wide Models

This model is simple and macroscopic approach. The model input is the total vehicle mileage (traffic volume integrated over the total link length of the network and over time) in the investigated region in terms of vehicle kilometers travelled (VKT). As output, these models deliver the global fuel consumption and emissions in the investigated area. This will be usually disaggregated at least with respect to passenger cars and heavy-duty vehicles (trucks). Each of these categories may be

V. Kanagaraj (✉) · M. Treiber
Technical University of Dresden, Dresden, Germany
e-mail: vkanagaraj.iitm@gmail.com

M. Palocz-Andresen et al. (eds.), *International Climate Protection*,
https://doi.org/10.1007/978-3-030-03816-8_20

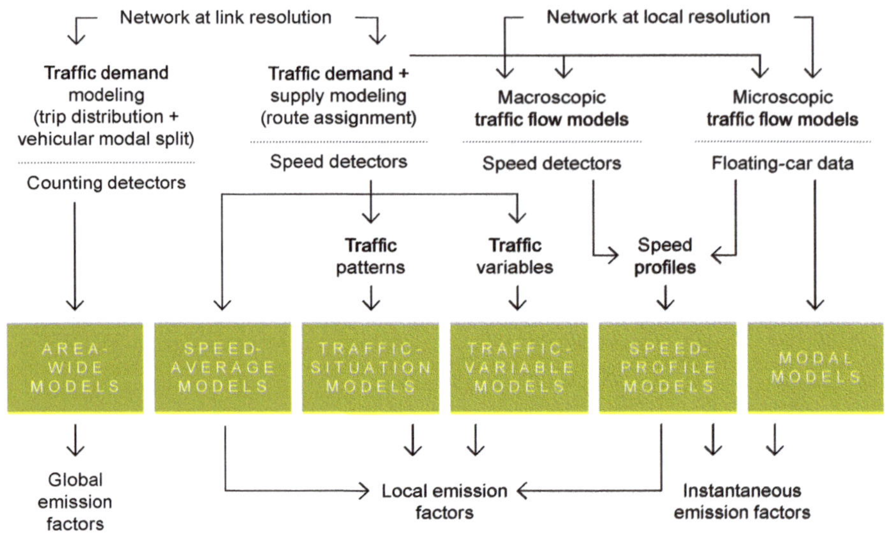

Fig. 1 Overview of fuel consumption and emission models

further disaggregated into several vehicle classes. The model input can be estimated by traffic demand models or by detector data since area-wide models are related to transportation planning rather than traffic flow modelling.

1.1.2 Average-Speed Models

The average-speed models use as input the average speed driven on a certain link of the considered network. In addition, some of these models include external factors such as the temperature. The standard tools to obtain the speed information obtained from the transportation planning models such as traffic demand and route assignment models. Other alternatives, one can directly measure speed and traffic volume by double-loop detectors or other stationary speed-detecting devices. The model output are local emission factors, i.e., volume or mass of consumed fuel or emitted pollutant per kilometer and per vehicle, on average. To date, the majority of fuel consumption and emission software use this model class, e.g., COPERT, MOBILE, MOVES or EMFAC.

But this kind of model is more related to transportation planning than to traffic flow modeling and cannot determine the effect of jams.

1.1.3 Traffic-Situation Models

The input of this model input consists of several distinct driving patterns. Most traffic-situation models define the set of driving patterns as a combination of the set of traffic-flow patterns (e.g., free, congested, stop-and-go), and the set of driving facility (e.g., highway, rural road, arterial road, residential street).

The traffic-flow situation may also be defined in terms of the level of service (LOS) assessing the traffic flow quality on an ordinal scale from 1 (best) to 6 (worst, i.e., completely congested). LOS is estimated based on average speed and ratio of volume/capacity of the particular link to be considered. But it is more tricky if stop-and-go traffic should be distinguished from stationary situations. Software implementing this class includes HBEFA, ARTEMIS, and some versions of MOBILE.

1.1.4 Traffic-Variable Models

In contrast to the traffic-situation models depends on a finite set of qualitative (categorically scaled) traffic patterns, this model class takes as input quantitative (i.e., metrically scaled) macroscopic factors related to traffic flow such as traffic density, traffic volume relative to capacity, queue length, and speed. These variables depends on the road category, design speed, signal cycles, link length, number of lanes, and type of intersection, etc. The models from transportation planning are no longer sufficient to provide this input but they have to be complemented by microscopic or macroscopic traffic flow models. The output of this model are local emission factors, usually related to a single vehicle. Representatives of this class include TEE (Traffic Emissions and Energetics) and the queue-based Matzoros Model.

1.2 Microscopic Models

The microscopic consumption and emission models need speed profiles of single vehicles at a high temporal resolution (of a few seconds or less) which can only provided by microscopic traffic flow models or by floating-car or trajectory data. As output, these models deliver local or instantaneous emission factors of single vehicles of a certain vehicle class. While traffic-situation and traffic-variable models evaluate coarse assessment affect to which degree congestions influence consumptions/emissions, only microscopic consumption/emission models allow us to answer questions related to individual vehicles and drivers such as.

How much fuel/emissions can be saved by a fuel-efficient driving style? How does this saving potential depend on different traffic conditions? Is it possible to implement a fuel-efficient behavior into driver-assistance systems for the longitudinal driving task (adaptive cruise control)? What saves more fuel/emissions: Avoiding high accelerations/decelerations or driving at low engine speeds? Are roundabouts or signalized intersections more fuel-efficient? Does it depend on the type of roundabout, or on the origin-destination (OD) matrix characterizing traffic demand and the topology of the intersection? Is the savings potential and/or the optimal driving style different when switching from traditional combustion engines to modern developments such as hybrid or all-electric cars?

What is the savings potential of recent ITS (Intelligent Transportation Systems) such as vehicle-to-vehicle or infrastructure-to-vehicle communication (e.g., a traffic light communicating its switching times to equipped cars)? How do the effects depend on the penetration rate of such ITS implementations? The microscopic consumption/emission models are strongly related to traffic flow models and traffic flow dynamics in general, and principally, they classify as two classes.

1.2.1 Speed-Profile Emission Models

This model class does not use the instantaneous information provided by the simulated or measured trajectories directly. Rather, it is aggregated to several speed profile factors of a driving cycle which, in turn, determine the instantaneous consumption and emission factors. speed profiles of single vehicles at a high temporal resolution which are obtained by floating-car data, trajectory data, or by a microscopic traffic flow simulation. At this level of detail, less road geometry data is needed since the speed profiles implicitly contain most of this information (except in road gradients). The outputs of speed-profile models are either local or instantaneous emission factors which are related to a single vehicle. Most models of this class (e.g. MEASURE or PKE) assume a linear multivariate mapping between the speed profile factors x and the estimates e of the instantaneous emission factors:

$$e = \mathrm{L} \cdot \mathrm{x} \tag{1}$$

Here, the components of the instantaneous emissions vector e may contain the CO_2, CO, HC, NO_x, PM (particulate matter), and others emission rate. The n m matrix L represents the linear relations between the m speed profile factors and the n components of the instantaneous emissions vector.

Table 1 displays typical speed profile factors that are used by many models of this class and their influence on consumption and emission varying from strongly negative (– –) to strongly positive (++) with respect to the reference. The coefficient of this factors are estimated by a multivariate linear regression.

1.2.2 Modal Emission Models

Modal emission models use the instantaneous information directly from the trajectory information (of floating-car data or microscopic simulations) (Fig.1). At any more advanced modal models, the vehicle operation mode is expressed by engine speed f(t) (including the idling mode), power demand (or torque), and possibly other history-

Table 1 A selection of common speed profile factors and their effect on CO_2 emissions (– – , strongly negative, –, negative, +, positive, ++, strongly positive)

Factor	Effect on CO_2 emissions
Constant of value 1	Reference (+++)
Fraction of time in speed range 0–25 km/h	++
Fraction of time in speed range 50–75 km/h	– –
Fraction of time in speed range 75–100 km/h	–
Fraction of time in speed range > 125 km/h	++
Standard deviation of speed	+
Average and standard deviation of acceleration	+
Average and standard deviation of deceleration	–
Frequency of acceleration-deceleration cycles	+
Fraction of time the vehicle is standing	+
Fraction of time the vehicle needs power near its maximum power	++
Fraction of road gradients greater than 5%	+
Engine speed (crankshaft revolution rate) 1,000–2,000 rpm	– –
Engine speed (crankshaft revolution rate) > 3,500 rpm	++

related factors such as engine age and temperature. At this microscopic level, the only road geometry related information that is used directly are road gradients and possibly the road surface quality. Depending on the situation and model complexity, further input is also included such as local road-related variables (e.g., uphill grade), external variables (e.g., altitude, air temperature), and variables related to the engine history (e.g., engine temperature). Models of this class are perfectly suited to be used in conjunction with time continuous microscopic traffic flow models. The models are linked such that the endogenous variables of the traffic flow models (speed, acceleration) are exactly the main exogenous variables of the modal emission models. The modal models gave have the potential to give the most precise description, but they have the highest demand on data for calibration, and validation. Particularly, it is extremely difficult to measure the instantaneous emission rates on a continuous basis at a time resolution of seconds.

Acknowledgements The first author is supported by a fellowship of the Alexander Von Humboldt Research Foundation, Germany at the Technical University of Dresden, Germany.

References and Further Reading

1. Treiber, M., Kesting, A.: Traffic flow dynamics: data, models and simulation. Springer, Berlin (2013)
2. Barth, M., An, F., Norbeck, J., Ross, M.: Modal emissions modeling: a physical approach. Transp. Res. Record J. Transp. Res. Board **1520**, 81–88 (1996)
3. Panis, L., Broekx, S., Liu, R.: Modelling instantaneous traffic emission and the influence of traffic speed limits. Sci. Total Environ. **371**, 270–285 (2006)
4. Cappiello, A., Chabini, I., Nam, E., Lue, A., Abou Zeid, M.: A statistical model of vehicle emissions and fuel consumption. In: Proceedings of the IEEE 5th international conference on intelligent transportation systems, pp. 801–809 (2002)
5. Ahn, K., Rakha, H., Trani, A., Aerde, M.V.: Estimating vehicle fuel consumption and emissions based on instantaneous speed and acceleration levels. J. Transp. Eng. **128**, 182–190 (2002)
6. Smit, R., Brown, A.L., Chan, Y.C.: Do air pollution emissions and fuel consumption models for roadways include the effects of congestion in the roadway traffic flow? Environ. Modell. Softw. **23**, 1262–1270 (2008)
7. Ericsson, E.: Independent driving pattern factors and their influence on fuel-use and exhaust emission factors. Transp. Res. Part D **6**, 325–345 (2001)
8. Treiber, M., Kesting, A., Thiemann, C.: How much does traffic congestion increase fuel consumption and emissions? Applying a fuel consumption model to the NGSIM trajectory data. In: TRB Annual Meeting 2008 CD-ROM, Washington, D.C., Transportation Research Board of the National Academies (2008)

Life Cycle Assessment of Conventional and Electric Vehicles

Gowri Asaithambi, Martin Treiber and Venkatesan Kanagaraj

1 Introduction

The transportation sector is one of the major contributors for global climate warming and greenhouse gas emissions. In the last 10 years, the global CO_2 emission increased by 13%, with 25% of the increase coming from transportation sector. Furthermore, by 2050, the global CO_2 emission is still expected to increase by 30–50% [1]. Above two-thirds of global emissions for 2014 originated from just ten countries, with the shares of China (28%) and the United States (16%) far surpassing those of all others. Combined, these two countries alone produced 14.3 $GtCO_2$ out of world total 32.4 $GtCO_2$ [2]. Figure 1 shows the CO_2 emissions from fuel combustion in 2014 by transportation sector and all sectors for top six emitting countries. The proportion of the CO_2 emissions by transportation sector is highest in United States (33.4%) followed by Germany (21.4%) and Japan (17.5%) countries. However, the total CO_2 emissions from all sectors is higher for China (9,134.9 $MtCO_2$) followed by United States (5,176.2 $MtCO_2$) and India (2,019.7 $MtCO_2$). In all the top six emitting countries, CO_2 emission in transportation sector has increased in 2014 compared to 2011 except Japan which shows 1% reduction [2, 3] which may be due to increase in use of alternative fuel vehicles.

Internal combustion engines (ICE) are the main power type of vehicles and contribute most of the total CO_2 emission in the transport sector. Electric vehicles (EVs) may provide a more promising way to solve the problem of CO_2 emissions and air pollution. The advantage of EVs mainly include high tank-to-wheel efficiency, zero/low local emissions and a quiet operation [4]. Therefore, many countries, especially those facing severe energy and environmental problems, have paid great atten-

G. Asaithambi (✉)
Indian Institute of Technology Tirupati, Chittoor, India
e-mail: gowri@iittp.ac.in

M. Treiber · V. Kanagaraj
Technical University of Dresden, Dresden, Germany

M. Palocz-Andresen et al. (eds.), *International Climate Protection*,
https://doi.org/10.1007/978-3-030-03816-8_21

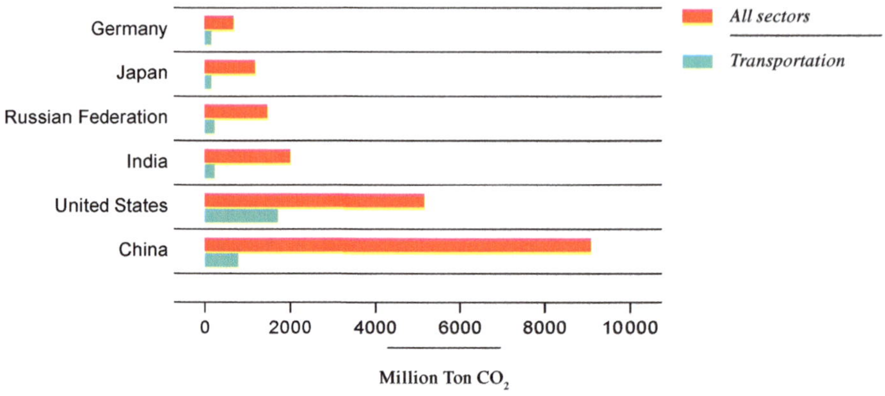

Fig. 1 CO_2 emissions from fuel combustion in 2014 for top six emitting countries (*Source* IEA [2])

tion to develop EVs to partially replace the conventional ICE vehicles. China has the largest market for electric cars in 2015, with over 225,720 registered vehicles whereas United States has a stock of about 210,330 [5]. In Germany, there are 25,502 EVs on the road by 2016 [6].

However, in order to compare EV and ICE vehicles based on energy, emission and economic effects, a comprehensive approach has to be considered. This approach is called life cycle assessment (LCA), which includes: (1) fuel life cycle and (2) vehicle life cycle. Figure 2 shows the life cycle phases for a vehicle. Life cycle of a vehicle is divided into two categories:

Fuel life cycle includes the following processes [7]:

Feedstock production—production of the raw materials in order to obtain the fuel needed;
Feedstock transportation—the raw material has to be transported to the refineries or processing plants;
Fuel production—refining/processing of the raw materials into standard fuel;
Fuel distribution—distribution of the fuels to fuel stations;
Fuel consumption—consumption of fuel during *vehicle operation* (sometimes assessed as part of vehicle cycle).

Vehicle life cycle includes the following processes:

Material production—the materials used include steel, plastics, non-ferrous metals such as aluminium, glass, rubber and composites such as glass fibre;
Vehicle assembly—energy is required to assemble components and operate manufacturing plant;
Vehicle distribution—transport of a vehicle from the assembly line to the dealerships;
Vehicle maintenance—maintenance and repair over the lifetime of the vehicle;
Vehicle disposal—end-of-life vehicles are shredded and a proportion of some materials are recycled for further use.

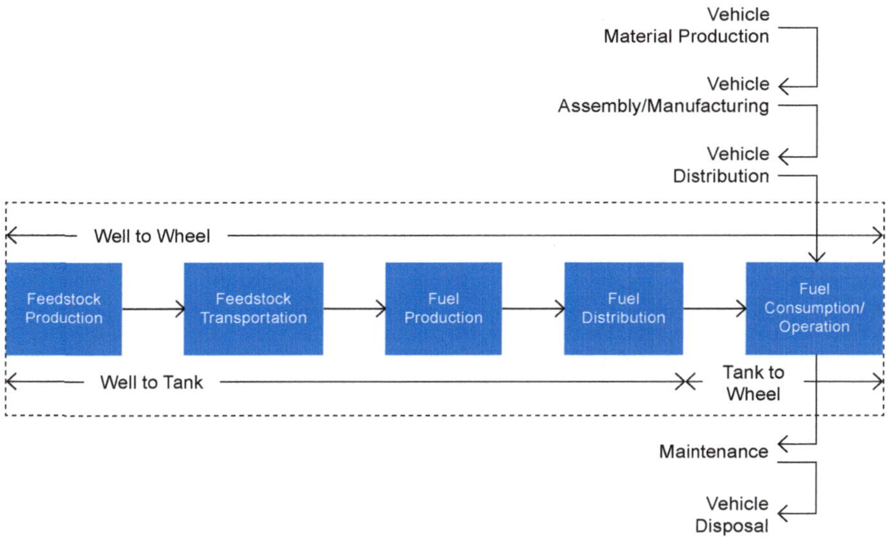

Fig. 2 Life cycle phases

2 Fuel Life Cycle Emissions

The fuel life cycle, also called Well-to-Wheel cycle, includes feedstock production, feedstock transportation, fuel production, fuel distribution, and fuel combustion. This section discusses various studies which focused on calculation of emission in fuel life cycle stage. In most of the studies, the GREET (greenhouse gases regulated emissions and energy consumption in transportation) model developed by the U.S. Department of Energy at the Argonne National Laboratory [8] was used for calculating the emissions of the fuel life cycle. This model first estimates energy use and then the emissions of fuel throughput for all the stages. In fuel cycle, emissions of pollutant (e.g., CO_2, NO_x) for a particular stage of the life cycle is calculated for each fuel type and combustion technology. GREET model includes more than 30 fuel types, involving 13 types of fuel feedstocks (e.g., petroleum, natural gas, coal, hydropower, solar energy, and wind) and 14 fuels (e.g., conventional gasoline conventional diesel, liquefied petroleum gas, compressed natural gas, methanol, biodiesel, and electricity). Because virtually no emissions are associated with electricity generated from hydropower, solar energy, and wind, these cycles are treated together as zero emission cycles in GREET. Combustion technology for a given process fuel are influenced by technology performance, technology costs, and emission regulations.

In the GREET model, combustion CO_2 emission factors in g/J of fuel throughput are calculated by using a carbon balance approach. In this approach, the carbon contained in a process fuel burned minus the carbon contained in combustion emissions for volatile organic compounds (VOCs), carbon monoxide (CO), and methane

(CH_4) is assumed to convert to CO_2. The following formula is used to calculate CO_2 emissions:

$$EF_{CO_2 jk} = \left[\frac{\rho_j}{LHV_j} \times C_{ratio\ j} - \left(VOC_{jk} \times 0.85 + CO_{jk} \times 0.43 + CH_{4,jk} \times 0.75 \right) \right] \times (44 \div 12) \tag{1}$$

where

$EF_{CO_2 jk}$ = Emission factor for CO_2 for fuel j and combustion technology k [g/J]
ρ_j = Density of fuel j [g/l]
LHV_j = Low heating value of fuel j [J/l]
$C_{ratio\ j}$ = Carbon ratio of fuel j
VOC_{jk} = VOC emission factor for combustion technology k, burning fuel j [g/J]
CO_{jk} = CO emission factor for combustion technology k, burning fuel j [g/J]
$CH_{4,jk}$ = CH_4 emission factor for combustion technology k, burning fuel j [g/J]
Carbon ratio of VOC = 0.85
Carbon ratio of CO = 0.43
Carbon ratio of CH_4 = 0.75
Molecular weight of CO_2 = 44
Molecular weight of elemental carbon = 12

Calculations involved in Eq. 1 require fuel specifications such as low heating value, fuel density, weight ratio of carbon which are given in the GREET manual. The above formula shows the calculation method for combustion CO_2 emissions by which carbon contained in VOC, CO, and CH_4 is subtracted. In the GREET model, the indirect CO_2 emissions from VOCs and CO decay in the atmosphere are considered. For example, VOCs and CO reside in the atmosphere for less than 10 days before decay into CO_2. In contrast, Methane has a larger decay time of about 12 years in the atmosphere. Moreover, its greenhouse effect is about 84 times stronger than that of CO_2. We emphasise that this model is general enough to include wind and solar energy and other renewable energy sources.

3 Vehicle Life Cycle Emissions

The following section describes the carbon emissions involved in each stage of the vehicle life cycle stage.

3.1 Material Production

This process considers the materials such as fuels, metals, chemicals, and other resources that make up the vehicle (body and doors, brakes, chassis, interior and exterior, tires, and wheels) and that must be extracted and processed. For the case of EVs, the material used for battery production [e.g., lithium iron phosphate ($LiFePO_4$), lithium nickel cobalt manganese (LiNCM), lithium manganese oxide ($LiMn_2O_4$)] has to be considered as well. The choice of battery depends on cost, lifetime, performance characteristics and behaviour under high and low temperature energy density, etc. Most of the studies calculate the emission of carbon dioxide for material production based on the mass of the material and energy consumption. It also depends on two emission factors: one for thermal energy by fuel and one for electrical energy generation. During production, the steel and aluminium is considered to have noticeable carbon emission effect, and the percentage values of thermal energy and electrical energy are 85 and 15% for steel, and 25 and 75% for aluminium, respectively [9]. The carbon emission due to material production of typical ICE and electrical vehicles is about 5 ton. The value of electrical energy depends strongly on the "energy mix". In Germany (2015), it is about 550 g/kWh.

Notice that the CO_2 emission for the material production and battery production are significantly higher than that for vehicle assembly, distribution and disposal. The material production step is responsible for almost 75% of the energy consumption and emissions during the vehicle life cycle [10]. The materials can be transported to the vehicle manufacturing industry in many different ways such as through the use of trucks, rail cars, ocean liners, barges, air freight, or pipelines. Material transportation stage also contribute to CO_2 emission.

3.2 *Vehicle Assembly and Distribution*

During the vehicle assembly and manufacturing stage, production-ready materials such as billets, ingots, sheet stock, rods, pellets, etc., are delivered to industries where parts are fabricated and assembled into a vehicle. Transformation processes as metal stamping, casting, forging, machining, and extrusion as well as polymer extrusion, injection, compression and blow molding, and calendaring are required to form the shape of parts of vehicles. Typical vehicle assembly processes include painting, heating, ventilation, air conditioning, material handling, welding and supply of compressed air. The carbon emission for assembly can be obtained by the product of the energy consumption for assembly and carbon emission factor. The energy consumption for assembly can be calculated based on vehicle mass and required energy per mass for assembling a vehicle. Typically CO_2 emissions for assembly is of the order of one ton. Sullivan et al. [11] estimated that CO_2 emissions for the vehicle manufacturing and assembly stage is about 2 ton. The distribution depends strongly on the distance and transportation mode of shipping the vehicle to the customer.

3.3 Vehicle Operation

The emissions of this stage are often called "tank-to-wheel" emissions and considered to be the most carbon and energy intensive phase in the life cycle of both ICE vehicles and EVs. The calculation of emissions for the vehicle operation stage depends on the life time of vehicle. The lifetime is generally measured in distance or sometimes years and this metric is important because it is the amount of miles or time by the emissions are normalized. Battery lifetimes are assumed to be same as the vehicle lifetimes in most of the previous studies. In a study by Wang et al. [12], the vehicle life time is considered as 300,000 km. In another study by Aguirre et al. [13], they considered vehicle lifetime as 180,000 miles. They concluded that the operation phase was attributed to 96% of ICE vehicle emissions, and 69% of EV emissions.

The emissions from EVs is dependent on electricity production. Electricity production is often classified using a specific type of grid and these grids are classified based on the type of natural resources which are used to produce electricity. If electricity is obtained from renewable energy sources such as wind turbines (7 g CO_2e/MJ), nuclear (9 g CO_2e/MJ), solar power, (30 g CO_2e/MJ), and hydropower (11 g CO_2e/MJ), electric vehicles have significantly lower emissions. Generating electricity from non-renewable energy sources such as coal (300–350 g CO_2e/MJ) or natural gas (100–120 g CO_2e/MJ) produces more emissions.

The impacts of EVs are highly dependent on vehicle operation energy consumption and the electricity mix used for charging. As examples, we compare some real-life cases considering a vehicle life time of 250,000 km, an ICE fuel consumption of 6 l/100 km, and a typical EV energy consumption of 20 kW/100 km. The following five cases are estimated based on CO_2 emission factor and energy mix for different countries.

Case 1: ICE vehicle (CO_2 emission factor for gasoline: 2.38 kg/l): $250{,}000 \times 6/100 \times 2.38 = 35{,}700$ kg
Case 2: EV vehicle (China CO_2 intensity 1100 g/kWh): $250{,}000 \times 20/100 \times 1.10 = 55{,}000$ kg
Case 3: EV vehicle (USA CO_2 intensity 650 g/kWh): $250{,}000 \times 20/100 \times 0.65 = 32{,}500$ kg
Case 4: EV vehicle (Germany CO_2 intensity 550 g/kWh): $250{,}000 \times 20/100 \times 0.55 = 27{,}500$ kg
Case 5: EV vehicle (Japan CO_2 intensity 400 g/kWh): $250{,}000 \times 20/100 \times 0.4 = 20{,}000$ kg

Notice that the ICE vehicle of this example produces 143 g CO_2 per km which is just above the current German fleet limit of 140 g/km. From the above calculation, it is observed that EVs in China produce more CO_2 emissions compared to ordinary ICE vehicles whereas that in Germany, USA, and Japan produce less emissions. This shows that the total EV emissions highly depend on the context of electricity mix in the region. Notice that the above calculations do not assume the need of spare EV batteries which would tend to make the balance less favourable for EVs.

Moreover, the emissions also depend on the vehicle lifetime, battery lifetime and driver behaviour. For example, in a study by Hawkins et al. [14], an assumption of a vehicle lifetime of 200,000 km exaggerates the benefits of EVs in the range of 27–29% relative to gasoline vehicles or 17–20% relative to diesel. Whereas assuming a vehicle lifetime of 100,000 km decreases the benefit of EVs in the range of 9–14% with respect to gasoline vehicles and results in impacts indistinguishable from those of a diesel vehicle. Also, assumption on driving behaviour also greatly impact the resulting lifecycle energy consumption and emissions. Aggressive driving (faster acceleration and braking), results in more energy consumption and emissions than moderate acceleration and braking. Similarly, more urban driving results in more energy consumption and emissions than highway driving [15].

3.4 Vehicle Maintenance and Disposal

This stage includes vehicle maintenance and repair over the vehicle life time. This phase makes a relatively small contribution of less than 10% of the emissions during operation both in terms of material and fuel. Onat et al. [16] reported that vehicle maintenance phase for ICE vehicle produces 12.19 g CO_2 equivalents/km and for EV it is 8.53 g CO_2 equivalents/km. Vehicle disposal is the final stage of a vehicle's life cycle. Generally, recycling is already being taken care of in production stage. Otherwise it will give a negative footprint.

Acknowledgements The third author is supported by a fellowship of the Alexander Von Humboldt Research Foundation, Germany at the Technical University of Dresden, Germany.

References

1. Lin, B.: China Energy Outlook 2011, 1st edn. Tsinghua University Press (2011)
2. International Energy Agency: Global EV Outlook: Beyond One Million Electric Cars. OECD/IEA, Paris, France (2016)
3. International Energy Agency.: CO_2 Emissions from Fuel Combustion Highlights (2013 edition), Paris, France (2013)
4. Zhang, X., Xie, J., Rao, R., Liang, Y.: Policy incentives for the adoption of electric vehicles across countries. Sustainability **6**, 8056–8078 (2014)
5. International Energy Agency.: CO_2 Emissions from Fuel Combustion Highlights (2016 edition), Paris, France (2016)
6. Statista: Anzahl der Elektroautos in Deutschland. https://de.statista.com/statistik/daten/studie/265995/umfrage/anzahl-der-elektroautos-in-deutschland/
7. Lane, B.: Life Cycle Assessment of Vehicle Fuels and Technologies, Final Report, London Borough of Camden, Mar 2006
8. Wang, M.: GREET 1.5—Transportation Fuel-Cycle Model. Volume 1: Methodology, Development, Use, and Results, Technical Report ANL/ESD/TM-22, Argonne National Laboratory (1999)
9. Schucker, M., Saur, K., Florin, H., Eyerer, P., Beddies, H.: Life cycle analysis: getting the total picture on vehicle engineering alternatives. Autom. Eng. **104**, 49–52 (1996)

10. Zamel, N., Li, X.: Life cycle analysis of vehicles powered by a fuel cell and by internal combustion engine for Canada. J. Power Sources **155**, 297–310 (2006)
11. Sullivan, J.L., Burnham, A., Wang, M.: Energy Consumption and Carbon Emission Analysis of Vehicle and Component Manufacturing, Technical Report ANL/ESD/10–6, Argonne National Lab, 2010
12. Wang, D., Zamel, N., Jiao, K., Zhou, Y., Yu, S., Du, Q., Yin, Y.: Life cycle analysis of internal combustion engine, electric and fuel cell vehicles for China. Energy **59**, 402–412 (2013)
13. Aguirre, K., Eisenhardt, L., Lim, C., Nelson, B., Norring, A., Slowik, P., Tu, N.: Lifecycle Analysis Comparison of a Battery Electric Vehicle and a Conventional Gasoline Vehicle. University of California, Los Angeles (2012)
14. Hawkins, T.R., Singh, B., Majeau-Bettez, G., Stromman, A.H.: Comparative environmental life cycle assessment of conventional and electric vehicles. Int. J. Ind. Ecol. **17**, 53–64 (2013)
15. Nealer, R., Hendrickson, T.P.: Review of recent lifecycle assessments of energy and greenhouse gas emissions for electric vehicles. Curr. Sustain. Renew. Energy Rep. **2**, 66–73 (2015)
16. Onat, N.C., Kucukvar, M., Tatari, O.: Conventional, hybrid, plug-in hybrid or electric vehicles? State-based comparative carbon and energy footprint analysis in the United States. Appl. Energy **150**, 36–49 (2015)

Mitigation in the Industrial Sector, CO_2 Trade

László Rácz

1 Prologue

According to the UN Intergovernmental Panel on Climate Change (IPCC) fifth assessment report, global industrial GHG emissions accounted for just over 30% of global GHG emissions and reached 15 Gt CO_{2eq} in 2010 (2005: 13 Gt CO_{2eq}). GHG emissions of the industrial sector can be divided into two categories: direct emissions produced at the industrial facility (e.g. due to its fossil energy consumption) and indirect emissions that occur off site but are connected to the industrial facility's operation (e.g. off-site energy production for the given industrial facility). Over half (52%) of global direct GHG emissions from industry and waste/wastewater are from ASIA region followed by OECD in 1990 (25%). In 2010 industrial emissions were comprised of direct energy-related CO_2 emissions (5.3 Gt CO_{2eq}), indirect CO_2 emissions (5.2 Gt CO_{2eq} from production of electricity and heat for industry), process CO_2 emissions (2.6 Gt CO_{2eq}), non-CO_2 emissions (0.9 Gt CO_{2eq}), waste and wastewater CO_2 emissions (1.4 Gt CO_{2eq}). In 2010 direct and indirect emissions were dominated by CO_2 (85.1), followed by CH_4 (8.6%), HFC (3.5%), N_2O (2%), PFC (0.5%), SF_6 (0.4%) [1].

The energy intensity of the industrial sector could be reduced by approximately up to 25% through widescale upgrading, replacement and deployment of best available technologies, and additionally up to 20% through innovation. "Barriers to implementing energy efficiency improvements relate largely to the initial investment costs and lack of information" [1]. Therefore, usually, the 'low-hanging fruits are picked for the first', i.e. the simplest and easiest works are done first.

The IPCC has identified six options for climate change mitigation in the industry (see Table 1).

L. Rácz (✉)
Department of Chemical Technology and Biotechnology, Budapest University of Technology and Economics, Budapest, Hungary
e-mail: liracz@gmail.com

M. Palocz-Andresen et al. (eds.), *International Climate Protection*,
https://doi.org/10.1007/978-3-030-03816-8_22

Table 1 Six options for climate change mitigation in the industry [1]

No.	Options	Specific actions
1.	Energy efficiency improvement	Regulations, voluntary actions and agreements, economic intruments, information programmes, government provisions
2.	Emissions efficiency improvement	Switch to lower carbon fuels, change of process parameters, change of process gases to ones with lower or zero global warming potential (GWP), carbon capture and storage (CCS)
3.	Material efficiency improvement in production	Reduce yield of losses, reuse of old materials
4.	Material efficiency improvement in product design	Use of lighter products, target-oriented material substitutions
5.	More intensive use of products	'Collaborative consumption', longer use of products
6.	Reducing overal demand	

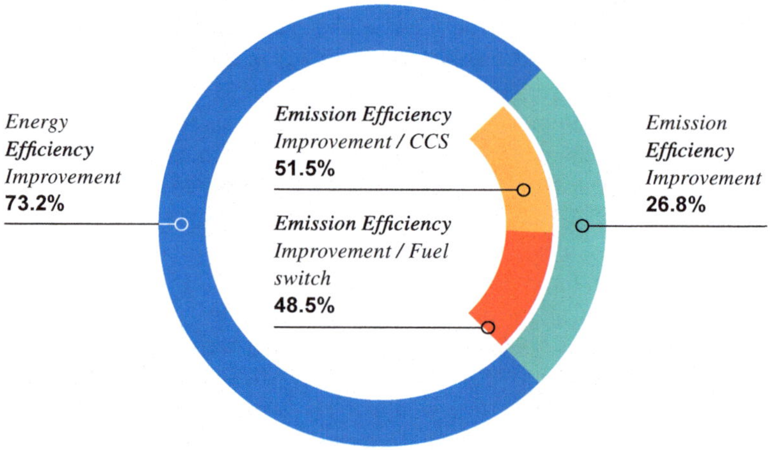

Fig. 1 The expected impact of the most important CO_2 emissions reduction actions in the industrial sector by 2050 (compared to the 2012 levels) [2] (CCS—carbon capture and storage)

The most important CO_2 emissions reduction actions in the industrial sector and their expected contributions by 2050 (compared to the 2012 levels) are shown on Fig. 1 [2].

In the following, climate change mitigation options of the industrial sector and specifically CO_2 trade are discussed in more details.

2 Energy Efficiency Improvement

According to the IEA 2015 Energy Technology Perspectives, in the two degree centigrade scenario and in the industrial sector approximately 73% of total CO_2 emissions reduction by 2050 should be achieved by energy efficiency improvements (see Fig. 1) [2]. Actions to improve energy efficiency can be facilitated by

- Regulations (e.g. performance standard)
- Voluntary actions, agreements
- Economic instruments (e.g. taxes, subsidies, cap and trade schemes)
- Information programmes (e.g. audits) and
- Government provisions [1].

Energy efficiency in the industry can be improved by heat and energy recovery from process streams (products, waste), cogeneration, thermal insulation, advanced control and optimisation, reducing losses (e.g. gas flaring), improving the yield of raw materials of the existing processes, commercialising of new processes with lower energy consumption, reducing wastes, using renewables, and creating products that enable energy savings.

The EU energy efficiency best available technique reference document (BREF) lists energy efficiency management (incl. commitment of the top management, benchmarking, taking corrective actions), continuous environmental improvement, identification of opportunities for energy saving, establishing and reviewing energy efficiency objectives and indicators, energy efficient design, increased process integration, maintaining the impetus of energy efficiency initiatives, maintaining expertise, effective control of processes, maintenance, monitoring and measurement, heat recovery, cogeneration, electrical power supply, electric motor driven sub-systems as the main elements for achieving energy efficiency at installation level [3].

For example a '**regulation**', the EU directive on industrial emissions (integrated pollution prevention and control) (IED) requires the application of the best available techniques (BAT) by the operator. Best available techniques "means the most effective and advanced stage in the development of activities and their methods of operation which indicates the practical suitability of particular techniques for providing the basis for emission limit values and other permit conditions" [4]. Best available techniques reference documents (BREFs) have been developed under IPCC (2008/1/EC) and IED directives (and are available at http://eippcb.jrc.ec.europa.eu/reference/). They outline the used techniques and processes, the actual emission and consumption levels and the new developments in a given sector or for a given topic. When applying for permits, operators should justify that their installations comply with some basic obligations in the prevention or minimalisation of pollution. BAT conclusions are the basics for permit conditions. The document that may result from this BAT analysis, the environmental permit, consolidates all emission level values for each environmental component and gives a comparison of the existing techniques with best available ones to recognise those that have not been installed yet and should be implemented due to the environmental aspects improvement.

Voluntary actions are good-faith actions, initiated by governments, professional associations or a group of individual companies. For example as a *governmental initiative* in the USA from 1992 Environmental Protection Agency's Energy Star programme helps to improve energy efficiency of buildings and facilities, presently covering industrial subsectors such as automotive manufacturing, cement, glass and paper production. "Energy Star provides energy performance indicators (EPIs) to help benchmark industrial plant energy performance. EPIs are external, industry-specific benchmarking tools that score a plant's energy performance and compare it to that of similar plants in its industry within the U.S. These statistical models generate an energy performance score … on a scale of 1–100 using actual plant data—not engineering projections, and evaluate a plant in terms of energy per unit of production at the whole plant level. … Plants that score a 75 or higher are eligible to earn Energy Star certification—recognition of the plant's superior energy performance within its industry—provided the plant satisfies an environmental compliance screen and its energy performance data is verified by a Professional Engineer" [5]. *Professional associations* may require their members to commit themselves to continuously improve energy efficiency, e.g. in the frame of the worldwide product stewardship initiative of the International Council of Chemical Associations—ICCA. On the eve of the Paris climate summit in December 2015, a *group of* ten leading oil and gas *companies* accounting for almost one fifth of the world hydrocarbon production, vowed to help climate change mitigation by increasing share of natural gas in the energy mix, eliminate routine flaring of natural gas (methane) and investing in carbon capture and storage (CCS) systems [6].

Economic instruments, including discriminatory taxes, discriminatory subsidies and certificates (green in the supply side, white in the demand side and brown as cap and trade systems) are used on the basis that climate change is (also) an economic issue. These tools should help to give price signals thus orienting industrial players in the long term, prevent relocation of investments into countries with less challenging climate change mitigation policy, involve entrepreneurs in the fight against climate change [7]. For example, in the USA, president Obama's 2017 fiscal budget, revealed in February, 2016 introduced a new plan for building a '21st Century Clean Transportation System' by public investments and incentives for private sector innovations to be funded by a new fee (10.25 USD/bbl crude oil) payable by oil companies [8]. The European Union from 2020 onwards, has set a target of 95 g CO_2/km for the average emissions of the new car fleet with time schedule accompanied with penalties for the laggers and super-credits for the front-runners [9]. Cap and trade tool will be introduced in more details in a next part of this subchapter.

Information programmes can cover governmental or company level *training and awareness increasing actions* to reduce GHG emissions. For example, the USA EPA supports programmes for aluminium, semiconductor and magnesium industries for reductions of occurrences of accidental release of flour containing GHGs from containers and equipments [10]. Article 8 of the EU Energy efficiency directive (EED, 2012/27/EU) prescribes that EU member states shall oblige 'large enterprises' (which are not small or medium enterprises—SMEs) to undergo an *energy audit* by December 5, 2015 and every 4 year onwards. The energy audit has to be carried out in an independent and cost-effective manner by qualified and/or accredited experts or implemented and supervised by independent authorities under national legislation. Large enterprises implementing energy management systems (e.g. ISO 50001) are exempted from the obligation.

Government provisions may arise from international or regional obligations (e.g. Kyoto Protocol, Paris Climate Agreement, EU Energy efficiency directive), international and country-wide initiatives. For example, Article 7 of the *EU* Energy efficiency *directive* (EED, 2012/27/EU) obliges national governments to set up national energy savings targets for 2020 and energy efficiency obligation schemes. Energy distributors and/or retailers of energy shall achieve a cumulative end-use energy savings target of 1.5%/annum "averaged over the most recent three-year period prior to 1 January 2013" in 2014–2020 amongst final energy consumers. Thus national governments are obliged to calculate volumes to be saved, define the obliged suppliers or retailers and distribute savings amongst them and set project types of savings by sectors of final consumers or distributors [11]. On the eve of the Paris Climate Summit in December 2015, an International Zero-Emission Vehicle *Alliance* (ZEV Alliance) has been formed as a collaboration of national (Germany, Netherlands, Norway, UK) and subnational governments (9 states of the USA and Canada) and announced that "we will strive to make all passenger vehicle sales in our jurisdictions ZEVs as fast as possible, and no later than 2050… Accelerating ZEV deployment will achieve greenhouse gas emissions reductions of more than 1 billion tons per year by 2050" [12]. *Individual countries* may decide to act individually to speed up the transition to the carbon-free economy. Specifically, in Germany by 2020, GHG emissions should be cut by at least 40% and by 2050,—by 80 to 95% (closer to the upper figure) both compared to the 1990 levels; in parallel primary energy consumption by 2020 should be reduced by 20% and by 2050—by 50% both compared to the 2008 levels [13].

In the iron and steel sector energy efficiency can be improved by heat and energy recovery from product and waste heat, better fuel delivery (e.g. pulverised coal injection), better furnace design and process controls, reduced number of temperature cycles and use of new technologies (e.g. coke dry quenching, high pressure recovery turbine). In the non-ferrous (aluminium and other) metal sector where indirect GHG emissions dominate (in the aluminium production they account for 80% of the total GHG emissions) due to the high electricity demand, energy efficiency can be improved by new technologies (e.g. multipolar electrolysis cells, inert anodes and carbothermic reactions) [1].

In the chemical industry, responsible for 7% of the global GHG emissions [14] exemplary actions for GHG emissions reduction are: thermal insulation, use of renewable energy (wind, biogas production), use of combined heat and power turbines, waste incineration for steam production, and waste heat boyler [1]. In the frame of the CEFIC's Sectoral Platform in Chemicals for Energy Efficiency Excellence (SPiCE) launched in 2013 with the aim to boost energy efficiency across the European chemical industry technology-specific 'Best practices' have been issued on at least 20 platforms [15]. Besides these direct actions, the International Chemical Councils Associations has launched in December 2012 a technology roadmap focusing on building efficiency. "The roadmap explores the potential energy and

GHG savings from five chemically-derived building technologies: insulation, pipe and pipe insulation, air sealing, reflective roof coatings and pigments, and windows. It estimates that combining ambitious building efficiency improvements with lower-carbon fuels could lead to a 41% reduction in energy use and a 70% reduction in GHG emissions by 2050" [16].

3 Emissions Efficiency Improvement

Emissions efficiency can be improved by reduction of GHG (largely carbon) emissions through switching to lower carbon fuels, changing process parameters, changing of the process gases to ones with lower or zero global warming potential and carbon capture and storage (CCS).

Switching to lower carbon fuels should result in lower direct CO_2 emissions at the same amount of combusted energy. IEA's 2015 Energy Technology Perspectives estimates that in the 2DC scenario and in the industrial sector roughly 13% of the total CO_2 emissions reduction by 2050 should be achieved by fuel switching [2]. Generally, fuels are used to convert feedstocks into final products but in some subsectors fuels can be used also as feedstocks, and some byproducts of the manufacturing processes can be utilised as fuels. Among the fossil energy sources occurring in the nature, natural gas has the lowest default emissions efficiency (56.1 kg CO_2/GJ) and lignite the highest (101 kg CO_2/GJ) [17]. Thus natural gas could ease the way to the future carbon-free economy. Substitution of fossil energy sources by renewable non-fossils (wind, solar, geothermal, aerothermal, hydrothermal and ocean energy, hydropower, biomass, landfill gas, sewage treatment plant gas and biogases according to Art. 2 of the EU RES—2009/28/EC—directive) and nuclear is the other way to improve GHG emissions efficiency.

Monitoring and controlling of the process parameters can result in significant emissions efficiency improvements. *Tunable diode laser spectrometer* (TDLS) is increasingly used for combustion control of fired heaters through measurement of oxygen, carbon monoxide, methane and water steam contents of the flue gas. Their real-time monitoring allows continuous adjustment of fuel-air ratio thus minimising methane emissions and lowering fuel consumption. *Low-NO_x burners* can be used to limit thermal NO_x production on fired heaters by reduction of peak flame temperature and restriction of oxygen availability through steam injection and flue gas recirculation (FGR). *Prevention of drop in alumina concentration* in the aluminium production by process control can lead to PFC emissions reduction (resulting from carbon of the anode and fluorine in the criolite). *High temperature NO_2 decomposition* can be applied in the ammonia production [1].

Changing of the process gases to ones with lower global warming potentials would have a great mitigation impact. Flourinated gases are regarded as the most potent and longest lasting anthropogenic GHGs [18]. Hydrofluorocarbons (HFCs) are mainly used as refrigerants (for substitution of ozone-depleting substances to be phased out under the Montreal Protocol). Perfluorocarbons (PFCs) are received in the production of aluminium and semiconductors. Sulphur hexafluoride (SF_6) is used in magnesium processing and semiconductor manufacturing [19]. Fugitive HFC emissions from leakages can be reduced by routine maintenance and proper disposal of the HFC refrigeration units, substitution of HFC with ammonia or hydrocarbon gases (properly addressing their flammability and toxicity) and application of alternative cooling technologies (e.g. geothermal). HFC-23 emissions can be thermally oxidised into products with lower GWPs. HFC used in foam blowing can be substituted by other blowing agents with lower GWPs. Fire extinguishers with HFC should be substituted by water mist and inert gas systems. SF_6 emissions from electricity transmission and distribution equipment can be reduced by proper disposal of the equipment, application of equipment with greater structural integrity, use of alternative insulating gases with lower GWPs, and leak detection systems. SF_6 emissions from magnesium production can be cut by its substitution with gases of lower GWPs, improving the efficiency of SF_6 usage and installing alternative production and casting systems. In semiconductor manufacturing fluorinated gases can be captured and directed to thermal or catalytic destroyment or substituted by other gases having lower GWPs [20].

Carbon capture and storage (CCS) shall play a very important role in the future climate change mitigation, since Article 4 of the 2015 Paris Agreement sets up a goal of net greenhouse gas neutrality (defined as a "balance between anthropogenic emissions by sources and removals by sinks") to be achieved in the second half of this century. Removal by sinks means GHG emissions reduction via forests and carbon capture and storage. According to IEA, CCS contributes one sixth of total CO_2 emissions reduction required by 2050, and near 14% in the industrial sector [2, 21].

In 2015 around 15 large-scale CCS projects were in operation around the world (incl. Canada, USA, Saudi Arabia), with capacity to capture up to 28 million tonnes of CO_2 per year. Further 7 large-scale projects are expected to be operational in 2016–2017 in Canada, USA, Abu Dhabi and Australia and another 11 are in advanced planning phase. Thus "the total number of large-scale CCS projects currently sits at 45, with a total CO_2 capture capacity of 80 Mtpa. This level of capture capacity is dwarfed by the amount of CCS deployment required in the next 20–30 years to meet climate targets, estimated at approximately 4,000 million tonnes of CO_2 captured and stored per annum by 2040" [22].

A study by the Sustainable Gas Institute (SGI) at Imperial College London states that the estimated available total global underground storage capacity for CO_2 is around 10,450 to 33,153 Gigatonnes which would equate to around three centuries of storage capacity for the world. Additionally, the study suggests that if the present 85–90% CO_2 capture rates were increased to 95–99%, "more of the world's fossil fuel reserves could be used while still remaining within the limits of a 2 °C climate-change scenario." While near-term costs of CCS represent a barrier, "beyond 2050, the costs of capturing and storing CO_2 are likely to be significantly less than the marginal cost of alternative ways to mitigate CO_2", the study says. Average marginal costs for emissions abatement in models used to support the IPCC's Fifth Assessment Report are USD 473–1,100/tonne of CO_{2eq} by 2050, rising further by 2100. The estimated cost for CCS, including transport, is USD 155–160/tonne [23].

The European Commission operates a CCS Project Network to support the early large-scale demonstration projects in the EU.

The captured carbon dioxide can also be used for enhanced hydrocarbon (oil, gas, coalbed methane) recovery (EOR, EGR, ECBM) thus 'killing two birds with one shot'. Currently 10 commercial EOR projects using anthropogenic CO_2 are reported (located in the USA, Canada, Saudi Arabia and Brazil) and 15 others are announced [24]. INA (Croatia) EOR CO_2 injection project Ivanić and Žutica represents one of the largest on-shore EOR projects in Europe: CO_2 is recovered from natural gases containing 23–26% CO_2 and injected into the reservoir. INA EOR project would reduce Croatia's yearly CO_2 emissions by more than 1% [25].

4 Material Efficiency Improvement in Production

Material efficiency in manufacturing of existing products can be improved by reducing yield of losses, and reusing of old materials. Reportedly approximately 20% of all paper, 25% of all steel and 41% of all aluminium is scrapped mainly during the downstream processes, and internally recycled. These *losses* should be reduced by process innovations and application of new design approaches. For example, in the aluminium industry near-net shape casting should be improved and the blanking and stamping process should be renewed by innovations. *Reusing of old materials* also cuts GHG emissions. For example, structural steel can be reused in construction without unsurmountable technical barriers [1]. Collection and recycling of certain used products could also help to save non-renewable sources (e.g. metals) with limited availability thus improving security of their supply, minimising environmental impact of the used products' incineration or dumping in landfills and lowering production cost of new products.

5 Material Efficiency Improvement in Product Design

Many products manufactured from the same material (e.g. steel) could be *lighter without loss of performance*, and some materials can be substituted by less GHG-intensive ones. For example, improvement of material efficiency in the car manufacturing (i.e. production of lighter cars) could lower both direct (caused by the car manufacturer) and indirect (caused by the car driver) GHG emissions. Presently options for *target-oriented material substitution* are limited: frequently comparable production quantities with the same functions are not available (e.g. in the steel and cement productions), or the new materials have higher embedded energy than the originals (e.g. epoxy-based composites and magnesium alloys versus steel and aluminium). However, "blast furnace slack and fly ash from coal-fired power stations can substitute to some extent for cement klinker" [1].

6 More Intensive Use of Products

A part of 'consumable' industrial products becomes waste without any service provision. In most cases the same service level (e.g. in the transport) could be delivered with fewer products, thus embracing *collaborative consumption* (the idea of Nobel memorial prize winner in economic sciences Elinor Ostrom, e.g. car, bike or flat sharing) instead of individual consumption would result in GHG emissions cut. *Longer use of industrial products* could reduce demand for replacement. Postponement of new product purchases through more intensive use of products, i.e. 'sustainable consumption' would reduce direct and in some cases indirect GHG emissions. For example, demand for iron and steel products can be reduced through meeting safety codes accurately (without exaggerations) and replacing buildings not as frequently. More efficient use of fertilizers in the agriculture can lead to lower GHG emissions in the chemical industry [1].

7 Reducing Overall Demand for Product Services

Beyond some threshold (e.g. in the developed countries) reduced consumption might not be harmful but can contribute to industrial emissions reduction [1]. By changing the way we live GHG gas emissions caused directly or indirectly by the industrial sector can be reduced.

8 CO_2 Trade

CO_2 trade is a mechanism to improve energy efficiency by economic instruments (see subsection 'Emissions efficiency improvement'). CO_2 trade under 'emissions trading' (ET) is one of the three flexibility mechanisms established by the Kyoto Protocol (1997), providing possibility for developed ('Annex 1') countries to acquire assigned amount units (AAUs), removal units (RMUs), emission reduction units (ERUs), certified emission reductions (CERs) from other Annex 1 Parties. Each unit is equal to one metric tonne of emissions (in CO_{2eq}) [26].

Presently, at the national level legislated carbon trading schemes exist in the European Union (28 countries), Norway, Iceland, Liechtenstein, Switzerland, New Zealand, Australia, South Korea, Kazakhstan and Australia ('carbon pricing scheme'). Some subnational schemes are legislated in the US (Regional Greenhouse Gas Initiative—RGGI in 9 states, and Califirnia's cap-and-trade scheme), Canada (Quebec's cap-and-trade scheme), and Japan (Tokyo and Saitama regional schemes) [27, 28].

The EU emissions trading scheme (ETS) is the world's longest-standing and largest emission trading system, launched in 2005. It covers around 45% of the total

GHG emissions of the EU and limits CO_2 (e.g. from power and heat generation and oil refineries), N_2O (from production of nitric, adipic, glyoxal and glyoxlic acids) and perfluorocarbons PFCs (from aluminium production) emissions of larger industrial emitters and aircraft operators in the present 2013–2020 (third) period. EU ETS is based on the 'cap and trade' principle. Cap (limitation) is made by allocating decreasing quantity of tradeable emission allowances to the covered companies/facilities. Under threat of heavy penalty companies/facilities must surrender allowances equal to their past year's emissions. Excess allowances can be used up or traded by the given company in the following period. Appropriately high prices of EU allowance (EUA; 1 EUA = 1 tonne CO_{2eq}) prices could facilitate investments in low-carbon technologies [29].

EU ETS has been developing since its establishment through '*trading periods*'. EU allowance (EUA) allocation in the first trading period (2005–2007) was done based on the estimated needs for allowances in the Member States which turned out to be excessive and resulted in carbon price falling to near to zero in 2007. The European Commission cut the total number of allowances by 6.5% in the second period (2008–2012) compared to the 2005 level but the economic crisis resulted in emissions reduction leading to the carbon price drop to a non-motivating level. In the present third trading period (2013–2020) an EU-wide cap on emissions (to be reduced each year by 1.74%) and allowance auctioning (with progressively increasing share) for allocating the EUAs are introduced. Total volume of emissions allowances is planned to be cut by an annual rate of 2.2% and further changes are on the agenda for the fourth trading period (2021–2030) [29].

The carbon price is expected to increase as EU reduction goals become stricter and stricter. In the past few years, the price has been relatively low due to the economic crisis. Therefore, the industry still manages to perform its activities without the need to implement low-carbon technologies.

Certain energy-intensive industries, covered by EU ETS may tend to transfer their production to non-EU countries with less stringent regulations in the field of GHG emissions which would lead to global emission increase (this situation is described by term 'carbon leakage', CL). In order to avoid such production transfers, EU sectors deemed to be exposed to a significant risk of carbon leakage can get higher proportion of free allowances starting from 2009 [30].

Free EUA is allocated on the basis of product-specific benchmark (BM) levels (product, heat, fuel or process emissions benchmarks) reflecting the most efficient EU installations performance. Volume of free allocation is calculated by multiplying the historical activity level (HAL) of the chosen historical period (e.g. TJ/year of 2005–2008 period) with the specific benchmark of the facility (e.g. 56.1 EUA/TJ for fuel benchmark) and taking into account some other factors. Only the most efficient installations of a given sector would get enough free allowances to cover all their needs.

9 Epilogue

Industrial sector is both a cause of problems (of global warming) and a source of solutions (to mitigate climate change). It is a significant indirect source of global warming. At the same time it is a (in some areas outstanding) solution-provider for the whole economy, being a developer or investor of cogeneration technology, next energy sources (algae) and new generation of biofuels, wind, solar and hydrogen economies and carbon capture and storage. All these technologies are also listed in IEA Blue Map scenario among key technologies to cut carbon dioxide emissions.

References

1. Fischedick, M., Roy, J., Abdel-Aziz, A., Acquaye, A., Allwood, J.M., Ceron, J.-P., Geng, Y., Kheshgi, H., Lanza, A., Perczyk, D., Price, L., Santalla, E., Sheinbaum, C., Tanaka, K.: Industry. In: Edenhofer, O., Pichs-Madruga, R., Sokona, Y., Farahani, E., Kadner, S., Seyboth, K., Adler, A., Baum, I., Brunner, S., Eickemeier, P., Kriemann, B., Savolainen, J., Schlömer, S., von Stechow, C., Zwickel, T., Minx, J.C. (eds.) Climate Change 2014: Mitigation of Climate Change. Contribution of Working Group III to the Fifth Assessment Report of the Intergovernmental Panel on Climate Change. Cambridge University Press, Cambridge, United Kingdom, New York, NY, USA (2014)
2. International Energy Agency: Energy technology perspectives 2015, executive summary. Viewed 25 May 2016. https://www.iea.org/publications/freepublications/publication/EnergyTechnologyPerspectives2015ExecutiveSummaryEnglishversion.pdf (2015)
3. European Commission: Reference document on best available techniques for energy efficiency. Viewed 25 May 2016. http://eippcb.jrc.ec.europa.eu/reference/BREF/ENE_Adopted_02-2009.pdf (2009)
4. European Commission: The industrial emissions directive. Viewed 25 May 2016. http://eur-lex.europa.eu/legal-content/EN/TXT/HTML/?uri=CELEX:32010L0075&from=EN (2010)
5. Energy Star: Energy performance indicators for plants. Viewed 29 May 2016. https://www.energystar.gov/buildings/facility-owners-and-managers/industrial-plants/measure-track-and-benchmark/energy-star-energy?s=mega (2016)
6. PhysOrg: Ten big energy forms vow to fight climate change. Viewed 29 May 2016. http://phys.org/news/2015-10-ten-big-energy-firms-vow.html (2015)
7. Lagarde, C: Economic instruments in the fight against climate change. OECD Observer, no. 267. Viewed 29 May 2016. http://www.oecdobserver.org/news/fullstory.php/aid/2603/Economic_instruments_in_the_fight_against_climate_change.html (2008)
8. The White House: Fact sheet: president Obama's 21st century clean transportation system. Viewed 29 May 2016. https://www.whitehouse.gov/the-press-office/2016/02/04/fact-sheet-president-obamas-21st-century-clean-transportation-system (2016)
9. European Commission: Reducing CO_2 emissions from passenger cars. Viewed 29 May 2016. http://eur-lex.europa.eu/legal-content/EN/TXT/HTML/?uri=CELEX:32014R0333&from=EN (2016)
10. United States Environmental Protection Agency: Sources of greenhouse gas emissions: industry sector emissions: reducing emissions from industry. Viewed 25 May 2016. https://www3.epa.gov/climatechange/ghgemissions/sources/industry.html (2016)
11. European Commission: Energy efficiency directive. Viewed 29 May 2016. http://eur-lex.europa.eu/LexUriServ/LexUriServ.do?uri=OJ:L:2012:315:0001:0056:en:PDF (2012)

12. Scribd.: International ZEV alliance announcement. Viewed 29 May 2016. https://www.scribd.com/doc/292065952/ZEV-Alliance-COP21-Announcement-3-Dec-2015 (2015)
13. Federal Ministry for the Environment: Nature conservation, building, and nuclear safety: climate protection in figures: facts, trends and incentives for German climate policy. Viewed 29 May 2016. <http://www.bmub.bund.de/fileadmin/Daten_BMU/Pools/Broschueren/klimaschutz_in_zahlen_2014_broschuere_en_bf.pdf (2014)
14. International Council of Chemical Associations: Chemical industry contributions to energy efficiency and mitigating climate change, p. 4. Viewed 25 May 2016. https://www.nikkakyo.org/sites/default/files/SummaryofTechnologyRoadmaps.pdf (2013)
15. CEFIC: SPiCE^3 sectorial platform in chemicals for energy efficiency excellence. Final report, Viewed 25 May 2016. http://www.cefic.org/Documents/PolicyCentre/SPiCE3-Sectoral-Platform-for-Energy-Efficiency-Excellence-Report.pdf (2015)
16. American Chemistry Council: ICCA highlights energy efficiency solutions made possible by chemistry at COP-18. Viewed 25 May 2016. https://www.americanchemistry.com/Media/PressReleasesTranscripts/ACC-news-releases/ICCA-Highlights-Energy-Efficiency-Solutions-Made-Possible-by-Chemistry-at-COP-18.htm (2015)
17. IPPC: 2006 IPPC Guidelines for national greenhouse gas inventories, Chapter 2, Stationary combustion, p. 2.18. Viewed 26 May 2016. http://www.ipcc-nggip.iges.or.jp/public/2006gl/pdf/2_Volume2/V2_2_Ch2_Stationary_Combustion.pdf (2006)
18. Myhre, G., Shindell, D., Bréon, F.-M., Collins, W., Fuglestvedt, J., Huang, J., Koch, D., Lamarque, J.F., Lee, D., Mendoza, B., Nakajima, T., Robock, A., Stephens, G., Takemura, T., Zhang, H.: Anthropogenic and natural radiative forcing. In: Stocker, T.F., Qin, D., Plattner, G.-K., Tignor, M., Allen, S.K., Boschung, J., Nauels, A., Xia, Y., Bex, V., Midgley, P.M. (eds.) Climate change 2013: The physical science basis. contribution of working group Fifth assessment report of the Intergovernmental Panel on Climate Change, Cambridge University Press, Cambridge, United Kingdom and New York, NY, USA. Viewed 29 May 2016. http://ghgprotocol.org/sites/default/files/ghgp/Global-Warming-Potential-Values%20(Feb%2016%202016).pdf (2013)
19. United States Environmental Protection Agency: Overview of greenhouse gases: reducing fluorinated gas emissions. Viewed 29 May 2016. https://www3.epa.gov/climatechange/ghgemissions/gases/fgases.html (2016)
20. Center for Climate and Energy Solutions: High global warming potential gas abatement. Viewed 29 May 2016. http://www.c2es.org/technology/factsheet/high-global-warming-potential-gas-abatement%20 (2016)
21. International Energy Agency: Carbon capture and storage. Viewed 29 May 2016. http://www.iea.org/topics/ccs/ (2016)
22. Global CCS Institute: The global status of CCS: 2015 summary report, Melbourne, Australia. Viewed 29 May 2016. http://hub.globalccsinstitute.com/sites/default/files/publications/196843/global-status-ccs-2015-summary.pdf (2015)
23. Sustainable Gas Institute, Imperial College London: Carbon capture and storage (CCS) key to unlocking unburnable carbon. Viewed 25 May 2016. http://www.sustainablegasinstitute.org/press-release-carbon-capture-and-storage-ccs-key-to-unlocking-unburnable-carbon/ (2016)
24. Carbon Capture & Sequestration Technologies @ MIT: Commercial EOR technologies using anthropogenic carbon dioxide; announced carbon dioxide capture and storage projects. Viewed 29 May 2016. https://sequestration.mit.edu/tools/projects/index_eor.html (2016)
25. Mavar, N.K.: EOR CO_2 project Croatia, environmental aspects. MOL Group Prof. J. **2**, 28 (2016)
26. United Nations Framework Convention on Climate Change: Kyoto mechanisms—background. Viewed 29 May 2016. http://unfccc.int/kyoto_protocol/mechanisms/items/2998.php (2014)
27. Carbon Market Data: World carbon market database. Viewed 12 May 2016. https://www.carbonmarketdata.com/en/products/world-ets-database/presentation (2016)

28. Parliament of Australia: Emissions trading schemes around the world. Viewed 12 May 2016. http://www.aph.gov.au/About_Parliament/Parliamentary_Departments/Parliamentary_Library/pubs/BN/2012-2013/EmissionsTradingSchemes (2013)
29. European Commission: The EU emissions trading system (EU ETS). Viewed 12 May 2016. http://ec.europa.eu/clima/policies/ets/index_en.htm (2016)
30. European Commissions: Carbon leakage. Viewed 12 May 2016. http://ec.europa.eu/clima/policies/ets/cap/leakage/index_en.htm (2016)

Impacts of Shipping on Environment and Climate

Shipbuilding and Maritime Transportation Industries

Eggo Bracker

1 In a Stress Field of International Trading and Globalisation

Ocean shipping is actually caught between the existing of too much vessels and overcapacity in total on one side and low transportation costs demands on the other.

From point of market economy, a solution for all prescribed minimum obligations and companies they will try to meet this at minimal cost. Of course there are always entrepreneurs who abstain for ethical reasons but in average they will gain lower profit than others.

On one hand must exists limits of growing in closed systems as the Earth is, (e.g. "Club of Rome" and books of Dennis Meadows et al. "The limits of growth") and on the other hand there are more and more individuals that have a right to reach out for a better life and living conditions.

How long can the planet (and the oceans) compensate the littering and pollution?

Globalisation effect is not only because of low overseas production costs. It is more a result of low transportation costs (low wages, tax free fuels and no follow up costs for the polluter) and of course scale effects.

The more container on one vessel the lower transportation cost and pollution for each container (Fig. 1).

As an example: The actual costs for a 20 t TEU container transport from China to USA cost less than €300,–. This transportation pricing allows companies to treat the whole world as a single manufacturing site and it allows them to select according lowest production costs.

E. Bracker (✉)
Ship-Automation and VDR, INTERSCHALT Maritime Systems GmbH, Part of Cargotec Oyj Group, Hamburg, Germany
e-mail: e.bracker@protonmail.com

M. Palocz-Andresen et al. (eds.), *International Climate Protection*,
https://doi.org/10.1007/978-3-030-03816-8_23

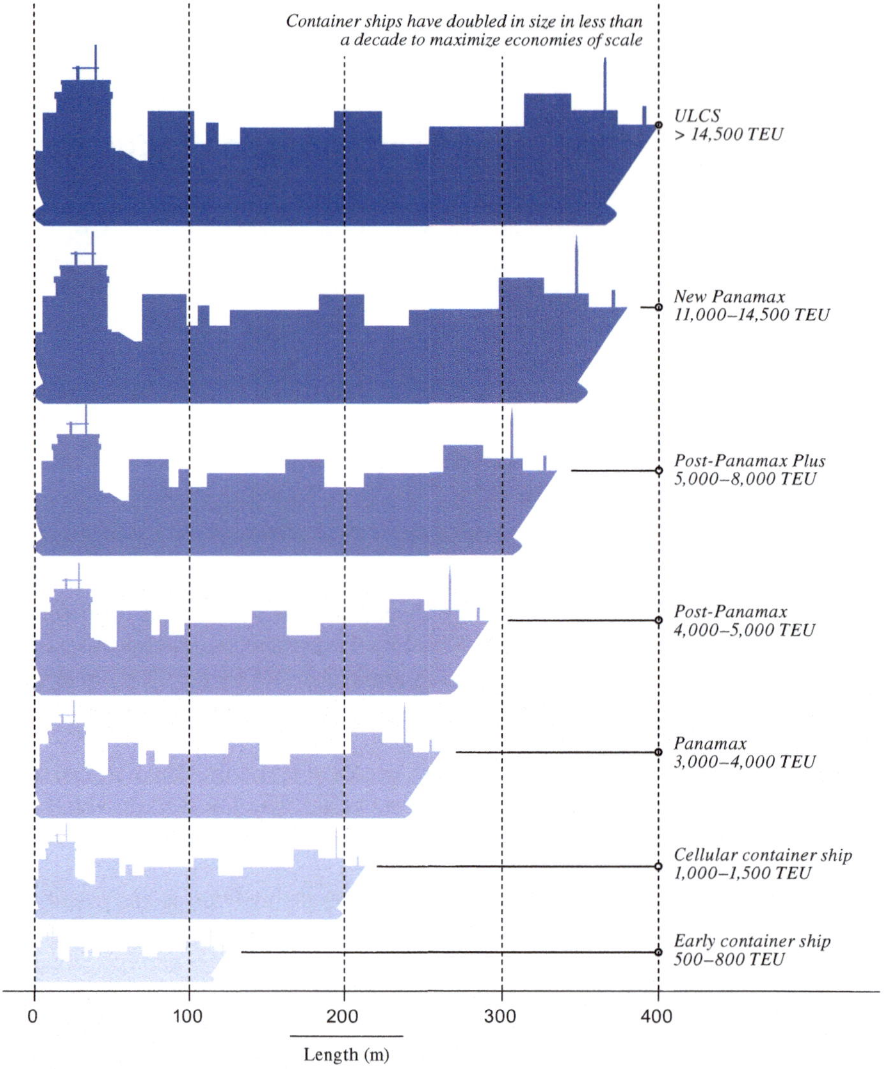

Fig. 1 Increasing sizes of container ships [1]

Distances no longer matters in this economical and profit oriented structures. In the following graphic we can see the growing of container transportation from 2006 to 2012 (remark: in between was 2008 the world financial crises) (Fig. 2).

This worldwide distributed production sites plus transportation are more profitable than local production including local minimum wages and short transportation distances.

This is obviously an internal objection…

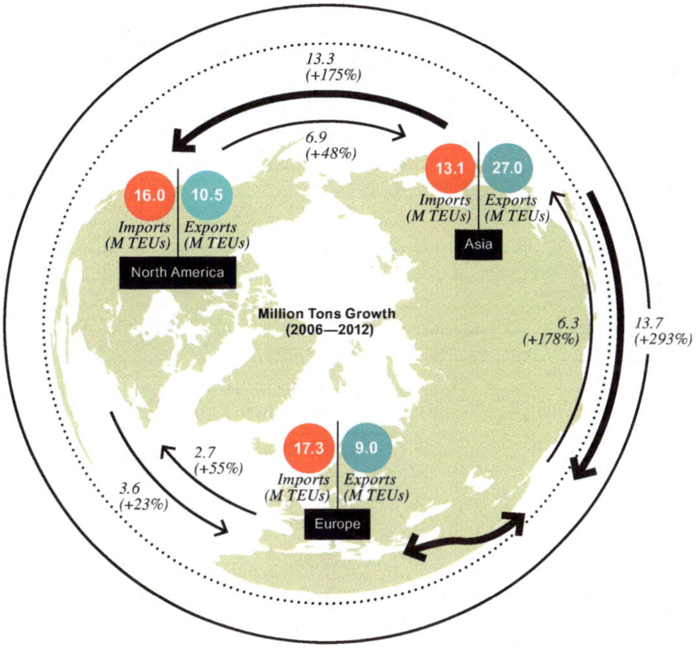

Fig. 2 Review of Maritime Transport [2]

… or somewhere are hidden costs that will occur …
… at least later on and in other places or circumstances.

As a matter of fact is meanwhile 90% of our consuming a result of globalisation.

2 Conventions and Organizations for (Environmental) Protection

2.1 United Nations Framework Convention on Climate Change, 21st Conference of the Parties, (COP 21) (for Example)

The Paris Agreement:
Global heating should be less than two degrees.
No more greenhouse gases (almost).
Exhaust goals should be constantly tightened.
Mitigate risks associated with warming.
Distribution of responsibility, obligation to provide information.

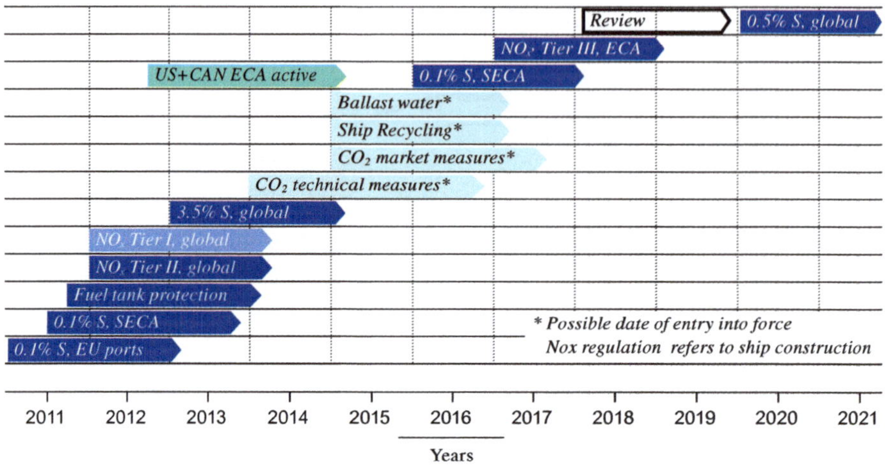

Fig. 3 Future Ship GmbH (GL Group) estimated dates for environmental regulations entering into force

Damage and losses, as far as money for the poor.
establishing: Severe Weather Insurances

All 196 countries signed—even OPEC and Russia
It contains no mechanism to make it binding
Figures have declared that all pledges are just self-interest
And they say emphatically, they are "not to save the planet."
(code words for, "do not take the climate into account")
The 2 °C goal is disconnected from any proposed mechanism
It got the US and China talking about climate
It got countries to think about climate change for a while

2.2 *International Maritime Organization (IMO)*

The International Maritime Organization (IMO) with MARPOL (International Convention for the Prevention of Maritime Pollution), is lagging behind in the issue of environmental protection and only makes compromises on the lowest international level (Fig. 3 and Table 1).

The table shows the great gap in efficiency (indirectly by the CO_2 output) of sea transportation. The opposite to other transportation types is clearly visible (Fig. 4).

From point of CO_2 emissioning the international shipping industry is only a smaller part in the complete lineup of greenhouse gas emitting. Regarding overall the vessels are much more efficient than all other transportation technologies but nevertheless they are emitting huge amounts of exhaust emissions.

Table 1 Exhausting of CO_2 per transported ton per kilometer

CO2 emissions in g CO_2/t km	Minimum	Maximum	Average
Tanker (Crude, LNG)	3	15	9
Bulker	2	29	16
Container vessel	12	36	24
Railway	14	120	74
Road	80	182	126

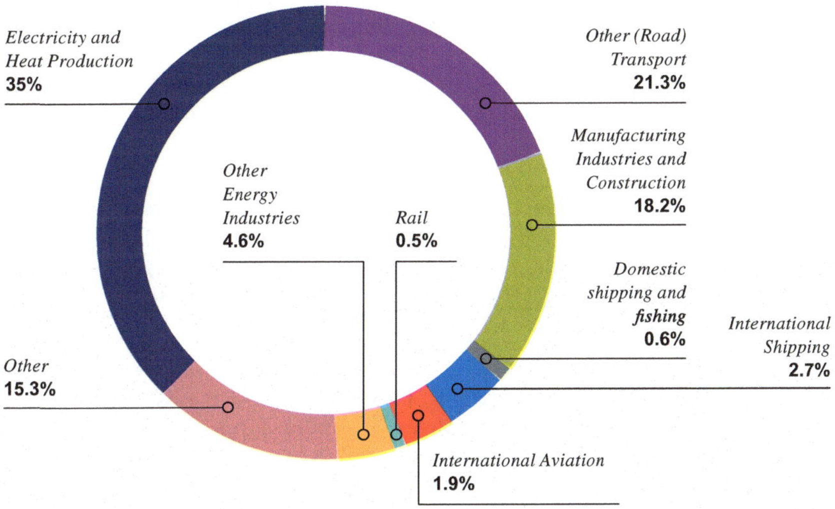

Fig. 4 CO_2 emissions worldwide [3]

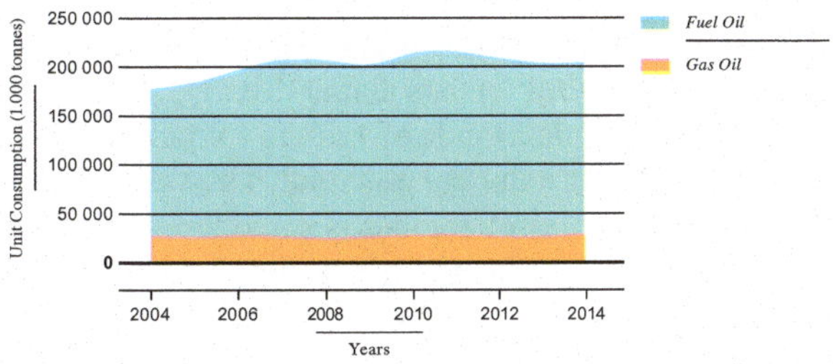

Fig. 5 Marine Fuel Consumption Worldwide in 2014 [4]

Marine fuels for propulsion are mainly heavy fuel oil (HFO) approx. 85% and marine diesel oil (MDO or gas oil) approx. 15% of total usage of approx. 220 million tons per year as shown in the next graphics. Alternative fuels like liquefied natural or petroleum gases (LNG and LPG) are nearly not used in merchant shipping, only some ferries and the newest cruise liner besides the LNG and LPG carriers starting to use this fuel (Fig. 5).

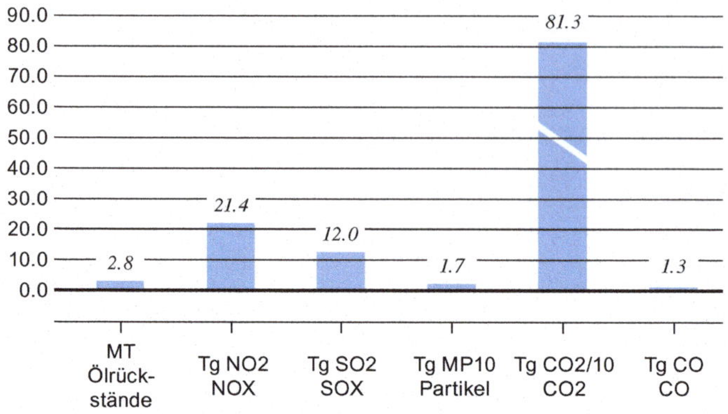

Fig. 6 Emissions from international shipping per year (Eyring, V. et al. 2004, DLR Institute for Physics of the Atmosphere) (1 Tg (Tera gram) is equivalent to 1 million ton (Mt))

While currently strong exhaust cleaning systems are installed in all cars or trucks, ships not even achieve the old car exhaust limits, which were valid decades ago. All exhaust techniques for ships already exists, there is no need for new development.

But if no regulation organization from itself forces the shipping companies to rethink and incorporate these techniques, it behoves—sad but true—finally the environmental protection organizations to draw attention on this problem (Fig. 6).

Heavy Fuel Oil (HFO) is the residues of oil refinery companies outputs. During the burning process it generates a lot of CO_2 and SO_x and much worser pollutants like the harmful soot particles and nitrogen oxides.

HFO (contains approximately 4.5% sulfur) is also the main human made sulfur(oxide) source and it is polluting the air and environment. For figuring out we can say that one big vessel emits the SO_x of approx. 50 Mio cars (Fuel accord. DIN EN 590 with actual maximum 0.001% sulfur).

That leads to the result that alone the 20 biggest vessels discharge more sulfur oxides than all 1 billion cars actual driving on all roads on our planet.

And there are thousands of vessels using HFO!

Here is a lot of reducing potential. The way we handle it in the past with burning heavy fuel oil in vessel engines is some kind of burning waste at sea.

Reduction of CO_2 and SO_x can simply be achieved by stronger worldwide regulations of the International Maritime Organisation (IMO) and the requirements from all flag states and its classification societies.

3 Fuel Efficiency by Technological Development

Several techniques are meanwhile available to reduce the total consumption of energy during the vessels operation and way across the seas.

Improved hull design with experimental and numerical hydrodynamic research can achieve reductions in emissions and fuel consumption.

A bulbous bow increases a ship's fuel efficiency by reducing its wave resistance.

A stern flap, a small plate that extends behind a ship's transom, lengthening the bottom surface of the hull, and reduces a ship's flow resistance and thus increase fuel efficiency by a few to several percent.

Special coatings (e.g. sharkskin-design) applied to propellers and hull may reduce fuel use up 4 to 5%, while simultaneously reducing maintenance requirements. This practice could likely compensate the costs for itself within a year.

Sail or kite-assisted propulsion can provide zero-emissions wind power, and the first experimental vessels are already underway which employs such technologies.

DK Group has developed an Air Cavity System (ACS), compressed air bubbles out of thousands little drilled holes from the vessel double bottom into the water and generates a water-air-bubble-mixture which reduces the frictional resistance, this reduces the ship emissions up to 15% during voyages.

The use of "cold ironing" at ports, where ships shut off their diesel generators and are connected to shore-based power grid for their electrical needs reduces the direct emissions in port areas and allow energy needs to be met by zero-emission sources, such as wind or solar energy.

References

DNV GL Expert Maritime Impact Issue; Bergmann, J. (2016). https://issuu.com/dnvgl/docs/maritime_impact_issue_01-2016_w

UNCTAD: Review of Maritime Transport (2013). https://unctad.org/en/publicationslibrary/rmt2013_en.pdf

IMO Second IMO GHG Study (2009). http://www.imo.org/en/OurWork/Environment/PollutionPrevention/AirPollution/Documents/SecondIMOGHGStudy2009.pdf

FuelsEurope: Statistical report (2015). https://www.fuelseurope.eu/wp-content/uploads/2017/05/fuelseurope-statistical-report-2015.pdf

'City Air Makes You Free'. Cultural Dimensions and Application of Urban Development Projects in Western Trans-Danubia

Béla Bakó

1 Prologue

'*Cities are the world that man builds for himself*', writes W. Schneider in his book *Überall ist Babylon. Die Stadt als Schicksal d. Menschen von Ur bis Utopia [The City as Man's Destiny. Cities from Ur to Utopia].* Carrying the idea a little further, we might as well say that a place equals the people who live in it. There are no people without a place to live, and there is no city without people [12]. At the same time, the basic question for urban development in our age is what *sort* of city man is building for himself and what kind of content is used to fill up the built environment of the city—the urban space. 'Our present age,' claims Michel Foucault, 'is best described as an era of spatiality. We are living in an age of synchronicity, the collateral, of near and far, right and left, as well as of scattering [4].' This concept of space entails the principle whereby the social world may be captured as a relational concept. The totality of the reality it describes has its roots in the mutual relations of its components [10]. It is in the spirit of this thought that we shall inventory the kind of social characteristics that urban space can stage for us and explore what sort of clues it offers for understanding the nature of the given geographic unit (city, area, region etc.). This way, through the concept of urban micro-spaces it becomes plausible that the space we view as a special manifestation of systems of social relations is not necessarily homogenous in character even among the circumstances of the metropolis [9]. As regards the different kinds of systems of relations which become accessible to the interested onlooker, the best witness is the collective memory of cities. This can lead us on to questions of space and identity, which Jonas Frykman summarises by claiming that 'everyone's place of birth outlines their collective and individual identity' [6]. On this basis it seems justified to investigate deeper into the urban tissue and the related social milieu which can serve as the basis of determining the content

B. Bakó (✉)
Vas County Scientific Educational Association, Szombathely, Hungary
e-mail: bakob.nco@gmail.com

M. Palocz-Andresen et al. (eds.), *International Climate Protection*,
https://doi.org/10.1007/978-3-030-03816-8_24

for later 'city building'. I am talking about the kind of content which constitutes a coherent unit and takes into account the characteristics of the urban tissue and the needs and desires of those who inhabit it.

One very good reason to talk about coherence in this context is that the concept is one of the crucial and widely known components of the toolkit and financial support policy used by the European Union, the 'chief sponsor' of urban development projects in Hungary and indeed probably in the whole of Central and Eastern Europe. At the same time the exact area involved and the exact nature of this cohesion are heavily debated in the theoretical writings and political practice of the member states (for more on this see [11]).

Exploring the nature of this cohesion from one particular angle could well serve as the central goal of this paper and could enrich a model for a viable (or, to use a currently trendy term, a creative) city defined along economic, social and cultural issues, while placing the mind-set and identity of the locality and local society into the urban space including the transformations of this identity in parallel with the changing of functions. This could become the cementing force which would mean the basis of territorial cohesion in the given urban space. Key concepts of this train of thought are the identity forming function of the urban space; preservation and management of the local heritage, or the genius loci; regional identity and network building; innovation, talent and knowledge society, as well as the sustainable way of life and social innovation.

This leads on to understanding the identity of the urban tissue which demonstrates not the identity of the place but, to quote Jonas Frykman, 'its mood, its soul, its ability to comprise memories and dreams' [6] in other words the environment which serves as the soil for a certain model of urban development and social existence which can be clearly outlined in space. This is a model which I like to define with the help of three key concepts—attraction, innovation, recreation—and this triangle creates the frame for a supported and viable urban space shaped by local identity.

2 Challenges of Integrated Urban Development

The urban tissue and the city's social and economic life create a coherent system where each component mutually influences the other and their totality and their interactions determine the quality of life in the city, its competitive potential, attractiveness, inclination for innovation or its recreational capacity.

Physical interventions taking place in the framework of urban development and urban rehabilitation aim at physically transforming the urban tissue in order to improve the physical conditions of urban life. In order to make sure that urban development yields substantial and lasting results which are sustainable in both the social and the economic sense, it is usually necessary to combine physical intervention closely with planning and executing activities which aim to transform the non-physical (social, economic, cultural etc.) conditions of urban life.

What we can declare with certainty on the basis of European research and the experiences of regional development projects is that social and relationship capital play a huge role in the innovative and economic performance of regions and city agglomeration economies. Besides, face to face relationships among actors of productive processes and their proximity in the geographic, social and cultural sense can trigger creative processes that go beyond innovation in the narrow sense.

This is particularly true in the case of interventions where the aim is urban rehabilitation and where, due to various interactions of the kind described above that had emerged over a long period of time, transforming the mere physical conditions brings little if any effective improvement in the quality of life or the competitive edge. Even though the built environment is up to the criteria of the 21st century, the local society barely profits from this, if at all.

Thus it seems justified to inventory, beyond the infrastructural aspects of contemporary urban development, the various possibilities for utilising the urban space. This idea is based on the conviction that in order to bring life into the urban space it is not sufficient to rehabilitate the built environment. Filling the space with content is far more important, as it can result in economic, social and cultural sustainability, create jobs or bring about wide-ranging long term opportunities for investment. From this point of departure it takes only one more step to connect local identity with the way in which it influences the urban environment—a connection which highlights the complexity of the transformation of space and society in the direction of development.

3 Conclusions Regarding Projects Completed and Currently Underway

Improving the physical and mental well-being of a population as a community can be the primary goal of shaping a city. Improving physical comfort becomes plausible through improving and renovation of the built environment—by upgrading public spaces, modernising the energy supply etc. Actions aimed at improving the mental well-being of the community generally take the shape of soft project elements happening alongside the other development efforts. At the same time it is also visible that on the one hand, these development are always limited to a narrow area, the action area designated for the purpose, while on the other hand, and this is particularly true of the so-called soft project elements, they do not constitute a coherent, identity shaping unit, much rather do they exude a sense of contingency. This sense of contingency could be resolved through two major subject areas. One is related to the idea of expanding development in space, while the other has to do with creating structures.

The present frames do not allow for a more extensive discussion of expansion in space, therefore I only offer a brief summary here. The problem of focussing on a narrow space is outlined by David A. McGranahan and Timothy R. Wojan who believe that the attitude of the local population is crucial to the development of the locality and the improvement of tourism (for more on this see [15]). However, just as the human eye spends a while dazzled before it can accommodate itself to a vast space which opens up suddenly, so the society of the present age is slow and cautious about accepting that its existential options and freedom of action have become radically expanded. At the same time, placing development into a global context and drawing on the synergies of local development should, indeed must, point in the direction of longer term, strategic thinking. It is in this sense that the nascent 'creative city, sustainable region' programme represents a model developmental strategy which extends the urban space, places it in a wider context and is thus able to realise a new strategy and way of thinking regarding small and medium sized towns or cities (for more on this see [17]).

From the perspective of our subject matter the second area deserves more attention. This concerns setting structural frames for development constituting a unified configuration, and doing this by placing the built environment, as well as the local community and, more broadly, the human factor, into the centre of our efforts. This is an intellectual and emotional surplus which points beyond a mere stark functionalist approach—something we might refer to as 'added value', were we talking about a product aimed at a market. It is an extra which defines the social demand for the product, and its potential success, beyond the specific, describable elements of matter, function and technology. This extra can alone determine the 'market position' of the 'product'—in our case the urban space itself, defining whether it is seen as lower, medium or top category.

I would like to define and categorise this extra with the help of three concepts—the previously mentioned triad of attraction, innovation and recreation.

4 Attraction

The concept of attraction manifests itself most commonly through the development of tourism in any locality. Most of the literature on tourism agrees that the existence of genuine and potential tourist attractions is a precondition for the development of tourism in an area. Tourist attractions represent the basis (raw material) of the partial tourist product and the complete destination tourist product. This is why such attractions need to be identified, registered, valued, classified and protected from damage and irrational use. Therefore, it would be more productive to go beyond functionalist, infrastructure-oriented, mid-term developmental strategies and move toward a culturally based urban development project which takes advantage of the characteristics of the built environment and tries to mark out a progressive direction both for the local society and the outside world (tourism, construction, investors etc.).

In spite of all this it is worth highlighting that already existing or newly created attractions play a part, beyond tourism, in encouraging the creativity of the local society. It can be generally stated that creativity is more emphatically present in places where we find a stimulating architectural and intellectual environment, where the city has characteristic quarters, an effervescent night life, cultural versatility and a wide range of sports activities. If people engaged in innovative work are present in great numbers, this, in turn, has a positive effect on the economic performance of the city. The essence of creativity in this case is that attractions form a very broad scale from technology through the arts all the way to the entertainment industry where the only cohesive element, very strong in its own right, is the value system itself: creativity, individuality, non-conformism and the mutual appreciation of individual merit. For representatives of activities with a high added value, post-materialist values such as the quality of life and a colourful, varied, intellectually inspiring environment are at least as important as the quantity of money received for work (for more on the motivation of employees see [16]). Accordingly, the cities best suited to attract the creative stratum are those who are able to provide people engaged in intellectual work with the conditions described above.

Prague, Barcelona or Venice are basically selling, besides their local historical and cultural values, a way of life or a myth. Salzburg is selling the myth of Mozart and of music. It suggests to its visitors that through the various events held in the town, particularly if they buy a ticket, they become participants of this mythology. Dublin is basing itself on the cult of James Joyce and the culture of beer. The name of New Orleans awakens the images and ambiance of its jazz pubs. The attraction of Bangkok, in turn, is based on Southeast Asian young women, even if Thai people are not awfully happy about this. Prague is selling its baroque historical milieu, with a certain amount of jealousy also Mozart, but mostly Kafka and Hrabal, but most of all a food which, compared to the amazing culinary spread of the great world cuisines, is something surprisingly plain and bland—the dumplings called *knedliky*. And, of course, they sell beer—something that you will find in a number of other points of the world. Its local flavours, its folklore which is not play-acting but genuine, and the very living cult of pubs in Prague authenticate this otherwise not very original, fundamentally German culinary element. Budapest, by contrast, is still lacking a mythology, nor does it offer a unique, loveable way of life which would tempt people to follow it. The reason for this is to be found in the mental condition of its society, the contemporary urban population which essentially functions according to pre-set patterns [2].

The job waiting to be done at this point is not only to make sure that old public functions, the agora or the institutions of the forum become once more embedded in the network and conditions of the community—it is also crucial to lay the foundations for or to conceive of public spaces which confirm the new identity. We must overcome the current vulgar approach of commerce which is driven purely by the principle of profit maximisation and create a new balance between the needs and goals of the community and commercial technologies; a state balance based on a new social contract written in line with their clearly defined interests.

5 Innovation

From the perspective of urban development, innovation plays an outstanding, but not exclusive role in enlivening the economy. The reason why it deserves to be discussed separately is the fact that creating, utilising and disseminating knowledge is one of the foundations of economic growth, development and the unified existence of the nation. This is why it is crucial to explore the questions related to innovation and place all of this into a new context. The more so, as in the recent times the nature and appearance of innovation have changed and this way it has become crucial to align the concept with the new contents. Today, the possibility of progress for a developed county (and in a worldwide comparison Hungary is clearly in this camp) is encoded in the sectors which are able to create something new using the power of innovation and intellectual capital. Companies and large corporations value innovation highly because it improves their results; employees appreciate it because they see it as one of the best ways for self-expression and for achieving fulfilment in their work.

If we accept the view widely held in contemporary sociology whereby in today's globalised information-based society no one exists unless they are part of the network, based on this consideration we could compile an inventory of the qualities and capacities of Hungary's urban spaces as they were designed and still exist today, and of their flaws and shortcomings accrued as a result of half a century's delay in progress. We lack the occasions to organise the local community in ways which could at the same time contribute to that local population forming a cultural community with the inhabitants of other cities and countries. It is often said that Europe ends at Vienna, referring to the fact that great itinerant cultural events usually appoint Vienna as their last European station. This is contrasted to the fact that once the Hungarian Opera House was directed by Mahler, and if we look at the most prominent institutions of the international musical scene we see that they are headed by grand personalities who originally have nothing to do with the national culture of the place [2].

As a result, the role of urban space becomes a dominant strategy here in connection with the concept of incubation. Incubation, in line with its original concept, is suited to help nascent innovative enterprises of any scale from one individual to a company to get through their initial, most vulnerable phase of life in the hope that this support will somehow pay dividends to the society as a result of later growth [1].

This may present a foundation for smaller cities which, due to their complex set of features, can deploy customary strategies of competition and thus become successfully integrated in the hierarchy of production and service systems or combine their inner resources, unique or not easily replicable qualities into a new system and thus attain specialisation [14], thus reducing their competitive disadvantage compared to large cities.

6 Recreation

Our third concept, that of recreation has mainly become known in this country as a term referring to the use of free time, but in fact, upon closer exploration, it turns out to have a more complex meaning. Taken more generally, recreation entails a sense of feeling good, of well-being, of creating the good life or a high quality of life, while recreation also results in the individual becoming refreshed, re-energised and entertained (for more on this see [5]).

Projecting this concept onto the topic of creative urban development is justified by the fact that contemporary man lives in a narrowed-down space (mostly dominated by the urban milieu) and his/her lifestyle is largely dominated by the pressure to achieve and by an exaggerated consumerism. Based on these it seems desirable to work consciously on developing a recreation culture which can be measured by the following four areas of 'consciousness'.

Free-time consciousness which refers to spending free time in a planned, conscious, mindfully attentive manner. From the perspective of urban development this means developing the spatial frames for a quality life, which appears to be a basic challenge in the case of small and medium-sized cities.

Health-consciousness—according to WHO's definition, health means not merely the absence of illness but also a state of physical and psychological balance with the external environment which entails a well-being in body, in mind and in the social sense. When examined in the context of urban life, this concept gravitates in the direction of a sustainable environment.

Life-style consciousness is partly related to the previous concept, but it also represents a more extensive or holistic approach. The quality or mode of a person's life is reflected by the kind of objects they use (objects turned objectifications and objectifications turned objects), what sort of environment they live in, what kind of values they profess and serve, what kind of relationship they have with other people, what kind of symbolic semiotic system they use and how they express themselves. In fact, these factors (objects, the environment, relationships, values, symbolic expression) together constitute culture and together they can form the basis of some sort of well-being in the urban space.

Environment consciousness—this is one of the most emphatic requirements of being civilised today: the intention to sustain the environment. Environmentally conscious action has come to constitute a philosophical approach, an entire system of thought, even though daily practice regularly contradicts rational action. This is why it is crucial that in every single aspect of the market economy we should seek for the opportunities of environment conscious behaviour and enforce them. In the urban space this could lead, as a long-term result, to the emergence of sustainable towns, cities and villages.

Regarding the criteria of urban development, in view of the above we can declare that in our contemporary social order without conscious, systematic development in the fields of fashion, design, architecture, film, music, book and record publication, high quality catering, art trade, cultural product industry or high level 'health and care', all in harmony with the goals of urban development, any urban object will remain only an isolated architectural or urban phenomenon. Unless the emerging structures are dictated by genuine, existing needs, this kind of hollow formalism can only mean an empty shell lacking authentic content.

7 'City Air Makes You Free'

'Stadtluft macht frei!', goes the old German proverb and this was indeed true of the mediaeval city—but how applicable is it in the urbanism of our day and age? And anyway—what makes a place into a city? As St. Augustine put it in a succinct and ingenious way, it is not the walls but the mentality of the people (*non muri sed mentes*) [13]. Based on this we could claim that instead of a functionalist urban development it would be more desirable if a culture-based way of thinking could take place regarding urban development.

Let us now review what are the greatest mistakes, based on our previous experiences, that should be avoided in the course of urban development.

Cultural development projects focusing their energy and cultural resources more and more on inner city businesses and culture zones, while the resources of the local society remain latent. Globalised culture becomes embodied in the world of the Shopping Mall and the Multiplex; spheres of artificial experience squeeze out the use of space into the suburban environment, while the inner city, the traditional core of the urban unit becomes empty.

The cultural wealth and infrastructure of the city do not fit the history and inner needs of the local society; typically it falls prey to global homogenisation.

Those in charge of 'organising' culture are not sufficiently well-prepared; they have a narrow political focus (primarily for financial reasons) and this narrow-mindedness has a negative influence on productivity and creativity.

There is a rift between the political environment, cultural policy decisions and the genuine cultural needs of the local societies, which has a negative influence on the retaining potential of the city, on the improvement of local identities and on exploiting latent potential.

What can be done? Economist David Throsby points out the following: cultural capital fuses the two most important criteria of sustainability—the environmental and the economic consideration. Sustainable economic and environmental development must move in a concerted and combined manner so that neither of them causes a loss of dynamism or stagnation in the other. They must generate processes which support the renewal of both the eco-system and the society [18]. One possible way of handling the problem is to reduce the number of short-term and temporary solutions, to identify and strengthen the main force-fields and to eliminate harmful self-generating processes before they become even more characteristic.

Throsby determines six criteria for creating the desired conditions.

Material and non-material well-being—the production of products and services which offer consumers both economic and cultural value.

Creating balance among the generations in the consumption of goods: we must act in such a way that different generations use and dispose of resources in a balanced and proportionate manner.

Fair and correct behaviour in the use of values and resources—the right of use is granted for all generations, but this right must be practised in a moderate and honourable manner. Thus, use of cultural capital is not equal to using up cultural capital.

Sustainable diversity, cultural plurality. The greater the cultural diversity and plurality, the broader the engagement of resources, the more opportunities there will be for renewing them and securing cultural 'multiplication'. A broad and colourful cultural palette can vouchsafe for the value system and richness of the future.

The principle of caution: we must only make decisions which strengthen the basic indicators and resources and do not generate irreversible processes.

The sustainability of the cultural system: it is necessary to understand the system and its variables and to explore synergistic energies. Between systems and subsystems, no matter what kind we are talking about, including cultural economy, there are always overlaps and co-operation, and understanding of these aspects is a basic criterion [18].

Based on all of this it may be declared, in line with the principles described by Patrick Geddes which are so crucial for the rebirth of cities, that planning is not a physical science but one of the humanities.

Geddes emphasises that every planning process must take into account the three co-ordinates of population—work—location. Therefore, planners need to be, besides draftsmen, anthropologists, economists and geographers. They need to understand the ways in which the local population work, have fun how and relate to their environment. Planning as a basically physical science must be superseded as regards the way in which it focuses on land use, infrastructure and transport. This practice is based on a two-dimensional city-planning and takes no notice of what goes on inside the coloured squares—these exemplary buildings.

Geddes emphasised that we must explore the strata of the city moving from the top down, all the way to the most distant past, and read from there up, constantly imagining history in front of our mind's eyes. It is important that we should be able to integrate and include the history, patterns and memories of the urban environment and its inhabitants, their history in the broadest sense, into the city planning process. Culture-based city planning must turn away from both the 'mentalist' and the 'aesthetic' aspect and establish far more robust ties with the forces that consume culture and those that create it [7].

8 Epilogue

Shaping the life of a city can have but one objective—to improve the quality of life of the people living in it. There are two ways to approach this.

According to one indispensably important technocratic school of urban development improving life conditions means a constant betterment of the city's infrastructure. At the same time we must bear in mind that cities were originally an intellectual product. In *Revolt of the Masses* Ortega y Gasset writes that when the 'urbs' or the 'polis' were created, this gave birth to a more revolutionary concept of space than Albert Einstein had done [3]. The actions associated with the city illustrate the life, the way of thinking and the memories of the people living in it, and urban development must always take this into account. This is so because a city is nothing other than memory cast in stone—every piece reports something from the past, remote or more recent, it shapes identity and points toward the future [8]. Once we manage to see beyond mere maps, ground plans, concrete and marble, the city will become reconstructed and along with its reconstruction the urban population will also reconstruct itself.

Literature

1. Bajmócy, Z.: Az üzleti inkubáció szerepe a vállalkozásfejlesztésben. Közgazdasági Szemle 12, pp. 1132–1150 (2004)
2. Bojár, I.A.: Budapest, a kreatív város – a lehetőségek kapujában. Demos Magyarország, Budapest (2005)
3. Czakó, G.: Azután. Agóra és lakógép. In: Lukovich Tamás – Csontos János szerk. A mi Budapestünk. Pallas Stúdió, Budapest (2002)
4. Foucault, M.: Eltérő terekről. In: Foucault, Michel: Nyelv a végtelenhez: tanulmányok, előadások, beszélgetések. Latin Betűk, Debrecen (2000)
5. Fritz, P.: Rekreáció mindenkinek I. Mozgásos rekreáció. Bába Kiadó, Szeged (2011)
6. Frykman, J.: Hely, valami másnak. Egy kulturális képzetrendszer elemzése. In: A kulturális örökség (eds.: Erdősi, Péter – Sonkoly, Gábor) L'Harmattan – Atelier, Budapest, 2004 (2002)
7. Geddes, P.: Town planning towards city development. A Report to the Durbar of Indore: Parts 1 & 2, Holkar State Printing Press, Indore (1918)
8. Granasztói, P.: Budapest arculatai. Szépirodalmi könyvkiadó, Budapest (1980)
9. Gyáni, G.: Az utca és a szalon. A társadalmi térhasználat Budapesten (1870-1940). Új Mandátum, Budapest (1998)
10. Gyáni, G.: „Térbeli fordulat" és a várostörténet. In: Korunk, 2007, Július (2007)
11. Illés, I.: A „területi kohézió" szerepe az EU és a tagországok politikájában. In: Kötőerők. Az identitás történetének térbeli keretei (ed. Cieger András). Atelier, Budapest, 2009, pp. 87–101 (2009)
12. Izsák, É.: A városfejlődés természeti és társadalmi tényezői. Budapest és környéke. Napvilág Kiadó, Budapest (2003)
13. Katus, L.: A középkor története. Pannonica – Rubicon, Budapest (2001)
14. Lux, G.: A gazdaság szerepe a városi térségek fejlesztésében: a globális kihívásoktól a fejlesztéspolitikáig. In: Somlyódyné Pfeil Edit (szerk.) (2012.) Az agglomerációk intézményesítésének sajátos kérdései: Három magyar nagyvárosi térség az átalakuló térben. Publikon Kiadó, Pécs, pp. 67–89 (2012)

15. McGranahan, D.A., Wojan, T.R.: The creative class: a key to rural growth. In: Amber Waves April 2007, pp. 17–21 (2007)
16. Miszlivetz, F. (szerk.): A magyar munkakultúra állapota és alakításának lehetőségei globális környezetben. Kutatási jelentés. Savaria University Press, Szombathely (2009)
17. Miszlivetz, F. (szerk.): Kreatív városok és fenntarthatóság. Javaslatok a Duna-stratégia megvalósítására Nyugat-Pannónia példáján. Savaria University Press, Kőszeg-Szombathely (2012)
18. Throsby, D.: The Economics of Cultural Policy. Cambridge University Press, Cambridge (2010)

Photocatalytic Conversion and Storage of Solar Radiation as a Renewable and Pure Energy

Ottó Horváth and Lajos Fodor

1 Hydrogen Production by Photocatalysis

The energy of the solar radiation reaching the surface of Earth during a period of just one hour is greater than the amount used up by the human world population in a year. Despite the several processes offering flexible conversion and storage of solar energy, only about 1% of the whole energy we produced is originated from this source. The average annual solar radiation energy in Central Europe is about 1,200 kWh/m^2. This intensity is acceptable for production of environmentally friendly fuel such as hydrogen from water by application of appropriate photocatalytic procedures utilizing the ultraviolet and visible range of the solar radiation. Depending on the phase of the photocatalyst, these methods can be homogeneous and heterogeneous. While in the previous case both the photocatalyst and the reactants exist in the same solution phase, in the latter case the catalyst is solid and either immobilized on the surface of a support or, as a colloid, suspended in the solvent, which is generally water also serving as a source of hydrogen. The discussion of these possibilities begins with the heterogeneous systems because their efficiencies and stabilities are mostly higher than those of the homogeneous ones.

1.1 Hydrogen Production by Heterogeneous Photocatalysis

1.1.1 General Concepts of Heterogeneous Photocatalysis

Heterogeneous photocatalytic systems are based on photoactive semiconductors, the function of which can be shortly described by the help of Fig. 1 [1].

O. Horváth (✉) · L. Fodor
Department of General and Inorganic Chemistry, Institute of Chemistry, University of Pannonia, Veszprém, Hungary
e-mail: otto@mk.uni-pannon.hu

M. Palocz-Andresen et al. (eds.), *International Climate Protection*,
https://doi.org/10.1007/978-3-030-03816-8_25

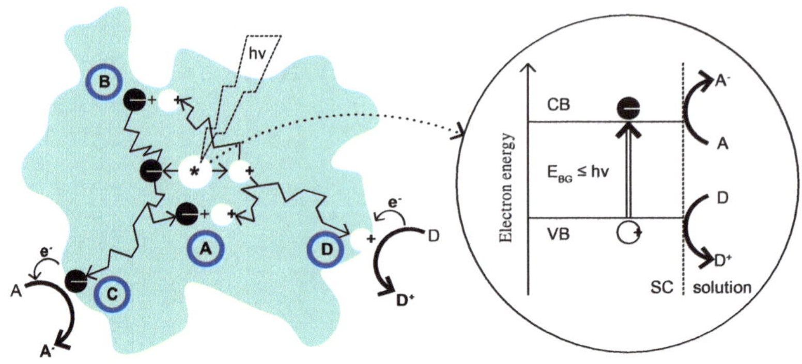

Fig. 1 The main processes in the function of a semiconductor (SC) as photocatalyst: **a** volume recombination, **b** surface recombination, diffusion of electron (**c**) and (hole) to the surface. * Designates the excited state resulting it charge separation

The most important three features of a semiconductor are the valence band (VB) containing electrons in its ground state, the conductance band (CB), to which electron can be promoted upon excitation, and the band gap (BG), i.e., the energy difference between the conductance and the valence bands. In the case of photoactive semiconductors, absorption of a photon the energy of which exceeds the band gap promotes an electron from the valance band to the conductance band, leaving a positively charged hole in the latter one. If the constituents of the electron–hole pair formed in this way reach the surface of the semiconductor particles, escaping the various recombination processes, they can participate in redox reactions. Thus, a light-driven electrolysis takes place on the surface of the semiconductor particle. Conduction–band electrons can reduce appropriate acceptors such as, in anaerobic systems, protons (H^+) from water, producing hydrogen (H_2) as an environmentally benign fuel, or, in air-saturated system, dissolved oxygen (O_2), generating superoxide anion radical ($O_2^{\bullet -}$) as a reactive intermediate for oxidation of various contaminants. Valence-band holes can oxidize suitable electron donors, e.g., a wide range of organic compounds as pollutants or oxygen in the water molecules, producing O_2 or hydroxyl radicals ($HO^{\bullet}$). The latter species are also very oxidative agents utilized for degradation of dissolved wastes in water.

In the case of heterogeneous photocatalytic hydrogen generation, there are 2 possibilities depending on the redox potentials of the photocatalysts applied; water splitting producing both H_2 and O_2 or only H_2 production at the cost of oxidation of some added sacrificial electron donor, which may be a pollutant to be degraded.

1.1.2 Hydrogen Production by Photocatalytic Water Splitting

Photocatalytic water splitting can be realized by application of two types of systems [2]. In the first one the production of both hydrogen and oxygen takes place on the surface of the same photocatalyst as indicated in Fig. 2. In this case the energy of a photon absorbed by the semiconductor is directly utilized in both the reduction of hydrogen and the oxidation of oxygen from water. Generally, a cocatalyst (mostly from the platinum-group metals such as palladium or rhodium) is deposited onto the surface of the semiconductor particles to overcome the overpotential of the reduction of proton [3–5].

The semiconductor photocatalyst of such a system meets two requirements as Fig. 2 demonstrates. The potential of the lower edge of the conductance band is more negative than the reduction potential of the H^+/H_2 system, while the potential of the upper edge of the valence band is more positive than the oxidation potential of the $O_2/H_2O(HO^-)$ system. The redox potentials of both systems depend on pH, according to the following equations:

$$\text{Cathode: } 2H^+ + 2e^- \rightarrow H_2 \quad E = 0.00\text{–}0.059\,\text{pH V}$$
$$\text{Anode: } 4OH^- \rightarrow O_2 + 4e^- + 2H_2O \quad E = 1.23\text{–}0.059\,\text{pH V}$$

If from the two requirements above only one is met, a partial water splitting can take place. In this case an added electron donor or acceptor species is needed as indicated in the previous section. The position of the band gap for some semiconductor photocatalysts are shown in Fig. 3, compared to the corresponding redox potentials of the water splitting. The values indicate that TiO_2 and $SrTiO_3$ can be most appropriate for water splitting because ZnO and sulfides (having narrower band gap) undergo photo-corrosion, while WO_3 and Fe_2O_3 can only be applied for O_2 production.

However, the band gaps of the promising semiconductors are rather wide, hence they can only be excited by UV light, strongly limiting the usable range of solar radiation. For an extension of the excitation wavelength toward the visible range, doping of the semiconductor (to decrease the band gap) or addition of sensitizers can be applied. The latter compounds absorb in the visible range and transfer electron

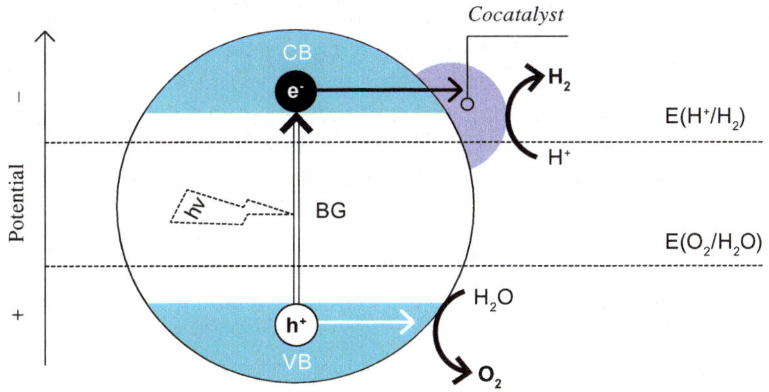

Fig. 2 Single-photocatalyst system for water splitting (adapted from Ref. [2])

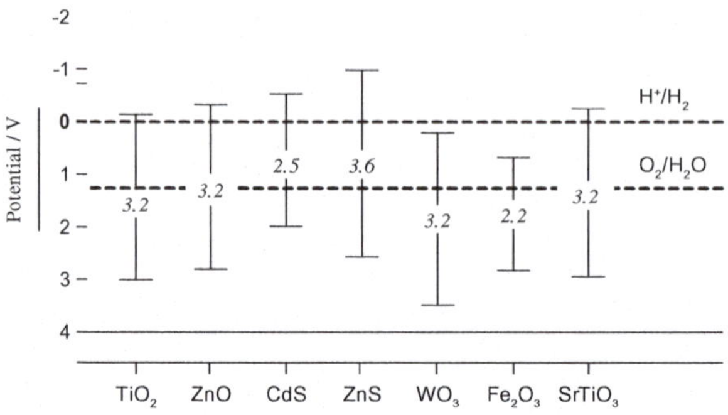

Fig. 3 The position of the band gap for some semiconductor photocatalysts, along with the redox potentials of the water splitting at pH = 0, with respect to normal hydrogen electrode (NHE)

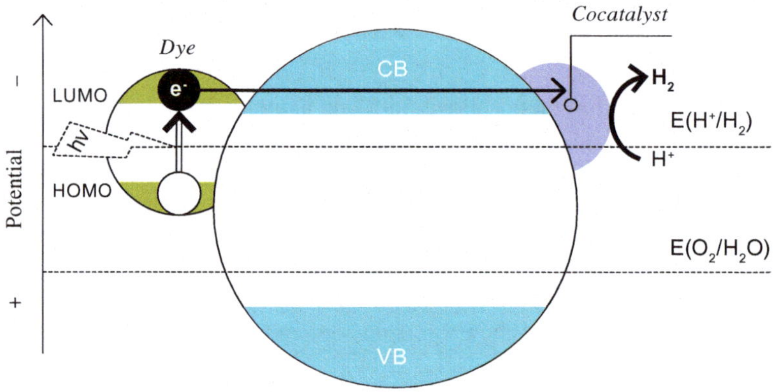

Fig. 4 Scheme of dye-sensitized photocatalyst (adapted from Ref. [2])

to the semiconductor (Fig. 4). The most frequently used doping materials (for TiO_2) are carbon, nitrogen and fluorine [6], while various metal complexes (with e.g. porphyrins, phthalocyanines and diimines) as well as organic dyes (such as Rhodamine B [7]) are primarily applied as sensitizers.

A combined system containing two semiconductors can also be applied for water splitting (Fig. 5). The band gap of one of them is favorable for hydrogen production, while that of the other one for oxygen generation. The two photocatalysts require individual excitation, and they are coupled through an appropriate donor-acceptor pair such as $IO_3^- - I^-$ or $Fe^{3+/2+}$. A typical semiconductor combination is $SrTiO_3$ for H_2 evolution and WO_3 for O_2 formation [2, 8]. The efficiency (i.e., the quantum yield) of such a system is close to unity at 420-nm excitation.

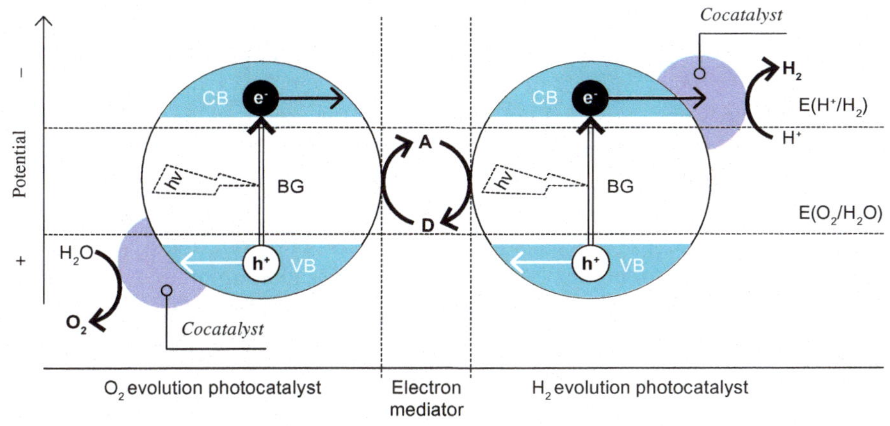

Fig. 5 Tandem photocatalyst system (Z scheme) for water splitting (adapted from Ref. [2])

1.1.3 Hydrogen Production with Sacrificial Electron Donors

As shown in the previous section, the total water splitting needs a semiconductor with a rather wide band gap (even if combined with an appropriate sensitizer) or a tandem system combining two photocatalysts. Of course, water splitting is the most reasonable solution for photocatalytic hydrogen generation. However, oxidation of water (i.e., its oxygen) requires a valence band of highly positive potential. For utilization of most of the visible photons in the solar radiation semiconductors with a rather narrow band gap are needed. In this case, if the position of the conductance band is suitable to the reduction of protons to hydrogen, the oxidation potential of the valence band is too low to oxidize water. A much more reducing compound is necessary for this anode reaction. Accordingly, in several hydrogen-producing aqueous photocatalytic systems contain such a sacrificial electron donor. The most widely used compounds for this purpose are triethanolamine (TEOA) and ethylenediaminetetraacetic acid (EDTA). Notably, some industrial wastes can also be applied, both organic and inorganic [9]. Thus, beside hydrogen generation, such systems can be utilized for oxidative transformation (degradation) of various pollutants into environmentally friendly, harmless end-products.

1.2 Hydrogen Production by Homogeneous Photocatalysis

Although heterogeneous catalysts have the advantage over homogeneous ones that they can be easily separated from the reaction mixture. However, their reactions with the corresponding electron donors and acceptors are limited on the surface of the catalyst particles, while in homogeneous systems the whole solution phase is utilized for this purpose. Additionally, they are more favorable from the viewpoint of the efficiency of light absorption, which is an important factor in photocatalysis.

Generally, the photocatalyst in homogeneous systems is a metal complex. In many cases the same coordination compounds can be used as both homogeneous catalysts and sensitizers in heterogeneous systems. Typical examples are the tris(bipyridine)ruthenium(II) complex ($[Ru(bpy)_3]^{2+}$) and its derivatives [10]. They were utilized as efficient sensitizers not only in heterogeneous photocatalysis, but in dye-sensitized solar cells (called also as Grätzel cells [11]), too. In the case of these photocatalysts, like most of the homogeneous systems, addition of a sacrificial electron donor (here EDTA) is needed. Besides, methylviologen (MV^{2+}) is also involved in the catalytic circle as an electron relay (accepting the electron from the excited catalyst and forwarding it towards proton, through a cocatalyst, colloidal platinum in this case) [10].

$$[Ru(bpy)_3]^{2+} + h\nu \rightarrow {}^*[Ru(bpy)_3]^{2+} \text{ (excitation)}$$

$${}^*[Ru(bpy)_3]^{2+} + MV^{2+} \rightarrow [Ru(bpy)_3]^{3+} + MV^{\cdot +} \text{ (oxidative electron transfer)}$$

$$MV^{\cdot +} + H^+ \rightarrow MV^{2+} + 1/2H_2 \text{ (proton reduction, cocatalyst: Pt)}$$

$$[Ru(bpy)_3]^{3+} + EDTA \rightarrow [Ru(bpy)_3]^{2+} + EDTA^+ \text{ (regeneration)}$$

The overall reaction is:

$$EDTA + H^+ + h\nu \rightarrow EDTA^+ + 1/2H_2$$

Nevertheless, photocatalytic water splitting was also realized with this system (in the absence of EDTA) if RuO_2 powder as cocatalyst for water oxidation was applied. In that case the regeneration of the photocatalyst is:

$$4[Ru(bpy)_3]^{3+} + 4HO^- \rightarrow 4[Ru(bpy)_3]^{2+} + O_2 + 2H_2O \text{ (cocatalyst: } RuO_2\text{)}$$

Beside $[Ru(bpy)_3]^{2+}$ and related compounds, various metalloporphyrins proved also to be suitable photocatalyst in homogeneous systems [12–15]. Due to their extremely high light-absorbing efficiencies in the visible range and favorable excited-state features, porphyrin derivatives such as chlorophylls play key role in natural photosynthesis providing the base of the food chain in the whole biosphere on Earth [16].

Instead of colloidal noble metals, like platinum, real homogeneous cocatalysts were also applied for hydrogen production in photocatalysis. Among these artificial hydrogenases, cobaloximes proved to be most efficient [17]. Their structures show similarity to that of cobalamin (B_{12} vitamin) having Co(II) as metal center in the coordination cavity of a corrin (porphyrin derivative) ligand. One of the most simple representatives of these compounds is bis(dimethylglyoxime)-cobalt(II). Reduction of these complexes generates Co(I) species that reacts with a proton source to produce Co(III)-hydride, from two of which, in a homolytic pathway, H_2 is eliminated.

2 Conclusion

In response to the current global energy crisis and challenges of climate change, scientists make considerable efforts to develop renewable resources capable of meeting projected energy demands. Solar energy conversion is an area of great promise; utilization of sunlight to make environmentally benign fuel such as H_2 from water offers a long-term solution of these problems. This short survey clearly demonstrates that both homogeneous and especially robust heterogeneous photocatalytic systems have been developed for efficient conversion and storage of solar radiation in the visible range.

Acknowledgements This work was supported by the Hungarian Scientific Research Fund (OTKA K101141 and NN107310).

References

1. Tahir, M., Amin, N.S.: Advances in visible light responsive titanium oxide-based photocatalysts for CO2 conversion to hydrocarbon fuels. Energy Conv. Manag. **76**, 194–2014 (2013)
2. Kudo, A., Miseki, Y.: Heterogeneous photocatalyst materials for water splitting. Chem. Soc. Rev. **38**, 253–278 (2009)
3. Fujishima, A., Honda, K.: Electrochemical photolysis of water at a semiconductor electrode. Nature **238**, 37–38 (1972)
4. Kitano, M., Tsujimaru, K., Anpo, M.: Decomposition of water in the separate evolution of hydrogen and oxygen using visible light-responsive TiO_2 thin film photocatalysts: effect of the work function of the substrates on the yield of the reaction. Appl. Catal. A **314**, 179–183 (2006)
5. Altomare, M., Pozzi, M., Allietta, M., Bettini, L.G., Selli, E.: H_2 and O_2 photocatalytic production on TiO_2 nanotube arrays: effect of the anodization time on structural features and photoactivity. Appl. Catal. B **136–137**, 81–88 (2013)
6. Anpo, M.: Photocatalytic reduction of CO_2 with H_2O on highly dispersed ti-oxide catalysts as a model artificial photosynthesis. J. CO_2 Utilization **1**, 8–17 (2013)
7. Le, T.T., Akhtar, M.S., Park, D.M., Lee, J.C., Yang, O.B.: Water splitting on Rhodamine-B dye sensitized Co-doped TiO_2 catalyst under visible light. Appl. Catal. B **111–112**, 397–401 (2012)
8. Abe, R., Shinmei, K., Hara, K., Ohtani, B.: Robust dye-sensitized overall water splitting system with two-step photoexcitation of coumarin dyes and metal oxide semiconductors. Chem. Commun. 3577–3579 (2009)
9. Preethi, V., Kanmani, S.: Photocatalytic hydrogen production. Mater. Sci. Semicond. Proc. **16**, 561–575 (2013)
10. Amouyal, E.: Photochemical production of hydrogen and oxygen from water: a review and state of the art. Sol. Energy Mater. Sol. Cells **38**, 249–276 (1995)
11. Kalyanasundaram, K., Grätzel, M.: Application of functionalized transition metal complexes in photonic and optoelectronic devices. Coord. Chem. Rev. **177**, 347–414 (1998)
12. Eng, M.P., Ljungdahl, T., Andreasson, J., Martensson, J., Albinsson, B.: Triplet photophysics of gold porphyrins. J. Phys. Chem. A **109**, 1776–1784 (2005)
13. Zhang, L., Lu, Y., Du, Y., Yang, P., Wang, X.: Synthesis and photocatalytic hydrogen evolution of meso-tetrakis(p-sulfonatophenyl)porphyrin functionalized platinum nanocomposites. J. Porph. Phthal. **14**, 540–546 (2010)

14. Horváth, O., Valicsek, Z., Harrach, G., Lendvay, G., Fodor, M.A.: Spectroscopic and photochemical properties of water-soluble metalloporphyrins of distorted structure. Coord. Chem. Rev. **256**, 1531–1545 (2012)
15. Horváth, O., Valicsek, Z., Fodor, M.A., Major, M.M., Imran, M., Grampp, G., Wankmüller, A.: Visible light-driven photophysics and photochemistry of water-soluble metalloporphyrins. Coord. Chem. Rev. **325**, 59–66 (2016)
16. Coyle, J.D., Hill, R.R., Roberts, D.R. (eds.): Light, Chemical Change and Life: A Source Book in Photochemistry. The Open University Press, Walton Hall (1982)
17. Dempsey, J.L., Brunschwig, B.S., Winkler, J.R., Gray, H.B.: Hydrogen evolution catalyzed by cobaloximes. Acc. Chem. Res. **42**, 1995–2004 (2009)

Green City—A Sustainable Energy Concept for a Climate Neutral University

Nikolai Strodel, Oliver Opel, Karl F. Werner and Wolfgang K. L. Ruck

1 Introduction

Universities have serious impacts on the environment as they can be seen as small cities with respect to size, population and energy consumption [1]. Moreover, environmental issues are becoming more complex and interconnected between direct and indirect effects. As living labs for research and education, universities are foreordained to implement innovative concepts. Although there are measures to reduce emissions and other adverse effects on the environment, a systematic approach is generally lacking [2]. Therefore the Leuphana University of Lueneburg decided to define an integral approach and strategic framework in order to achieve a sustainable and climate-neutral university. This chapter focuses on the energy concept as part of these endeavours.

2 District Heating System and Campus

District Heating Systems (DHS) provide heat for residental and commercial heating requirements such as space heating and water heating. With high energy efficiency, high reliability and the possibility to implement renewable energy sources (e.g. solar energy), DHS contribute to a more sustainable future energy supply and lower specific CO_2-emissions. These systems operate at different temperature and pressure levels. With respect to the temperature levels, DHS have been developed in the course of four generations [3]. The Leuphana University in Lueneburg is supplied by a low temperature DHS which can be assigned to the 3rd generation. Provided with two cogeneration units (Fig. 1) burning bio-methane and two peak load boilers

N. Strodel · O. Opel (✉) · K. F. Werner · W. K. L. Ruck
Sustainable Energy Research, Leuphana University of Lueneburg, Lüneburg, Germany
e-mail: opel@fh-westkueste.de

M. Palocz-Andresen et al. (eds.), *International Climate Protection*,
https://doi.org/10.1007/978-3-030-03816-8_26

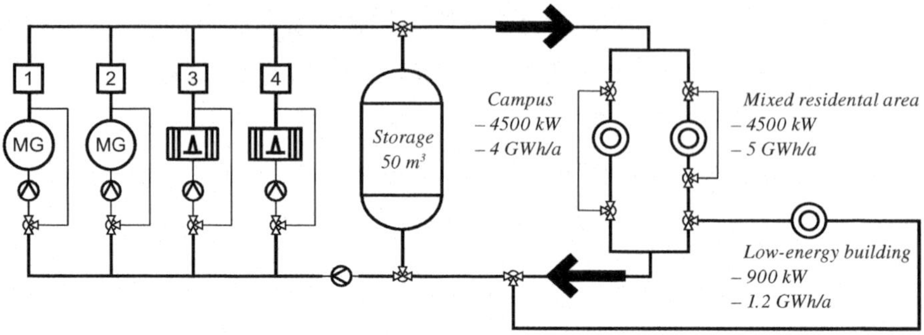

Fig. 1 Sketch of the existing district heating system (DHS)

(Fig. 1(3,4)), the system covers the heating demands of the university and a mixed residential and commercial area nearby. A more detailed description of the energy system and operating conditions is delivered in [4].

The existing campus was constructed in 1936 and formerly used as a casern. The casern was converted in the early 90s, when the barracks of the southern part were broken off. This part was rebuilt as a mixed residential (about 400 flats) and commercial area. The campus consists of 14 old casern buildings, 2 buildings used as dormitories and a new central building (low-exergy building) which is planned to be finished in 2017. The concept for the new building uses two internal, cascaded temperature levels at 55 °C/35 °C and 35 °C/25 °C, respectively. This low-exergy approach is realized by using the return flow of the DHS for the mid- and low-temperature-level within the new building (Fig. 1). Along with the conceptual design of the new building, new concepts of heat storages were thought about to refine the campus and the mixed area with a total of 0.25 km^2 in terms of efficiency and share of renewable energy.

2.1 Considered Storage Options

Water tanks used as short term heat stores are cheap and widespread. To avoid disadvantages connected with this technology, such as high specific heat losses and short storage periods, additional storage options for mid- and long-term storage were considered [5]. In this chapter, the integration of a long-term storage concept is presented.

Basically there are several techniques for seasonal storage of thermal energy. A very cost-effective and energy-saving technology associated with heating and cooling of buildings or districts is Aquifer Thermal Energy Storage (ATES). Since 2010 there are investigations on how to integrate ATES into the existing DHS in Lueneburg.

Depending on the season, ATES systems are generally operating bidirectional. The basic principle is to store surplus heat in the subsurface during the summer period and to recover the heat in the winter. Considering a common two well ATES system, surplus heat is stored by pumping cold water from the cold well through a heat exchanger and subsequently the hot water of the heat exchanger outlet is injected into the subsurface. During winter, when additional heat is needed, the storage serves as heat source by discharging hot water from the hot well to the cold well. The flow direction has switched.

3 Integration of Aquifer Thermal Energy Storage (ATES)

The hydro-geological requirements for ATES in Lueneburg were investigated and a simulation environment in TRNSYS (Transient Systems Simulation) was developed to simulate possible case studies. The annual mode of operation of the ATES system can be split up in a charging period and a discharging period.

Charging period (April until September)

As the cogeneration units produce renewable heat and electricity from biomethane, whereas the peak load boilers burn natural gas, the share of cogeneration should be as high as possible to achieve a maximum share of renewable heat. Hence, only surplus heat originated from the cogeneration units is used to charge the aquifer storage by transferring heat from the primary supply side (DHS) to the secondary side.

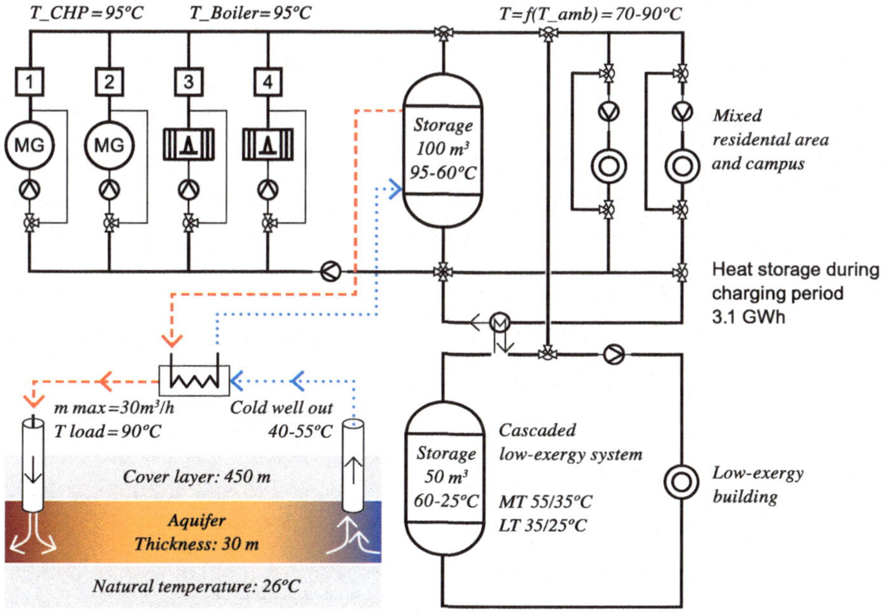

Fig. 2 ATES system operating mode from April until September

The surplus heat is calculated by comparing the amount of heat that is generated with the amount of demanded heat while a minimum state of charge in the buffer storage has to be guaranteed. The load temperature on the secondary side is limited to 90 °C due to water chemistry. During the charging period of six months, 3.1 GWh can be stored in the ATES (Fig. 2).

Discharging period (October until March)

For the optimal discharge of the aquifer during the discharging period, an additional 50 m^3 low-temperature buffer storage is used. The efficiency of the aquifer is dependend on the amount of heat recovery which is dependend on the return line temperatures and the supply temperature of the aquifer.

In the first months of the discharging period, when the supply temperature of the aquifer is at a temperature level in between 90 and 60 °C, the discharge flow passes the load valve through the outlet to the 100 m^3 storage. At the end of the discharging period, when the supply temperature is below 60 °C, the discharge flow is diverted to the low temperature storage. The lower the return line temperatures of the DHS are, the better it is for discharging the aquifer in order to get a maximum of heat recovery. During the discharging period 2.3 GWh can be recovered according to the model (Fig. 3) which corresponds to 75% of the charged heat energy (Fig. 2).

More detailed information about the considered ATES system can be retrieved from [6–9].

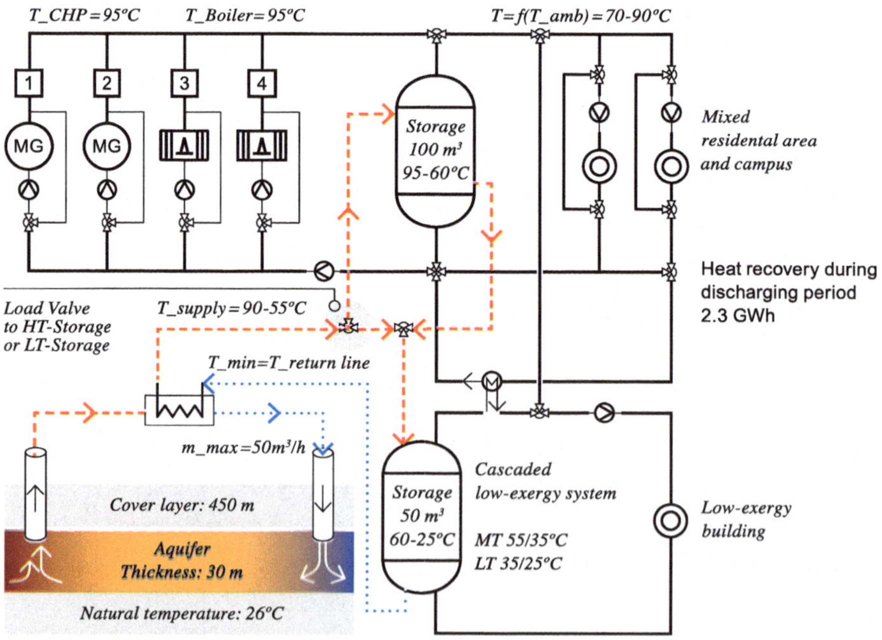

Fig. 3 ATES system operating mode from October until March

4 Climate Neutrality and Sustainability

In the course of seasonal heat storage in an aquifer, the share of renewable heat supply is increased by more than 20% (from 75 to 95%). By charging the aquifer storage with surplus heat originated from the cogeneration units, they are not switched off in the summer and therefore the two units are operating nearly 4,800 h longer (2,400 h each). With the cogeneration units' runtime extension, the generation of renewable electricity is increased by 2.8 GWh/a (from 6.4 to 9.2 GWh/a). With credits for surplus baseload electricity fed into the grid, which is valued with negative CO_2-emissions, the amount of CO_2-emissions of the considered system is affected positively by maximizing the runtime of each cogeneration unit. The system comparison between the existing system and a system with ATES highlights the environmental impact of ATES in such a system. Table 1 gives an overview over the yearly CO_2-emissions connected with heat, electricity as well as cars and business trips. The system enlargement with ATES results in additional savings of 2,400 t CO_2 per year.

5 Opportunities and Outlook

The considered energy system with a seasonal heat storage in an aquifer proved to be a sustainable energy concept on the district level. Further research is aiming at the total replacement of conventionally generated heat. As mentioned before, there are 5% of the heat supply left that are provided by conventionally driven peak load boilers. As the cogeneration units are continuously operating all over the year, the limitation of contribution is already reached. To rise the share of renewable heat, an additional renewable heat source is needed. The enlargement of the ATES system with a solar thermal system is one opportunity to achieve the goal setting of 100% renewable heat supply on the district level. 1,000 m^2 of evacuated tube collectors appears to be sufficient for a share of 98%.

Table 1 Allocation of annual CO_2-emissions of the Leuphana University (campus) based on CO_2-equivalents according to [5]

"Source" of emission	Existing system	System with ATES
Heat energy		
Cogeneration units	1304 t	1864 t
Boiler	838 t	184 t
Electricity		
Production (cogeneration)	−5254 t	−7545 t
Consumption (renewable)	10 t	10 t
Cars and business trips	1531 t	1531 t
Balance	−1571 t	−3956 t

References

1. Alshuwaikhat, H.M., Abubakar, I.: An integrated approach to achieving campus sustainability: assessment of the current campus environmental management practices. J. Clean. Prod. **16**(16), 1777–1785 (2008)
2. Seiffert, M.E.B., Loch, C.: Systemic thinking in environmental management: support for sustainable development. J. Clean. Prod. **13**(12), 1197–1202 (2005)
3. Lauenburg, P.: Temperature optimization in district heating systems. In: Advanced District Heating and Cooling (DHC) Systems. Elsevier, pp. 223–240 (2016)
4. Strodel, N., Opel, O., Kranz, S., Werner, K.F., Ruck, W.K.L.: Accepted paper: optimization of cogeneration by seasonal heat storage in an aquifer. In: Nachhaltige Energieversorgung und Integration von Speichern (NEIS) (2016)
5. Opel, O., Strodel, N., Geffken, J., Tribel, A., Ruck, W.K. L.: Climate-neutral and sustainable campus Leuphana University of Lueneburg (2016). https://doi.org/10.1016/j.energy.2017.08.039
6. Strodel, N., Opel, O., Kranz, S., Werner, K.F., Ruck, W.K.L.: Erhöhung des regenerativen Anteils der Wärmeversorgung eines Nahwärmenetzes durch die saisonale Wärmespeicherung in einem Aquifer. In: Ostbayerisches Technologie-Transfer-Institut e.V., editor, pp. 113–120 (2016)
7. Strodel, N., Opel, O., Werner, K.F., Ruck, W.K.: Ausbau und Energieeffizienzerhöhung des Energiesystems der Leuphana Universität Lüneburg. In: Nachhaltige Energieversorgung und Integration von Speichern (NEIS). Hamburg: Vieweg + Teubner Verlag (2015)
8. Strodel, N., Opel, O., Werner, K.F., Ruck, W.K.L.: Integrales Energiekonzept für eine klimaneutrale Universität: Projekt "klimaneutraler Campus". In: Ostbayerisches Technologie-Transfer-Institut e.V., editor. Fachforum Thermische Energiespeicher, pp. 89–92 (2015)
9. Strodel, N.: Wahrscheinlichkeitsbasierte Energiesystem- und Wirtschaftlichkeitsanalyse eines Energieverbundsystems unter Einbindung eines Aquiferwärmespeichers. Verbesserung der Investitionsplanung durch Erhöhung der Prognosefähigkeit und Prognosegenauigkeit. Dissertation, Leuphana Universität Lüneburg. Lüneburg (2018). http://opus.uni-lueneburg.de/opus/volltexte/2018/14506/

Dimensioning Method of the Thermal Comfort

László Bánhidi

It is known, that the subjective comfort of occupants of closed spaces is affected by numerous factors. These factors are listed by G. Blachere as follow:

- acoustic factors,
- smelling and breathing,
- touching,
- vision and colours,
- temperature, humidity and air flow,
- vibration and movement of building,
- special factors (e.g. sunshine, ionization).
- security factors,
- group behaviour (separation),
- factors related to daily life,
- impact of unexpected threats,
- economic factors.

It is obvious that these factors cannot be researched together from any technical or other point of view or their inspection would be highly difficult. Accordingly, that shows that the human body adapts to a given environment through a complex process with various factors working at the same time and in interaction. The body reacts to this joint impact.

In the dimensioning of closed spaces the key criterion is the so-called „pleasant thermal comfort", which is defined as:

… Pleasant thermal comfort is the condition of mind that expresses its satisfaction with the thermal environment. …

L. Bánhidi (✉)
Budapest University of Technology and Economics, Budapest, Hungary
e-mail: laszlo.banhidi@mailbox.hu

M. Palocz-Andresen et al. (eds.), *International Climate Protection*,
https://doi.org/10.1007/978-3-030-03816-8_27

Thermal comfort is influenced by six parameters:

- air temperature, its distribution in space and time, its changes,
- mean radiant temperature of boundary layers,
- relative air humidity and partial pressure of water vapour in air,
- relative air velocity,
- metabolic heat in humans, heat release, thermo-regulation,
- insulation of clothing, impact on evaporation,
- insulation of clothing.

The first four are physical parameters while the other two are linked to the adaptability of the human body; they are important to maintain the heat balance.

1 Key Factors for the Heat Balance of the Human Body

Heat generation of the human body, mainly depending on the activity performed but also influenced to a certain extent by age and gender etc., which cannot be changed by technical means. Heat loss of the human body, largely depending on clothing and the impact of the technical parameters mentioned above.

In accordance with standard MSZ EU 1752, the data included the following table (See Table 1) should be used in the thermal comfort dimensioning of closed spaces:

The factors included in the table are defined in the following chapters.

2 PMV-PPD

These two figures are defined as follows:

PMV is the predicted mean vote for thermal comfort,
PPD is the predicted percentage of dissatisfied with thermal comfort.

Table 1 Required values for the three categories of the thermal environment

Category	State of the whole body		Local discomfort			
	Predicted percentage of dissatisfied, PPD	Predicted mean vote, PMV	Percentage of dissatisfied in the case of draught, DR %	In the case of vertical temperature difference, %	In the case of warm or cold floor, %	In the case of asymmetrical radiation, %,
A	<6	-0,2<PMV<+0,2	<15	<3	<10	<5
B	<10	-0,5<PMV<+0,5	<20	<5	<10	<5
C	<15	-0,7<PMV<+0,7	<25	<10	<15	<10

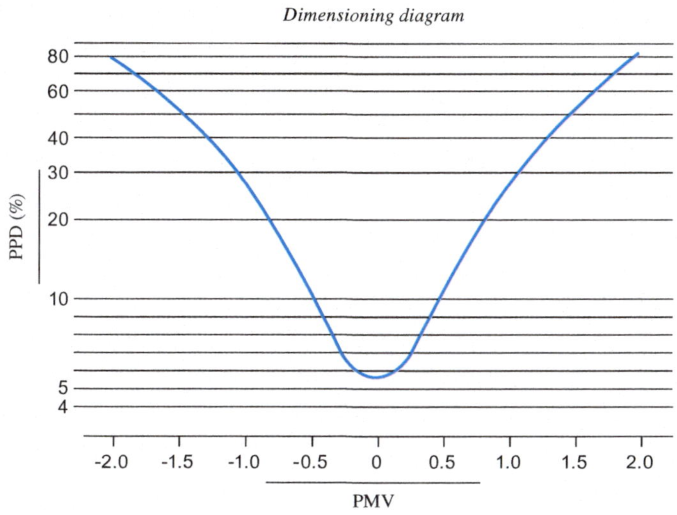

Fig. 1 Dimensioning diagram

Fig. 2 Vertical air temperature difference

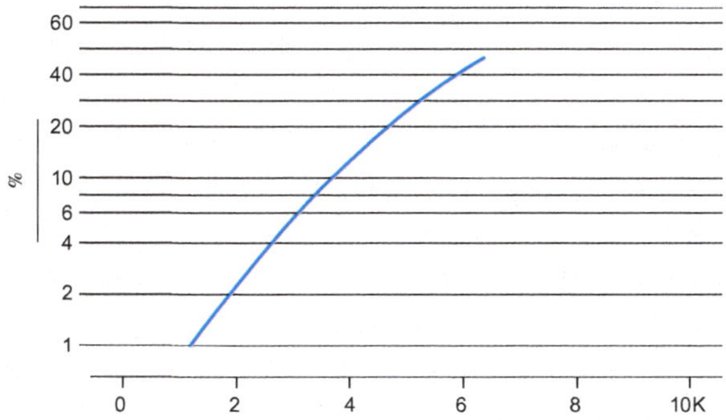

The dimensioning diagram you can be found on Fig. 1.

It is important to note that it is impossible to create a thermal environment that makes everyone satisfied. The optimal percentage is 95%.

The so-called local discomfort factors included in the table are interpreted as follows.

Dimensioning diagram for draught (mean air velocity, local air temperature, turbulence intensity, category).

(PPD, Air temperature difference between the head level and the feet level).

(PPD, Floor temperature).

Asymmetrical radiation:

(PPD, Asymmetrical radiation).

As seen above, the values currently used for dimensioning will change along with the expected climate changes.

Fig. 3 Warm of cold floor

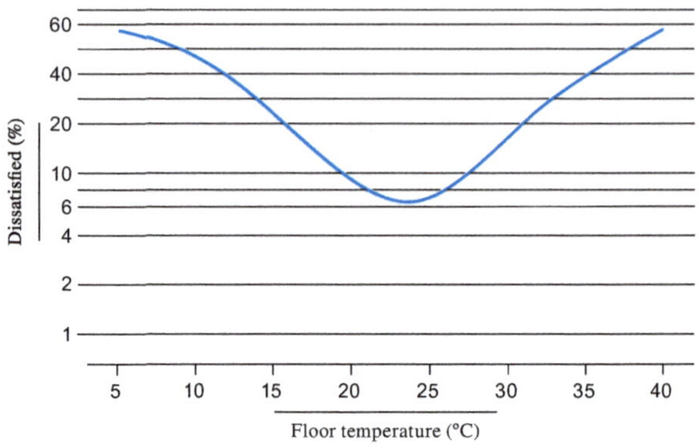

Fig. 4 PPD in the case of various levels of asymmetrical radiation

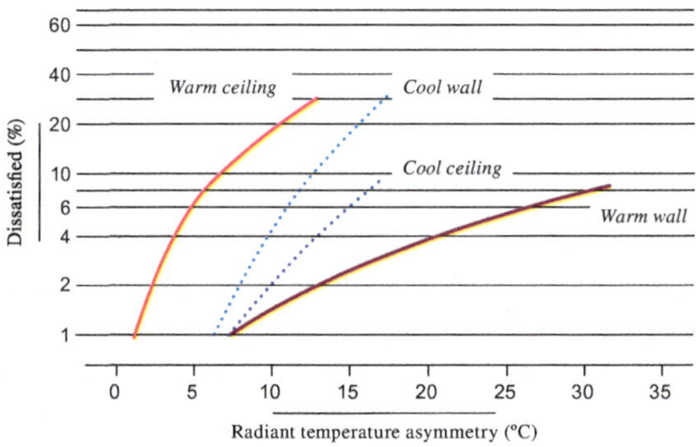

References

1. CR 1752.: Ventilation for Buildings—Design Criteria for the Indoor Environment, CEN Report (1998)
2. MSZ CR 1752.: Thermal Comfort Dimensioning of Indoor Spaces
3. Fanger, P.O.: Thermal Comfort. Robert E. Krieger Publishing Co., Malaba, Florida, p. 244
4. Bánhidi, L., Kajtár, L.: Comfort Theory. Műegyetemi Kiadó (2000)

Using Geothermal Water Resources in Hungary

Jenő Kontra and Zoltán Magyar

1 Definition of Geothermal Energy—Geologic Reserves

Heat content of solid and liquid components of rock within a defined volume between the initial undisturbed upper and the chosen lower temperature is called geological reserve of geothermal energy (Definition by Pál Szilas A.) (Fig. 1).

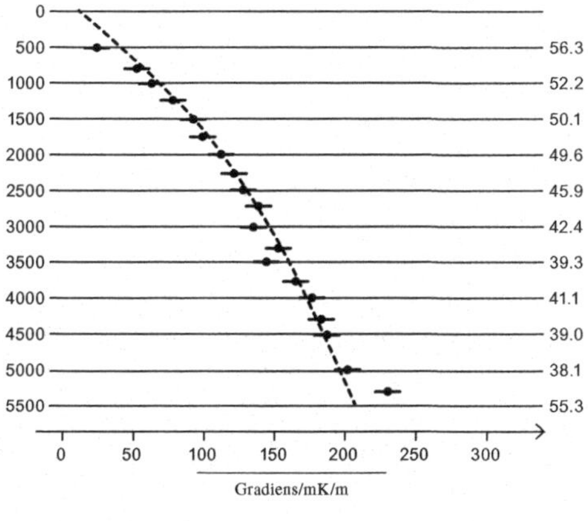

Fig. 1 Average temperature-depth curve in Hungary

J. Kontra (✉) · Z. Magyar
Budapest University of Technology and Economics, Budapest, Hungary
e-mail: kontra@egt.bme.hu

M. Palocz-Andresen et al. (eds.), *International Climate Protection*,
https://doi.org/10.1007/978-3-030-03816-8_28

2 Short Historical Review

Even people of antiquity thought about utilization of geothermal energy. The Romans used water of natural thermal springs mainly for supplying baths and heating the related buildings. They constructed the first thermal bath in Hungary in Aquincum. Utilization of hot water for bathing was renewed after an interruption of more than thousand years during the Turkish occupation.

The erection of the first hot water well by deep boring in Hungary is connected to the name of mining engineer Vilmos Zsigmondy. He countersank a 970 m deep well in the second half of the 19th century in Budapest in the city garden Városliget during about 10 years (being a record that time). Between the two world wars, borings mainly for hydrocarbons led to exploration of thermal water in the great Hungarian plain Alföld. Thus, under leadership of Ferenc Pávay Vajna, the first thermal well in Hajdúszoboszló, still working, was established in 1925. Although on the Margaret Island and in the Lipót city, some dwellings and hotels were heated by thermal water in the twenties of the 20th century, hot water extracted by deep wells built in a growing number in the Alföld between the two world wars was used dominantly for balneological purposes.

Intensive utilization of thermal water began in the late fifties, and from that time, hot water has been used increasingly for heating and warming agricultural plants, hospitals and housing estates. The first facility was thermal heating of the hospital in Szentes in 1957, followed by heating the housing estate in new Szeged by water of thermal wells in 1963. County Csongrád has become the Hungarian centre of production and utilization of geothermal energy. That time, wells were constructed, first of all, by the aid of the technological development office OMFB; more than 40 wells were erected this way. Scientific research began in parallel in Hungary, pioneered by Tibor Boldizsár. He published the results of the terrestrial thermal flow measurements carried out in Hungary in 1944, for the first time on the continent. According to the data of the Hungarian Central Statistical Office, there were more than 1,100 thermal wells in 2000. Outflow temperature of 286 wells was 30–40 °C, in 179 wells 41–60 °C and 121 wells produced 61–90 °C as outflow temperature. Calculating with 15 °C as final cooling temperature, heat quantity produced during one year amounts to 30.4 PJ corresponding to the calorific value of 740,000 tons of oil.

3 Origin of Geothermal Energy

Geothermal energy means the internal heat content of rocks in the earth crust. Concerning origin, this heat is coming from core upwards to the surface resulting from radioactive decay processes deep in the earth. This heat has been generating for billions of years and continues to generate without being bound to a geographic point.

On earth, there are active and passive geothermal zones. In the active zones, still active volcanoes can be found and tectonic activity can be observed (These zones include New Zealand, California, Japan, Kamchatka and Hawaii).

Hungary belongs to the passive geothermal zone with favourable circumstances.

Heat flows by deep convection and heat convection to the surface. This is called terrestrial heat flow. The terrestrial heat flow was first measured by Professor Boldizsár in 1943. Temperature of the earth's external surface is identical to the air temperature measured by meteorologists. The difference between this temperature being low over the year and the high temperature deep in the earth induces this continuous heat flow to the surface.

4 Definition of Geothermal Energy Stock and Geological Reserve

For heat extraction and planning heat exploitation as well as for heat management, it is important to know the amount of geothermal energy. There are different estimations, but the exact description of fundamental terms is unambiguous.

The term geological reserve of geothermal energy designates heat content (more correctly: energy) of solid and liquid components of rock within a defined volume between the initial undisturbed upper and chosen lower temperature limits. (Definition by Pál A. Szilas).

In general, the mean annual temperature on the surface is used as chosen lower temperature limit. Rock within a defined volume has a good permeability, its gaps contain water or are replenished by water from the surface.

Geothermal field is a continuous rock body from which thermal energy is transported to the surface by water. This means that water originating from deepness acts as heat carrier.

Specific internal energy content of a unit volume of porous and permeable rock body saturated with geothermal fluid is as follows:

$$E = \phi \cdot \rho_F \cdot c_F \cdot T_F + (1 - \phi)\rho_K \cdot c_K \cdot T_K$$

where:

φ porosity
ρ density
c specific heat
T absolute temperature
F index relates to fluid
K index to rock.

If theoretically we were able to use the geothermal energy content up to the ambient temperature (To), we could write it as follows:

$$E_{elm} = \phi \cdot \rho_F \cdot c_F \cdot (T_F - T_o) + (1 - \phi)\rho_K \cdot c_K \cdot (T_K - T_o)$$

In the aquifer layer deep in the earth, the temperature is in equilibrium:

$$T = T_F = T_K$$

therefore:

$$E_{elm} = [\phi \cdot \rho_F \cdot c_F + (1 - \phi)\rho_K \cdot c_K] \cdot (T - T_o).$$

Simplifying and using summation temperature and specific heat of rock:

$$E_{elm} = \rho \cdot c \cdot (T - T_o)$$

where:

$$\rho = (1 - \phi)\rho_K \cdot c_K + \phi\rho_F$$

and

$$c = \frac{\phi\rho_F c_F + (1 - \phi)\rho_K c_K}{\phi\rho_F + (1 - \phi)\rho_K}.$$

The theoretically exploitable energy amount is:

$$E_{elm} = \int_A \int_o^z \rho c(T - T_o) d_Z d_A$$

This is related to an area of A and deepness σ. This is the energy of the initial geological reserve.

With aquifer water of the Pannonian sediment complex—if the cooled water is injected back—not only the internal energy of aquifer water but also a part of the internal energy of the rock frame can be brought to the surface. In Hungary, Upper Pannonian sandy aquifer rocks, from where thermal water can be exploited economically, can be found on an area of 40,000 km^2.

5 Geothermal Conditions of the Pannonian Basin

A geothermal system is a restricted domain of earth crust being in heat transfer connection with its environment.

A geothermal reservoir is part of the system from where internal energy can be partially brought to the surface via a heat carrying medium. Accordingly, the reservoir is a geological formation of adequate porosity and storing thermal water or steam. This is called natural geothermal reservoir.

An artificial reservoir does not contain any aquifer fluid, from there heat can be extracted with a heat carrier injected from the surface. This system is called hot dry rock.

In deep-seated strata, heat is transferred by convection or heat conduction.

Geothermal reservoirs are continuously heated by terrestrial heat flow. Heat coming to the surface is replenished by the conductive heat flow being of relatively lower intensity than the convective heat transfer.

Average value of terrestrial heat flow is 90 mW/m^2 in Hungary. This is much higher than the value of 60 mW/m^2 measured on average in other areas, therefore, heat energy can be produced in Hungary from deepness economically if the other geological circumstances are also favourable:

- in sinking and sedimentary basins,
- geothermal gradient of 50–60 °C/km,
- presence of adequate aquifer strata.

In such porous, sandy and/or sandstone reservoirs, no convection flows can be generated. Temperature gradient measured along the deepness is fundamentally linear.

Maximum layer temperature of these reservoirs with low enthalpy does not surpass 150 °C. The largest reservoir vacuum heated by heat conduction is the Upper-Pannonian sandy reservoir in the sediment complex of the Hungarian Great Plain. Its average thickness is 200 m, its area is about 40,000 km^2, lying in the Carpathian basin disregarding frontiers (towards Romania, Slovakia from the Small Plain, Serbia).

Geothermal gradient and terrestrial heat flow.

Change of temperature along deepness is called geothermal gradient, designated by gg.

$$gg = \frac{dT}{dz}$$

Its value changes with strata as material and thermal conductivity coefficient of crossed layers are different. In a point, geothermal gradient is defined by physical and thermal properties of all the strata.

On the whole, geothermal gradient is a skew line consisting of straight lines that can be substituted by a straight line with a steepness shown in the figure.

Terrestrial heat flow is a property expressing geothermal conditions under a given area better than geothermal gradient with its approximate value. Deepness-dependant inhomogeneity of layer temperature induces terrestrial heat flow.

Its value is:

$$q = -k\frac{dT}{dz}$$

for a concrete stratum:

$$q = -\frac{k}{H}(T_1 - T_o)$$

where:

k heat transmission coefficient of the given stratum,
H stratum thickness,
$T_1 - T_o$ temperature difference between contacts.

Heat flow vector:

$$q = \frac{1}{\sum_{i=1}^{n} \frac{H_i}{k_i}}(T_n - T_o).$$

Magnitude of heat flow vector is defined by the heat transmission coefficient related to all layers as a whole. Therefore, it can be said that the magnitude of the terrestrial heat flow is the most characteristic property of geothermal conditions of an area (Boldizsár).

In Hungary, earth crust is thinner than the European average, therefore, our geothermal conditions are rather favourable in terms of heat flow. Here, the average value of the terrestrial heat flow is 90–100 mW/m^2 being almost double the continental average.

Geothermal gradient on the earth is 0.020–0.033 °C/m on average, while it makes in general 0.042–0.055 °C/m in Hungary. Thus, in Hungary, temperature of layers is about 60 °C in a deepness of 1,000 m, in some locations even higher.

6 Fluid Quantity Exploitable from Closed Reservoir

Volume of a given deep-seated reservoir is V, porosity of constituting rock is ϕ; and rock pores are filled by homogeneous fluid.

Deep-seated fluid is exploited by deep boring. By opening the reservoir, water body expands flexibly, and reservoir pressure decreases. This is an isothermal change, the bulk elasticity factor makes β.

Heat flow escaping through the well:

$$\dot{m} = V \cdot \phi \cdot \beta \cdot \rho \frac{dp}{dt}.$$

In the case of a constant production quantity, within time period $t_2 - t_1$, when initial p_1 pressure decreases to p_2, fluid of mass M can be exploited from the reservoir:

$$M = V \cdot \phi \cdot \beta \cdot \rho \cdot (p_1 \cdot p_2)$$

Pressure at well bottom:

$$p_2 = p_{kf} + \rho \cdot g(H + \frac{w^2}{2_g} + h^{'})$$

where:

p_{kf} well head pressure
H deepness of well bottom
w average velocity in well
$h^{'}$ head loss

$$h^{'} = \lambda \frac{H}{d} \frac{w^2}{2g}.$$

The reservoir can be exploited until pressure p_1 decreases to p_2. Exploitation through flexible expansion is so long possible.

If pressure p_2 is not low enough, i.e. there is no considerable pressure difference, thermal water can be brought to the surface by artificial exploitation only.

Pressure p_2 can be diminished by gas lift or drowned pump.

However, most reservoirs in Hungary are replenishing ones rather than closed reservoirs. As an example, reservoirs between clay layers are hydraulically connected. Due to exploitation, mass flows are generated even between the reservoirs based on a sophisticated hydraulic model. Fluid quantity exploitable from such complex systems depends on the exploitation intensity and also on water replenishment - often from the surface vicinity. A considerable part of Upper Pannonian sequences in the Hungarian Great Plain belongs to this type.

7 Thermal Water Re-Injection

7.1 Necessity of Thermal Water Re-injection

Thermal water re-injection can be explained by three reasons:

recipients on the surface are negatively impacted by waters with high salt content,
decrease of reservoir pressure due to thermal water exploitation shall be avoided,
deep-seated water reserves have to be managed economically.

We note here that all the three requirements can be fully met by injecting water back into the original layer rather than in other deep-seated layers.

7.2 Theoretical Background of Water Re-Injection

Water absorption capacity of a re-injecting well is an essential and dominant parameter of the re-injection process on long-term run. This depends on the layers' status, geologic conditions and quality of water to be re-injected. Absorption capacity is the water quantity injectable by overpressure on one surface unit. Water injection is successful if the sum of pressure of the re-injecting pump and pressure of the water column between cooled water level at rest and ground level is higher than reservoir pressure. Heat lift and gas lift are of little importance in this case as the temperature of the re-injected water is near to the temperature of the cooled well. Usually, the initial absorption capacity is identical to the yield of the well. The largest problem is that due to the suspended matter content of water, clogging generate in the vicinity of the well, and the well fills, thus closing the opened sector. A maximum reinjection pressure value of about 10–20 bars can be still regarded as economical, but in most cases, this pressure is increasing with time, resulting in prevention of the process. In this case, well cleaning is required. As important criteria, suspended matter content of water shall be below 1 mg/l, and water quality shall not be different from the original water quality. Re-injection of chemically treated water is not allowed.

7.3 Equipment and Process of Re-Injection

Closed, pressurized thermal systems are suitable for re-injection of cooled thermal waters. As major criterion, the whole process of exploitation—primary side of utilization—and re-injection shall be carried out with adequate overpressure. In this case, no scales and precipitants generate as a result of gas emission.

A simpler case of re-injection can be observed in karstic-carbonate and cracked aquifers. In general, their absorption capacity is good, thus, water can be returned to the reservoir layer by a minimum pump pressure.

The process operation is much more difficult in sandy and sandstone layers. In this case, the well bottom shall be specially designed for better absorption capacity (in Hódmezővásárhely, Kistelek). Productive re-injection can be started after pilot tests and measurements of about half a year.

Major operations include investigation of the absorbing well, its continuous hydrodynamic measurement, analysis of absorption capacity drop and regular well cleaning. Then, the well is to be investigated again, the whole system shall be monitored and reinjection shall be permanently operated.

In the past years, district heating systems with re-injection have been working successfully in Bóly, Hódmezővásárhely, Veresegyház and Kistelek.

Reference

1. Excepet, D.: Tem fugiass imusant et fuga. Pocket Books, New York, NY (2013)

Part IV
Climate and Environmental Law and Legislation

From Copenhagen to Paris: The Way Towards a New International Climate Change Agreement

Attila Pánovics

1 Introduction

Climate change is now commonly recognized as a serious threat to humanity. It is arguably the most urgent, complex long-term and systemic challenge of our time. We live in an age where we have to cope with a changing climate whilst continuing to mitigate. While mitigation is crucial, it is also necessary to adapt to already experienced changes in climate.

As climate change unfolds, severe impacts are anticipated for both ecosystems and human societies. In policy terms, these warnings have been translated into the 2 °C threshold: global mean temperatures must not increase by more than 2 °C above pre-industrial levels in order to avoid irreversible changes to the planet's climate.[1]

Tackling climate change requires incentives across the entire policy spectrum, and an effective and balanced international response. Greenhouse gas (GHG) emissions need to be considered globally as they affect the climate system no matter where they are released. While classic negotiation tactics views all policies as a zero sum game, limiting the temperature rise requires the widest possible cooperation and substantial reductions in carbon emissions by all countries. Even large groups of countries acting together (such as the European Union and its Member States) cannot resolve these problems on their own. The nationally determined level of effort by countries differs, with a risk of competitive disadvantage for industries if an uneven playing field will remain.

Although the impacts of both observed and projected climate change are not distributed equally, all members of the international community will benefit if climate change is prevented from reaching dangerous levels. Delaying action will close the

[1] The 2 degree target was discussed by a German climate researcher Wilfrid Bach in an interview with Spiegel in 1988 [1].

A. Pánovics (✉)
Department of International and European Law, University of Pécs, Pécs, Hungary
e-mail: panovics.attila@ajk.pte.hu

M. Palocz-Andresen et al. (eds.), *International Climate Protection*,
https://doi.org/10.1007/978-3-030-03816-8_29

window for effectively reducing emissions and preparing for the adverse impacts of climate change.

2 The Road to Paris

The 2015 Paris Agreement marks the latest step in the evolution of the UN climate change regime and builds on the work undertaken under the UN Framework Convention on Climate Change.[2] UNFCCC was signed in 1992, at the Earth Summit in Rio de Janeiro. 17 years after the famous UN Conference on Environment and Development (UNCED), the fifteenth Conference of the Parties (COP)[3] was the second time when the majority of the world's leaders gathered to seriously discuss climate change.

COP3 adopted a supplement to the UNFCCC in 1997, called the Kyoto Protocol. It had an aim to reduce the emissions of six GHGs, based only on 1990 emissions. These targets applied only to 35 industrialized countries, and covered the period 2008 to 2012, known as its first commitment period. After a long ratification period, the Protocol finally came into force in 2005.

The Kyoto Protocol is an old, top-down style agreement,[4] which has a clearly finite life to 2020. Its second commitment period started in 2013 and will run until 2020.[5] The United Nations is encouraging governments to speed up their acceptance of the second commitment period, in order to provide further momentum for the years leading up to 2020.

The Kyoto Protocol did not provide specific long-term targets to work forward. Discussions on how to continue after the commitment periods were initiated at the UNFCCC conference in Buenos Aires in 2004. In 2009, close to 115 world leaders attended the high-level segment at COP15 (making the Copenhagen Conference one of the largest gatherings of them ever outside UN headquarters in New York), but the international community could not agree on anything definite and left home empty handed.

The 'Copenhagen Accord' was "taken note of ", but not "adopted" by the COP. Non-binding nature of the document was disappointing to many negotiators and observers alike. Nevertheless, the key elements of the document included a reference to consider limiting the temperature increase to below 1.5 °C (a demand made by vulnerable developing countries), and the developed countries promised to provide $30 billion for the period 2010–2012, and to mobilize long-term finance of a further $100 billion a year by 2020—the Copenhagen 'Green Climate Fund' (GCF).

---

[2]There are now 197 Parties to the Framework Convention and 192 Parties to the Kyoto Protocol.

[3]The Conference included the fifteenth COP to the UNFCCC (COP15) and the fifth COP serving as the Meeting of the Parties to the Kyoto Protocol (COP/MOP5).

[4]In reality the Protocol always had a bottom up approach, in that countries put forward negotiating positions.

[5]With absolute reduction targets set for the EU, Australia, Norway and Switzerland.

When looking back to the history of the UNFCCC, the Paris Agreement marks an historic milestone in the history of UN climate talks as a result of decisions taken at previous COPs in Copenhagen (COP15), Cancún (COP16), Durban (COP17), Doha (COP18), Warsaw (COP19) and Lima (COP20). The annual climate conferences of the Parties achieved a range of important outcomes between 2009 and 2015, for example:

2010—The 'Cancún Agreements' included the most comprehensive adaptation package ever agreed to help developing countries deal with climate change, encompassing finance, technology and capacity-building support.

2011—A decision was adopted to launch a negotiation for form a "protocol, another legal instrument, or an agreed outcome with legal force" in 2015, to be implemented starting in 2020. Under the 'Ad Hoc Working Group under the Durban Platform for Enhanced Action' (ADP) all countries had to make a legally binding contribution, even if their commitments took different forms.

2012—At the 'intermediate' Doha Conference an ADP negotiation roadmap was developed to work out the elements of the new global climate agreement in 2014, to prepare the draft text in May 2015, and to conclude the negotiation by the end of 2015.

2013—The text adopted at the Warsaw Conference referred to all Parties, without differentiation, and used the word "contributions" (intended nationally determined contributions) instead of "commitments" to describe what countries would have to put forward.

2014—Parties adopted the 'Lima Call for Action' at COP20, which elaborated key elements of the Paris Agreement. Countries concluded that INDCs (Intended Nationally Determined Contributions) will form the foundation for climate action post 2020.

While no breakthroughs were achieved, the UNFCCC process was kept on track. In spite of modest results, all COPs have been important to make progress towards reaching a long-term and effective climate change agreement to replace the Kyoto Protocol. This would not have happened without the efforts of the international community in the 5 years beforehand.

3 The Paris Agreement

6 years after Copenhagen, the Paris Conference was seen as a make-or-break for the UN climate talks, and the delegates deservedly hailed the adoption of a historic pact. The international community could turn past failures into success, and COP21 adopted the Paris Agreement on 12 December 2015.[6]

The Agreement was adopted by 196 Parties (195 countries and the European Union) to the UNFCCC. In advance of the Conference, almost all Parties had presented their INDCs covering 5- or 10-year periods starting in 2020. Nevertheless, the

[6]See Decision 1/CP.21. The text of the Agreement is contained in the annex to this decision.

Paris Agreement is a concise document that sets a framework for action, but provides little guidance for how to achieve its goals. Flexibility is used as a priority to meet the differentiated responsibilities, capacities and needs of all Parties.

These are the main characteristics of the Agreement (and the accompanying Decision) that are based on moving away from previous failures in the COP process:

As opposed to the Copenhagen Accord and the outcomes of the following COPs, the Paris Agreement takes the form of a legally-binding international treaty.

It is the first international treaty on climate change covering almost all emissions of the world. Mitigation through emissions reduction is the core of the collective action, but with an equal role for adaptation, finance and technology.

The Agreement contains a long-term goal (to limit global temperature rise to 2 °C with a strong aim for limiting temperature rise to 1.5 °C above pre- industrial levels by the end of the century). In order to achieve the main target, the Agreement requires all Parties to the UNFCCC to put forward 'Intended Nationally Determined Contributions' (INDCs), and provides a dynamic mechanism to take stock and strengthen ambition over time. It also includes and a 5-years "ambition cycle",[7] and sets up an enhanced transparency and accountability framework. This ends the old division of countries in the UNFCCC and the Kyoto Protocol.

Developed countries confirmed that they will continue to take the lead in mobilizing climate finance from a variety of sources to support the poorest and most vulnerable countries. The Agreement provides an ambitious solidarity package with provisions on climate finance and on addressing needs linked to adaptation and loss and damage associated with adverse effects of climate change.

Developed countries will continue to provide $100 billion of climate finance per year for mitigation[8] and adaptation action, and a new, higher goal is to be set by 2025.

Several factors played a key role in brokering the first major multilateral deal of the 21st century, and shaped the successful outcome of the conference:

The global setting changed completely when compared to Copenhagen resulting in word-wide bottom up mobilization of all stakeholders (governments and non-state actors). Especially global civil society, academics[9] and business organizations applied huge pressure on governments to overcome all resistance.

Following the lack of agreement in Copenhagen, the European Union and its Member States has been building a broad coalition of developed and developing countries in favor of high ambition (the 'High Ambition Coalition'). The EU had the world's most ambitious commitments on climate change[10] and pushed for a new

[7]This is known as the „review and ratchet" process.

[8]The involvement of developing countries in mitigation actions started at the Montreal COP in 2005 (COP11) and was formalised at the Bali COP in 2007.

[9]The Fifth Assessment Report of the Intergovernmental Panel on Climate Change (IPCC) put the scientific evidence on climate change right back on the political agenda, and increased the urgency of action.

[10]The EU became the first major economy to present a climate plan (i.e. INDC) on 6 March 2015, reflecting its 2030 climate and energy policy framework. It has also signed the Paris Agreement in New York on 22 April 2016.

agreement to be finalized in Paris. Concerted EU diplomacy also helped secure a landmark agreement.

In June 2015 industrialized country leaders agreed at the G7 meeting that they should phase out GHG emissions by the end of the century. Joint statements of the two largest polluters and global powers—the US and China—also raised hopes that a new agreement could indeed be adopted.

A new grouping called the 'Climate Vulnerable Forum' has emerged (43 countries most vulnerable to climate change).

The decisive role played by the French government before and during the Paris Conference is hard to overstate, especially considering the environment it was operating in.[11] The French presidency focused on achieving realistic goals, and successfully facilitated the complex global negotiation process building trust and confidence among the participants.

In contrast to Copenhagen, the Paris Conference is a good example of how the international community can forge collective action on tackling climate change. The adoption of the Agreement is a clear signal of the global clean energy transition. The document sets out a global action plan to put the world on track to avoid dangerous climate change acknowledging that this will require a peaking of GHG emissions as soon as possible, and achieving climate neutrality in the second half of the 21st century.

4 Next Steps

COP21 promises to be a historically significant turning point in the fight against climate change, although there is still a very high chance of not meeting the main goal. The transition to a low carbon economy demands a fundamental shift in economics, energy, technology, business and investment behavior, and production and consumption patterns. This needs to be properly managed, and the international community has to identify clear processes for managing climate impacts.

The ramping up of ambitions must be seen against the broader context of the Addis Ababa Action Agenda from the Third UN Financing for Development Conference held in July 2015, and the Sustainable Development Goals (SDGs) that were adopted in September 2015 to support their implementation.[12] Mutually supportive actions will bring together and intensify the efforts of all actors (of course, the synergies between these actions need to be exploited fully).

In accordance with Article 20(1), the Agreement is open for signature until 21 April 2017. On 22 April 2016, 175 Parties signed the Paris Agreement including the US and China; this was the largest number ever to sign an international treaty on a

---

[11] The terrorist attacks on Paris raised concerns about whether the negotiations would go ahead at all.

[12] These decisions have been made in parallel with climate negotiating sessions.

single day. Now the signatories should ratify the Agreement as soon as possible to achieve its swift entry into force and show commitment to phase out fossil fuels.[13]

The Paris Agreement has left a wide range of technical issues open, but the clean energy transition is irreversible and non-negotiable. The outcomes of the Conference highlight that an enormous amount of work remains to be done on the implementation of the Agreement, such as detailed provisions on transparency, accountability and technology mechanisms. It is rather "a bridge between today's policies and climate-neutrality before the end of this century" [2].

In order to maintain the positive momentum from Paris, the implementation of COP21 commitments must remain a strategic priority in climate diplomacy at the UN and other international fora. The Agreement is a landmark achievement for multilateralism, and, when implemented, it will definitively accelerate the transition to a climate resilient global economy.

References

1. Von Kampe D., Schumann H.: Die Zeit läuft uns davon Professor Wilfrid Bach über Programme. Zur Verhinderung der drohenden Klimakatastrophe Der Spiegel 45/1988, p. 63. http://www.spiegel.de/spiegel/print/d-13530065.html
2. Paris Agreement. http://ec.europa.eu/clima/policies/international/negotiations/paris/index_en.htm

[13]The Agreement is due to enter into force in 2020. In accordance with Article 21(1), it will enter into force on the thirtieth day after the day on which at least 55 Parties representing at least 55% of global GHG emissions have deposited their respective instruments of ratification, acceptance, approval or accession.

Renewable Energy Sources Act 2017 in Germany—Auctions for Renewable Energy Transition

Henning Thomas

1 Never Change a Running System? Why Auctions for Renewable Energies Are Introduced

From the first Act that initiated a support for renewable energy sources for the generation of electricity in Germany—the *Stromeinspeisungsgesetz* (Electricity Feed-in Act) of 1990—the basic support scheme remained essentially the same. The Electricity Feed-in Act and all later amendments and new regulations with the Renewable Energy Sources Act from 2000, 2004, 2009, 2012 and 2014 have relied on feed-in tariffs granted for the operators of renewable energy installations for the electricity fed into the grid as well as on claims for priority grid access and priority take off of renewable energy.

The Renewable Energy Sources Acts without doubt have been largely successful in increasing the contribution of renewable energy in the electricity sector. The share of renewable energy sources in the gross electricity production reached 30.0% in 2015 [1]. However, since 2013 Germany experienced an intense debate on the electricity prices and, in the first line, on the "EEG allocation" (*EEG-Umlage*) that must be paid by all electricity supply companies (which demand this EEG allocation from the final consumers on a contractual basis). By this mechanism, the costs for the generation of renewable energy sources in effect are allocated to the final consumers of electricity in Germany. Due to the amount of the EEG allocation—6,354 cent/kWh in 2016, 6,170 cent/kWh in 2015 and 6,240 cent/kWh in 2014—this has been regarded as a main driver of the total costs of electricity for final consumers—although other factors like the market prices for electricity also play a major role in the level of the EEG allocation. In addition, the political intention to have better steerage in the expansion of renewable energy sources grew for several reasons that include the cost discussion, but also problems in the expansion of the electricity grids and the

H. Thomas (✉)
Hamburg, Germany
e-mail: henningthomas@gmx.de

M. Palocz-Andresen et al. (eds.), *International Climate Protection*,
https://doi.org/10.1007/978-3-030-03816-8_30

goal of a better adjustment between building of generation installations and new grid infrastructure.

While the before-mentioned are mainly the national reasons why Germany has been open for a system change in the renewable energy support, the final initiative and need to introduce auctions resulted from the European Commission Guidelines on State aid for environmental protection and energy 2014–2020. These Guidelines, as a principle, require that from 1 January 2017 aid shall be granted "in a competitive bidding process on the basis of clear, transparent and non-discriminatory criteria" [2].

It must be added, however, that these guidelines have no direct legal effect upon the national states so that Germany could have insisted on not introducing auctions. This might have led to the European Commission's interpretation of the state aid rules on the support schemes for renewable energy sources being challenged in front of the European Courts. But even more, the guidelines themselves give substantial leeway that could have been used. In the case of wind energy, the guidelines allow "for installations with an installed electricity capacity of up to 6 MW or 6 generation units" to grant aid without a competitive bidding process [2].

After massive debates with the wind energy industry and certain Federal states in particular, the German legislator nevertheless decided to introduce auctions for wind onshore, wind offshore, photovoltaics and biomass installations with few exemptions only. A main exemption applies for installations with an installed capacity of up to 750 kW, but this—in spite the endeavours of several proponents—is far below the threshold of 6 MW/6 generation units that could have been used in case of onshore wind. While several stakeholders demanded to make use of this exemption allowed by the Commission's guidelines, the Federal Ministry for Economic Affairs and Energy (BMWi) considered this rule to be "not unerring" and voted against a utilization of this exemption possibility [3].

This clearly demonstrates the strong intention of the German legislator and the BMWi in particular to shift to auctions for all sources of renewable energy with a major potential of expansion in the electricity sector. The BMWi summarizes the three main reasons for the introduction of auctions as follows: to increase cost efficiency of the support mechanism by introducing competition, to establish a reliable path of expansion for renewable energy sources and to maintain the diversity of actors within the renewable energy industry [4].

It must be noted that only the first and the second reason can be regarded as aims for the introduction of auctions. Also with a view on the difficulties of setting an appropriate price level by statute, the legislator wants a competitive bidding process to determine the support level for renewable energy sources and is of the view that this might raise cost efficiencies. With providing a path of expansion for the several renewable energy sources and by providing auctions only for the capacity that is within this path of expansion, the legislator limits the expansion of the several renewable energy sources to a maximum. This must be seen in relation to the aim of cost efficiency, because it at least allows the legislator to restrict the overall costs of the expansion of renewable energy sources to a total maximum (while this leaves open whether this is the most cost efficient mechanism to transform the electricity sector seen as a whole). The third reason that the BMWi provides—maintaining the diversity of actors—is not in itself a goal that is pursued with the introduction of auctions. Instead, there are concerns that the diversity of actors is threatened by the introduction of auctions (see closer below 4). The real intention is therefore to safeguard the diversity of actors existing today while auctions are introduced to some extent—the main goals of the auctions are to pursue cost efficiency and an expansion path for renewable energy sources.

2 Regulatory Mechanism: Auctions as a Precondition for Feed-in Support

The first step towards the introduction of auctions for all renewable energy sources were auctions for stand-alone photovoltaics. Based on an ordinance detailing the procedure, the first three auction rounds took place in 2015. The auctions for stand-alone photovoltaics were intended to work as a pilot project for the introduction of auctions for other renewable energy sources as well [5].

Already in 2015 as well, the drafting of the Renewable Energy Sources Act 2017 designated to introduce auctions as a general principle for renewable energy in the electricity sector had been started. The Federal Ministry for Economic Affairs and Energy drafted several papers outlining the basic points of the upcoming auctions and stakeholders could comment on these basic points for the new design of the Renewable Energy Sources Act [6]. Also, a working group on citizen energy and diversity of actors was initiated [7].

After further political debates and compromises, the new EEG 2017 was adopted by parliament in July 2016. In principle, the successful participation in an auction is now a precondition to claim any financial support granted under the Renewable Energy Sources Act, see Section 22 EEG 2017. If a bid has been accepted in an auction, the financial support is available by means of the so-called direct marketing of renewable energy. This means that the operators of the plants have to sell their electricity produced on the energy markets, but in addition to the market price they can claim a so-called "market premium" from the responsible grid operator [8]. It is therefore the amount of this market premium that is to be established in the auctions.

However, the auctions are technology-specific and also the preconditions, procedure and exemption possibilities differentiate between wind onshore and offshore, photovoltaics and biomass. In the case of onshore wind, which is the focus of this contribution, in general all plants with a capacity above 750 kW need to participate in an auction to be able to claim any support. An exception may apply under further conditions for plants commissioned prior to 1 January 2019 that have received a permit under the Federal Immissions Protection Act (BImSchG) before 1 January 2017 already, see Section 22 para. 2 number 2 EEG 2017. Up to a maximum of 125 MW installed capacity per year, an exception may apply as well in cases of pilot wind energy plants that are closer defined in Section 3 number 37 EEG 2017.

Summing up, this leads for all "normal" onshore wind energy plants after the before-mentioned dates to the necessity to participate in the auctions that are organized for this particular renewable energy source.

3 Preconditions for Auctions for Onshore Wind and Procedure

Following a long debate on the point in time in which the participation in an auction shall take place, the Federal Ministry for Economic Affairs and Energy opted for a "late auction" in the case of wind onshore. This view remained from the drafts to the final version of the new EEG 2017. The late auctions mean that only wind energy plants that already received a permit under the Federal Immissions Protection Act at the latest three weeks before the auction may participate in this auction, Section 36 para. 1 EEG 2017. This results in participation in an auction only being possible at a relatively late point in time in the course of planning a wind energy project. In turn, only a comparably low [9]—but still substantial—financial security of 30 Euro per kW installed capacity is necessary if a bid shall be made in the auction.

The necessity of a permit and a security as preconditions for participation in the auctions must be seen against the background that the projects which received an acceptance in the auctions shall be realized. By setting a rather late point in time during the project and requiring a permit for construction and operation already being in place, the legislator expects that most of the projects that win in an auction will also be constructed and go into operation. Also, the legislator wants to avoid speculative bidding which he considers to be more probable in case of earlier auctions.

For the specific auction round, all parties interested must offer their bid by providing a specified capacity and a price level in particular and complying with a number of formal requirements in addition. The auctions are administered by the Federal Network Agency (BNetzA) which publishes the total volume that shall be procured in every auction round. Up to this total volume, the BNetzA then accepts the bids that have been issued, starting from the cheapest price. To avoid windfall gains, a maximum level is set which is starting at 7 cents/kWh in the first auction in 2017.

It is important to note that the price values that are offered are corrected dependent on the location of the specific wind energy plant and a correction factor provided in Section 36 h EEG 2017. This mechanism continues the principle that the support level for onshore wind takes account of the specific wind conditions of the location of the wind energy plant. The regulatory purpose is to achieve an expansion of wind energy plants in all regions of Germany and not concentrated in the areas with best wind conditions only.

After the operator of a wind energy plant has received the acceptance of a bid for his project, he shall bring his plant into operation within 24 months after the publication of the acceptance (as a general principle). In this case, the operator may claim the support level that has been set in the acceptance of his bid. The market premium for the electricity fed into the grid is calculated on basis of the value determined in the auction. Thus, the legislator decided that the remuneration shall follow the "pay as bid" principle—in contrast to the "uniform pricing" principle (which had been tested in case of stand-alone photovoltaics as well).

4 Special Rules for Citizen Energy

The late point in time of the auctions leads to high efforts on planning and financing of a project being necessary before the achievement of a feed-in support and its amount being secure for the operator. For all onshore wind energy operators, the risks in the planning of a project are increased by an additional factor: There is a new insecurity whether a support by the Renewable Energy Sources Act is available at all and which support level can be claimed for the project in particular.

While this insecurity concerns all actors initially, several studies pointed out that small actors would be more affected by this insecurity. According to a study by trend:research and Leuphana University, "public energy" (in a wider sense) shall sum up to a share of about 46 per cent of the total installed capacity of renewable energy in Germany (2012) while institutional and strategic investors would have a share in the installed renewable capacity of about 41% and energy supply companies of about only 12% [10]. The main reason that smaller actors are more affected shall be that these small actors, unlike larger actors with several projects in their pipeline, cannot spread the risk of not receiving an acceptance within an auction (to the desired price level) to further projects [11]. The risk of "sunk costs" in case of no success in an auction therefore would concern these actors much more intense. In addition, it would be more complicated for these small actors to gain credit for the financial security necessary against the background of this new risk [12].

As a consequence of these considerations and again following heavy political debates, the EEG 2017 contains in Section 36g EEG 2017 special conditions under which "citizen energy associations" (*Bürgerenergiegesellschaften*) may participate in an auction for onshore wind plants. A main alleviation is that bids for up to six plants with a total maximum capacity of up to 18 MW may be done without a permit under the Federal Immissions Protection Act being granted (subject to further conditions). Also, only an initial financial security in an amount of 15 Euro per kW installed capacity must be made to be able to participate in an auction. Only in the case the bid is being accepted, a second security to the same amount needs to be provided thereafter. A further relief for these players is that—in short—uniform pricing applies for the price level set within an auction for these projects, so that they can benefit if the highest accepted price is above the bid they have made within the auction.

For being able to fall under these alleviations, it is crucial that a citizen energy association (*Bürgerenergiegesellschaft*) is participating in the auction. Important elements of the definition of this citizen energy association in Section 3 number 15 EEG 2017 involve that the association must consist of at least ten natural persons as members or shareholders being entitled to vote. Also, at least 51% of the voting rights must be with natural persons registered with their principal residence for at least one year within the city or region in which the wind energy plant shall be built. Also, no member or shareholder may have more than 10 percent of the voting rights in the association. In addition, a financial participation of the community in which the wind energy plant shall be built with at least 10 percent needs to be existent or must have been offered to the community according to Section 36g EEG 2017.

Thus, the special rules for citizen energy ease the participation in the auctions for these actors, but this is subject to a number of conditions that can only be fulfilled by a restricted group of players (which is also the clear intention).

5 Restriction of Acceptances in the Grid Expansion Area

At a late stage in the legislative procedure, the legislator took account of the problems with the expansion of the electricity grids that occur in certain grid areas. In particular in the north of Germany, extensive measures of grid expansion are necessary. Due to the grids not having sufficient capacity to date, substantial curtailments of renewable energy—and in particular wind energy—plants take place in the north (so-called feed-in management), leading to considerable amounts of electricity not being produced due to constraints in the energy network.

To avoid worsening this situation, Section 36c EEG 2017 restricts the further expansion of wind onshore in the "grid expansion area" to a certain maximum. That means that acceptances within an auction shall be granted only up to a total maximum per year in case of bids for plants within this grid expansion area. The grid expansion area has been defined by means of an Ordinance; it may cover 20 percent of the area of Germany as a maximum.

While it is meaningful to consider the grid abilities also in the expansion of the renewable energy plants, the rather short-term political compromise on this grid expansion area needs to be regarded critical. It is at least questionable whether the mechanism found now takes sufficient account of the period necessary for a further expansion of the grids (which has to be the first priority) and of means to use "surplus" electricity for a transitional period e.g. by industrial customers that can shift their demand, power to heat or energy storage. What is missing is an overall assessment on whether it is necessary to curtail the further expansion of wind energy within the grid expansion area or whether—on contrast—it would be more economic from an overall perspective to use part of the electricity for a second-best usage like heat generation within an interim period until the electricity grids are sufficiently expanded.

6 Outlook

This contribution revealed some of the political debates around the introduction of auctions for renewable energies. At the outset of this new support mechanism, there is still substantial skepticism on whether auctions can lead to cost reductions by an increased competition. However, the political decision has been taken and all players now have to change their planning and include auctions as one step within the project procedure, having effects on the other steps as well. It is to be hoped that the several safeguards the legislator foresees for his goals work well so that auctions can continue the success story of the Renewable Energy Sources Act. From the perspective of climate change mitigation, it is most important that the bids that have been accepted also lead to the renewable energy projects being realized and contributing to a low carbon energy industry.

References

1. Federal Ministry for Economic Affairs and Energy (BMWi), Erneuerbare Energien auf einen Blick, www.bmwi.de/DE/Themen/Energie/Erneuerbare-Energien/erneuerbare-energien-auf-einen-blick.html, last accessed 19 July 2016
2. European Commission, Guidelines on State aid for environmental protection and energy 2014–2020, (2014/C 200/01), paras. 126 et seq
3. BMWi, Ausschreibungen für die Förderung von Erneuerbare-Energien-Anlagen, Eckpunktepapier, July 2015, www.bmwi.de/BMWi/Redaktion/PDF/Publikationen/ausschreibungen-foerderung-erneuerbare-energien-anlage,property=pdf,bereich=bmwi2012,sprache=de,rwb=true.pdf, last accessed 19 July 2016, p. 6
4. BMWi, Ausschreibungen für die Förderung von Erneuerbare-Energien-Anlagen, Eckpunktepapier (above note 4), last accessed 19 July 2016
5. See closer *Malte Kohls/Guido Wustlich*, Die Pilot-Ausschreibung für Photovoltaikanlagen, Neue Zeitschrift für Verwaltungsrecht (NVwZ) 2015, 313 et seqq.; *Jens Vollprecht/Christoph Lamy*, Die Freiflächenausschreibungsverordnung – ein erster Überblick, Zeitschrift für neues Energierecht (ZNER) 2015, 93 et seqq
6. For further information see the consultation documents provided by the BMWi on https://www.erneuerbare-energien.de/EE/Redaktion/DE/Standardartikel/EEG/eeg-ausschreibungen.html, last accessed 19 July 2016
7. See the information by the BMWi provided on http://www.erneuerbare-energien.de/EE/Navigation/DE/Recht-Politik/EEG-Ausschreibungen/Akteursvielfalt-Buergerenergie/akteursvielfalt-buergerenergie.html, last accessed 19 July 2016
8. See closer *Guido Wustlich*, in: Martin Altrock/Volker Oschmann/Christian Theobald, EEG, 4. ed. 2013, § 33 g para. 14
9. See BMWi, Ausschreibungen für die Förderung von Erneuerbare-Energien-Anlagen, Eckpunktepapier (above note 4), last accessed 19 July 2016, p. 8
10. Trend:research, Leuphana Universität Lüneburg, Definition und Marktanalyse von Bürgerenergie in Deutschland, October 2012, https://www.buendnis-buergerenergie.de/fileadmin/user_upload/downloads/Studien/Studie_Definition_und_Marktanalyse_von_Buergerenergie_in_Deutschland_BBEn.pdf, last accessed 19 July 2016, p. 42
11. Fachagentur für Windenergie an Land, Charakterisierung und Chancen kleiner Akteure bei der Ausschreibung für Windenergie an Land, July 2015, www.fachagentur-windenergie.de/fileadmin/files/Veroeffentlichungen/FA-Wind_Studie_kleine_Akteure_in_Ausschreibungen_IZES_07-2015.pdf, last accessed 19 July 2016, p. 2, 16, 42; IZES, Bewertung von Ausschreibungsverfahren als Finanzierungsmodell für Anlagen erneuerbarer Energienutzung, Bericht im Auftrag von Bundesverband Erneuerbare Energie e.V., 26 June 2014, www.bee-ev.de/fileadmin/Publikationen/Studien/IZES20140627IZESBEE_EE-Ausschreibungen.pdf, last accessed 19 July 2016, p. 53

12. Fraunhofer ISI et al., Ausschreibungen für erneuerbare Energien, wissenschaftliche Empfehlungen im Auftrag des BMWi, 7 July 2015, www.bmwi.de/BMWi/Redaktion/PDF/Publikationen/ausschreibungen-eeg,property=pdf,bereich=bmwi2012,sprache=de,rwb=true.pdf, last accessed 19 July 2016, p. 2, 16, 42

Micro PEMS for the Control of Emissions in Cars

Diego Ernesto Contreras Domínguez, Stefan Lehmann, Virgilio Vásquez López and Michael Palocz-Andresen

Abbreviations

AAA	Applus Autologic Analyzer
AG	Aktiengesellschaft (corporation)
CO	Carbon monoxide
HC	Unburned hydrocarbon
λ	Lambda sensor
MIL	Malfunction indicator lamp
NO	Nitrogen monoxide
NO_2	Nitrogen dioxide
NO_x	Sum of nitrogen monoxide and nitrogen dioxide
OBD	On-board diagnosis
ppm	Parts per million
vol.%	Volume percent (1 vol.% = 10,000 ppm)

1 Comparison of the Traditional PEMS with the Micro PEMS Technology

To the traditional test bench test the measurement of emissions by cars with the PEMS, the Portable Emission Measuring System was introduced in the frame of the Real Driving Emission Test Procedure RDE in the EU on September 01, 2017.

D. E. C. Domínguez · V. V. López · M. Palocz-Andresen (✉)
ITESM Instituto Tecnológico de Monterrey, Mecatrónica, Estado de México, Mexico
e-mail: Palocz-andresen@t-online.de

S. Lehmann
Dräger Safety AG & Co., Lübeck, Germany

M. Palocz-Andresen et al. (eds.), *International Climate Protection*,
https://doi.org/10.1007/978-3-030-03816-8_31

Most important producers are SEMTECH, AVL and Horiba [1–3]. Current PEMS systems have a volume of ca. 0.5–1 m^3 and cost 113,000–226,000 dollars. Much smaller is the TNO Smart Analyzer [4]. However, similar to the larger analyzers, the structure of the analyzer is a typical PEMS technology and the place of the installation is in the luggage rack room of cars.

2 Installation of the Micro PEMs in Automobiles

The installation of the micro PEMS is contained in Fig. 1.

Micro PEMS means the installation of an independent micro measuring device in transient position at the end pipe which can be installed and deinstalled very easily. It has an own energy supply system and collects and transmits the data independently from the automobile's OBD system. A connection to the OBD system is possible for an interim time interval with a specific hard- and software system. The OBM1 system does not influence the operation of the OBD system. It could be used for temporary applications, e.g. for estimating the exhaust gas quality in the control procedure of the Periodical Inspection in Germany (TÜV), MoT in England and Inspection and Maintenance I/M in the USA.

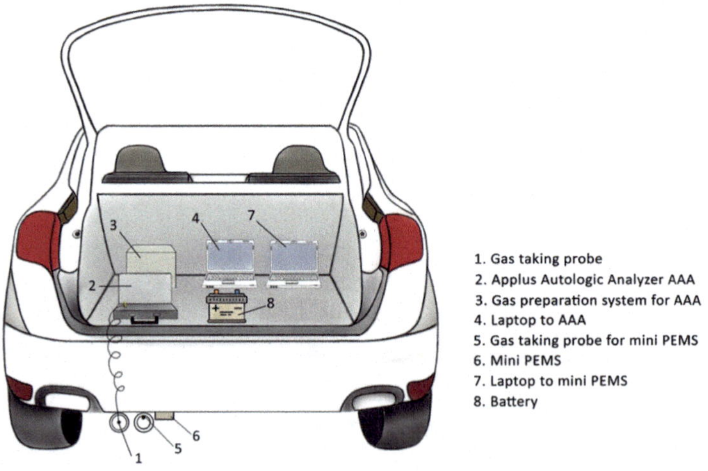

Fig. 1 Installation of the conventional PEMS and the mini PEMS in the automobile

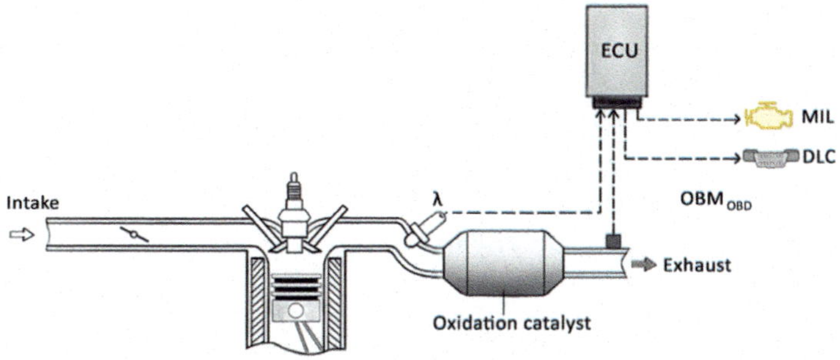

Fig. 2 Measuring real driving emissions with two deterioration causes due to a weaken λ sensor and a defect catalyst with reduced O_2 converting capacity

3 Importance of Long Life Controlling with Mini PEMS in the Automobile

The OBD system monitors only the signals of individual sensors and indicate a MIL signal if the limit of single sensors is exceeded. Smaller failure of several elements over a period of time caused by aging, can easily lead to a multiple deterioration. More than one error can lead to exceeding the emission limiting value without any warning signal in the MIL, because the individual sensor signals do not exceed the emission limiting values, see Fig. 2.

Mini PEMS with long-life measuring potential can discover higher emissions than allowed.

4 Experiments in an Experimental Car

The experiments were done at the test car of the ITESM, Universidad de Tecnologico de Monterrey in Mexico, Departmento Mecatronicam, model VW Jetta TSI 2014 Canada Edition, see Fig. 3.

The following devices were used in the first experiments, see Table 1:

The construction of the gas analyzing technology with a common gas preparation module and different gas analyzers is presented in Fig. 4.

In the future, the gas measuring devices will be placed parallel, not serial as shown here. Thus, we can avoid undesired influences of the exhaust gas by the first analyzer.

Fig. 3 Experimental car with the sampling probe in the exhaust pipe

Table 1 Devices used in the experiments

Element	Applus Autologic gas analyzer	Dräger X-am 8,000 analyzer	Dräger X-am 7,000 analyzer	Dräger analyzers type X-am 5,600 with V 7.1
O_2	0–25 vol.%			
CO_2	0–20 vol.%	0–5 vol.%	0–5 vol.% (range 1) 0–100 vol.% (range 2)	0–5 vol.%
CO	0–15 vol.%	0–2,000 ppm		0–2,000 ppm
HC	0–4,000 ppm	0–2,000 ppm[a]		
NO_x	0–4,000 ppm			
NO		0–200 ppm		0–200 ppm
NO_2		0–50 ppm		0–50 ppm

[a]calibration with iBut, no selective for single hydrocarbons

5 Comparison of the Measuring Results Gained by Dräger and the Applus Autologic 5 Analyzer

The blue color presents the CO distribution of the Applus Autologic Analyzer, the Dräger analyzer type X-am 8000 and X-am 5600 is displayed in orange. The Applus Autologic Analyzer presents the results as volume percentages and the Dräger gas analyzer as parts per million. Thus, the measurement of the Applus Autologic Analyzer was transformed into ppm, in order to enable the sensor comparison.

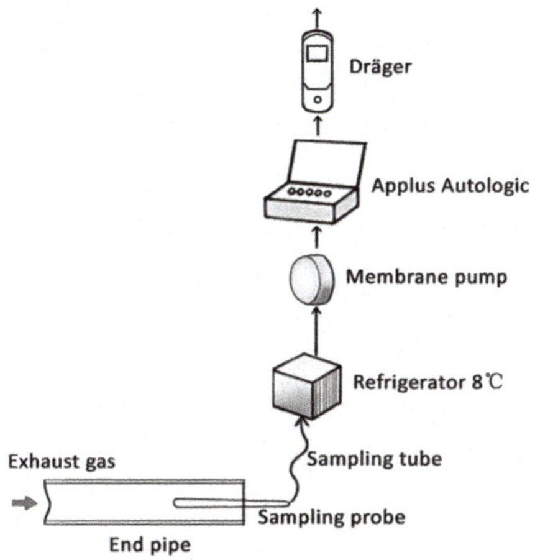

Fig. 4 Flow diagram of the gas measuring system

An internal combustion engine generates CO mainly when it is running on a rich mixture because of the higher fuel consumption than the stoichiometric value. For the present graph that condition is met at the cold start of the engine, where the engine control unit commands for a rich mixture to quickly warm up the engine and achieve the peak efficiency. After the cold start, most modern vehicles control the air-fuel mixture to a lean condition to reduce the fuel consumption and decrease the pollutant emissions, see Figs. 5 and 6.

Figure 6 clearly shows that the Dräger sensor is more sensitive and has a lower detection limit than the AAA. So, the Dräger sensor has a high quality and can detects

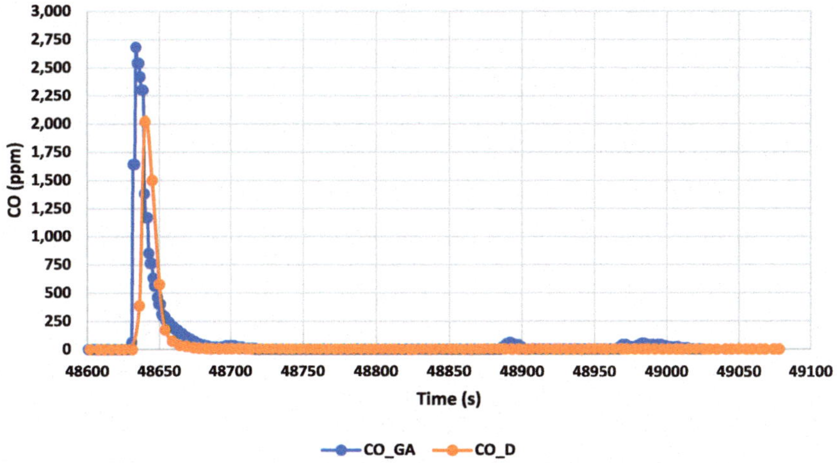

Fig. 5 CO distribution in the cold start measured with the AAA and the X-am 8000 Test 1

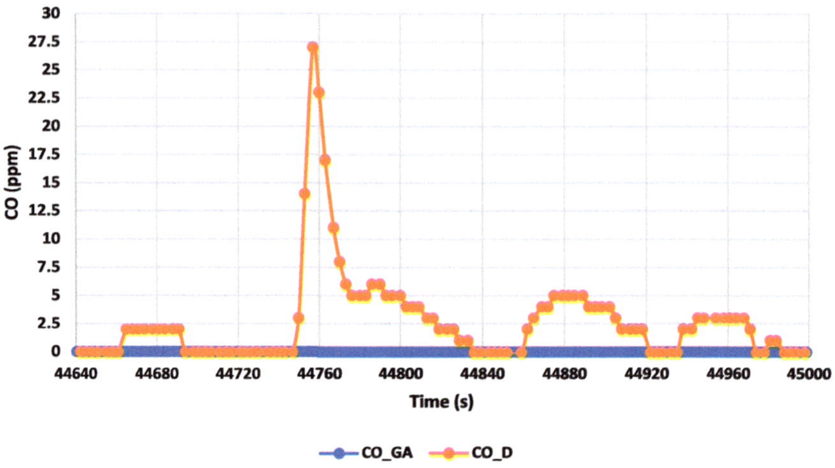

Fig. 6 CO distribution after the cold start in the slow travelling phase, gained by the AAA and the X-am 5600 Test 2

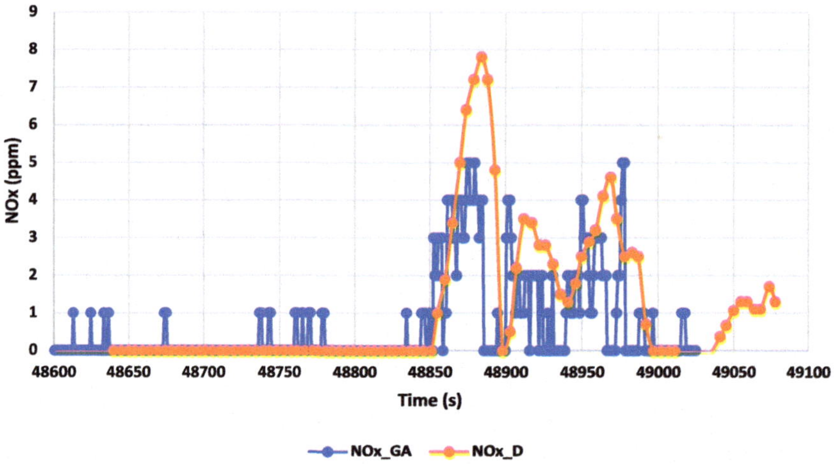

Fig. 7 Distribution of NO_x gained by the AAA and the X-am 8000 Test 1

small differences in the combustion and the exhaust gas after treatment process of the automobile.

Both analyzers, containing a photo-ionization and an electro chemical detector, measure more than two dozen different hydrocarbons without selectivity. The calibration is prepared with isobutane. Similar to CO, the internal combustion engine produces most of its unburned hydrocarbons when it is running on a rich air-fuel mixture, which can be seen in Fig. 7, as the largest emissions are presented in the cold start phase.

Both analyzers measure simultaneously NO and NO_2. Also, as it can be seen from the graph, the resolution and sensitivity for the Applus Autologic Analyzer is not quite that of the Dräger analyzer for the measurement of NO_x, see Fig. 7.

6 Outlooks

The predicted market introduction of the technology is 2021. The application is useful in controlling organizations, in automobile manufacturers and OEM companies, in schools and universities, and in research institutes.

Bibliography

1. Gaseous and Exhaust Flow Measurement. SEMTECH® LDV. http://www.sensors-inc.com/Products/SEMTECH/LDV
2. AVL M.O.V.E GAS PEMS iS. https://www.avl.com/vehicle-development/-/asset_publisher/gYjUpY19vEA8/content/avl-m-o-v-e-gas-pems-is
3. Horiba: Real Driving Emissions Test with On-Board System—RDE. http://www.horiba.com/en_en/applications/mobility-and-transportation/automotive-manufacturing/real-driving-emissions-test-with-on-board-system-rde/
4. Measuring Real-World Emissions with TNO's Smart Emissions Measurement System (SEMS). https://www.tno.nl/en/focus-areas/traffic-transport/roadmaps/mobility-logistics-digitalisation/sustainable-mobility/improving-air-quality-by-monitoring-real-world-emissions/measuring-real-world-emissions-with-tno-s-smart-emissions-measurement-system-sems/

Development Trends in Forest Economics

László Jáger

1 History of Forest Economics

Economics is a social science that describes the factors that determine the production, distribution and consumption of goods and services [1]. Using the given definition above, primary goal of forest economics is to optimise wood production and harvest in order to achieve the highest economic value and greatest production volume. In broader terms wood processing and wood trade can also be seen as an integral part of forest economics.

Wood was always a primary construction material and source of energy. History of forest economics goes back as long as 500 years, since the first evidences of systematic planning and management of forest are dated back to around 1500.

Wood played important role at two specific areas:

In central Europe, gold mining was the primary source of income of the king. The inner framework of mines were built of wood, therefore wood supply was a crucial factor to maintain gold supply [2].
In the Western countries ship building was the most important factor which led to reduction of forest cover. In some countries, as UK or Ireland, it went down as low as 1% of total area [3].

Wood had remained a key source of energy until the start of industrial revolution, when it was replaced by coal, but its importance as a raw material was kept high continuously.

L. Jáger (✉)
Budapest, Hungary
e-mail: dr.jager.laszlo.tamas@gmail.com

M. Palocz-Andresen et al. (eds.), *International Climate Protection*,
https://doi.org/10.1007/978-3-030-03816-8_32

2 Development of Theoretical Background of Forest Economics

During the XIX century the increasing demand of wood required better inventories and accurate future estimations. Development of mathematics provided a rapid progress within the field of forest measurement and empirical knowledge was replaced by accurate calculations. Mathematical and statistical models were developed to provide more and more accurate models and methods in order to measure current and to predict future harvest volumes. At that time natural procedures were in the centre of regulation: key concept of forest inventories was to keep harvest level equal or less than current growth rate.

Forest economics was an important but steady sector of local economy and rules were developed during the XIX century kept valid for the next hundred years. Most of these rules changed with the rapid economic development after the Second World War.

3 1960s, the Production Oriented Forest Economics

With an increasing demand of wood and rapid development of scientific knowledge, forestry also turned toward industrial solutions. A key issue in this transformation was species improvement and the application of exotic species. Industrial forestry was introduced, large bulldozers were applied to achieve efficient site preparations and large areas were converted into plantations in order to produce more and more wood. Natural factors were hardly considered and large areas of even aged forests were managed with clear-cuts.

Communist states played a key role in cultivation of exotic species. Only in Hungary, 500,000 ha of Robinia forests were planted between 1960 and 1980. This means that currently more than 25% of the forest cover is a plantation.

Aggressive industrialisation stared to show its drawbacks along the 1970s. The famous book of Rachel Carson: Silent Spring in 1962 was the first alarm signal of using aggressive chemicals [4]. The book documented the detrimental effects on the environment—particularly on birds—of the indiscriminate use of pesticides. Carson accused the chemical industry of spreading disinformation and public officials of accepting industry claims unquestioningly.

Negative effects were seen in the forests also: forest dieback due to acid rain became a significant risk factor, mostly in Germany and Check Republic. This is the point when a new factor shall be added to forestry economics: how to handle the risk factor. Large areas of monocultures, especially pines are exposed to fire risk, attack of insects and other damages. Level of fire risk should not be underestimated. Even in the European Union, where it is assumed to have an efficient protection against forest fire, about 65,000 fires occur in every year, burning approximately half a million hectares of wildland and forest area. Most of the burnt area, over 85%, is in the European Mediterranean region [5, 6].

While scientists started to understand the problems of application of exotic species, the global industry turned its attention toward tropical forests and aggressive loggings were started in this much more fragile ecosystem.

4 1980s, the Internationalisation of Forest Economics

Speed of tropical wood harvest [7] showed a significant increase between 1970 and 1980 (Table 1).

Where is the balance between economics and biodiversity? To what extent the loss of habitats is accepted due to increasing demand of wood? This is an important question even today and it gives us a lesson that forestry economics shall be assessed together with an environmental perspective. To give a direct priority to biodiversity would be an easy answer, but it must be noted that replacement of wood with plastic or steel cannot be regarded as a solution of the problem. On the contrary, it will increase the environmental risk even higher, enough to recall the famous Bhopal catastrophe in 1984.[1] The solution shall be found within a more environment oriented forestry where both environmental and economics features are considered somehow.

As a result of this perspective, the orientation moved toward ecological issues and level of loss of biodiversity grew an international attention.

Table 1 World output and exports of hardwood products

Year	Harvest (million m^3)	Export (million m^3)	Share of export
1946	113,700	1,310	1.2
1950	110,500	3,300	3.0
1955	112,100	6,620	6.0
1960	149,600	12,306	9.3
1965	160,000	21,200	11.6
1970	207,066	36,741	16.7
1975	209,365	36,379	17.4
1979	243,402	47,364	22.6

[1]The Bhopal disaster, also referred to as the Bhopal gas tragedy, was a gas leak incident in India, considered the world's worst industrial disaster. Over 500,000 people were exposed to methyl isocyanate gas and other chemicals.

5 1990s: The Focus on Biodiversity

Several attempts were made from 1990 to stop the biodiversity loss of the globe. The most famous was the Earth summit in 1992, Rio de Janeiro. Forest certification scheme system was started in 1997.

In Europe, Ministerial Conference of forestry affairs started in 1990 to initiate proposals and actions for the protection and sustainable management of forests in Europe. European Union developed the network of Natura 2000 to promote further protection of forests.

All problems are rooted in the same question: should we harvest more or less wood? The answer is not easy, especially under the shadow of the new enemy: the global warming.

6 2000s Energy and Biomass Production

When global warming is discussed, it is worth considering how energy price may interfere with forest economics. It is evident that trucks, tractors and chainsaws use petrol as fuel, but there is a much stronger relationship than fuel price of the production.

When oil price is increasing, consumers are keen to find alternatives in order to reduce their costs. One of the most obvious renewable energy sources is biomass. It is not without reason that EU commitments refer to 20% emissions reductions by 2020, and 20% use of renewables by the same date. There is a significant room for further biomass production as currently the oil is 20 times more important source of energy than total renewable [8] (Fig. 1).

The steadily increasing oil price resulted a strong incentive for bioenergy production between 2000 and 2008. As a result, large areas of agriculture were converted into biomass production and great investments are seen in the field of bioethanol and biodiesel production. However, the wold is more complex than just level of demand and production: One significant side effect was a rapid growth of agricultural price index. This increase occurred the "Arabic spring" which was a sudden political crisis in many Arabic countries during 2007. The explanation lays within economy again: these countries relies heavily on agricultural inputs and therefore harshly exposed to the sudden changes of price level of agricultural products.

All in all, increasing oil price trend seemed to be very promising from the point of biomass producers but the great economic crisis in 2008 pushed the oil price as low as 30 $ and a similar drop was seen again in 2016.[2]

[2]The absolute peak occurred in June 2008 with the highest inflation adjusted monthly average crude oil price of $136 for one barrel. From there we see one of the sharpest drops in history. Note that the fall from the 1979 peak took until 1986 (7 years) to fall as much (percentage wise) as it lost in only six months from 2008–2009.

It is pretty obvious that traditional forestry investments are hard to predict in such conditions where energy price index has such a great volatility. In broader context one may say that world is trembling with bigger and bigger waves like a trembling suspension bridge does just before collapsing.

During this decade the intensive exploitation of tropical timber was also continued and it became evident that traditional tools were not enough efficient to stop this process. The core root of the problem that tropical countries are rather poor and want to sell whatever they can to provide similar life standards that is available in developed countries. Developed countries—on the other hand—express their worries about destroying the environment and the reduction of the forest cover; a similar process had been completed around 300 years ago in the western world.

The solution is not so simple to put an overall ban on tropical timber since in the developing countries the land conversion is not driven by the wood as a sellable product but the primary aim is to gain more agricultural land as a place to agricultural production.[3]

The evidences of climate changes also underline the need to increase rainforest protection and to rely more on northern hemisphere wood supplies where forest regeneration is much easier, effectively supervised by the state authorities and nature protection considerations are much smaller.

One solution can be to increase the income of tropical countries in other form than relying on their natural resources and to distribute of the wealth has been collected

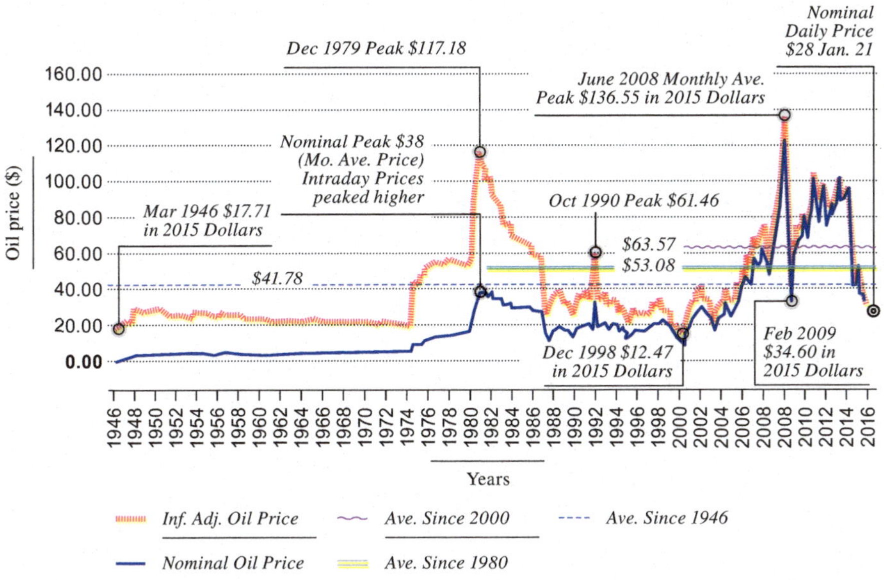

Fig. 1 Historical oil price index between 1946 and 2016 [9]

[3]There are examples when harvested wood was just fired up due to trade restrictions.

by developed countries along the centuries. In the field of forest economics this may lead to the specific idea. Preservation of the natural ecosystem should be regarded as a product which can be a subject of trading. Here one can find the origin of the theory of carbon exchange [10].

Will be the carbon stock exchange the universal tool to solve market turbulences? It is hard to say now, however it is evident that current price (5 euro/tonnes of CO_2) is not enough to put enough market pressure to achieve significant targets [11].

7 Forest Economics in an European Context

As previously international questions of forestry economics had been discussed, it is worth to have a deeper look of forestry in European context. Within Europe, policy implications are much stronger than in other sectors of economy:

forestry is a strange mixture of liberal and strongly regulated sector
forestry is subject of long term planning, there is a weak ability to adapt sudden market changes
forest economics is highly affected to natural, environmental factors
there is a strong public participation and emotional perspective of the society.

Additionally, geographical features affect forest economics:

From North to South there is a steady increase of biodiversity and production volume
From West to East there is an increase of relative wood price, compared to incomes and wages.

The result of these features is that there are no unique answers in forest economics. A solution which works in the west will not lead to a solution in the east. Challenges of the Mediterranean are different from problems of Nordic countries. What is common that society shall seek ways of sustainable forest management and wide scale use of wood and wood products.

References

1. Wikipedia: Economics. https://en.wikipedia.org/wiki/Economics
2. Magyar, E.: A feudalizmus kori erdőgazdálkodás az alsó-magyarországi bányavárosokban, vol. 101, pp. 1255–1747. Akadémiai Kiadó (1983)
3. Neeson, E.: A history of Irish forestry. Lilliput Press Ltd. (1991)
4. Carson, R.: Silent Spring. Mariner Books (1st. Pub. Houghton Mifflin, 1962). ISBN 0-618-24906-0. Silent Spring initially appeared serialized in three parts in the June 16, June 23, and June 30, 1962 issues of The New Yorker magazine (2002)
5. San-Miguel-Ayanz, J., Schulte, E., Schmuck, G., Camia, A., Strobl, P., Liberta, G. et al.: Comprehensive monitoring of wildfires in Europe: the European Forest Fire Information System (EFFIS) (2012)

6. Riera, P., Mogas, J.: Evaluation of a risk reduction in forest fires in a Mediterranean region. Forest Policy Econ. **6**(6), 521–528 (2004)
7. Kumar, R.: World tropical wood trade: economic overview. Resour. Policy **8**(3), 177–192 (1982)
8. Klemperer, W.D.: Forest resource economics and finance. McGraw-Hill Inc. (1996)
9. http://inflationdata.com/Inflation/Inflation_Rate/Historical_Oil_Prices_Table
10. Najam, A., Huq, S., Sokona, Y.: Climate negotiations beyond Kyoto: developing countries concerns and interests. Clim. Policy **3**(3), 221–231 (2003)
11. Dong, Y., Whalley, J.: Carbon, trade policy and carbon free trade areas. World Econ. **33**(9), 1073–1094 (2010)

Security Risks of the Climate Change

Ilona Bodonyi

1 Concept of Security

Speaking about the security concept we can find several different definitions on it. The classical sense of security—from the Latin securitas, refers to tranquility and freedom of care. "The English word "security" has a wide range of meaning including "to feel safe," and "to be protected" and is used to describe a situation without any risks or worries [1]. (…) the term security can be used in three meanings.

- "traditional meaning"—security as an attribute of state, absence of military conflict—"military security",
- security used in a broader sense yet still referring directly to the phenomena taking place in international relations, or directly/indirectly caused by inter-state relations
- security as a public good,
- security in a universal sense (of a unit and of a social entity)—human security" [1].[1]

The traditional concept focuses on the protection of citizens and national interests from an internal perspective: which according to Kennan (1948) is "*the continued ability of the country to pursue the development of its internal life without serious interference, or threat of interference, from foreign powers*".

In the UN concept of security already appeared the non-military aspects as well: "*with a view to the creation of conditions of stability and well-being, (…) the United Nations shall promote: a) higher standards of living, full employment, and conditions of economic and social progress and development; […] c) universal respect for, and*

I. Bodonyi (✉)
NUPS, Budapest, Hungary
e-mail: ibodonyi@gmail.com

[1] [1] p. 4.

M. Palocz-Andresen et al. (eds.), *International Climate Protection*,
https://doi.org/10.1007/978-3-030-03816-8_33

observance of, human rights and fundamental freedoms for all [...]".

In order to keep going into the concept of security in depth, two well differentiated trends can be distinguished. The first one is supported by those believing that the State is the essential object of security, even if they accept the inclusion of new factors. The concepts of Common Security, Integrated Security and Defensive Security are part of this first trend. The concept of Common Security which was first introduced by Olof Palme in the context of the Cold War, pointed out that "*... countries cannot gain security at the expense of others, nor through only military power. Common security requires peoples to live in dignity and peace, to have the needed food, find a job and live in a world without poverty and destruction*". In its postulates this concept is close to the one of Human Security.

The Integrated Security endorses a comprehensive approach to security, which recognizes a balance between its political, economic, social-cultural and environmental aspects.

This concept was made known by the UN in the framework of the "Study on Defensive Security Concepts and Policies" in 1993 which describes it as "*a condition of peace and security attained step-by-step and sustained through effective and concrete measures in the political and military field*".[2]

Human Security principle was gradually introduced, promoted by the UN since 1994. Criticized by some politicians and analysts due to its lack of precision, this concept includes within the questions to be considered: economy, food, sanity, politics, environment and people individual protection, as well as the communities in which this people live. Its implementation is related to the controversial principal of the Responsibility to Protect. This principle points out that each State is responsible to protect its population, but the international community, through the UN, is also responsible of that protection using, in case it was necessary, coercive instruments. Recently Defence, Diplomacy and Development—3 D model as a new trend is intended for serving the security both in national and international level—that means the global security.[3]

"Security is an elastic and diverse concept that can be understood in different forms, depending on its objects: the perception of threats, the protected values, and the means through which these values can be protected.[4] Despite the people-centred Human security[5] the new concept has paved the way for a shift in the focus of various security issues, from the domain of national security to a much greater spectrum. The shift is significant, for example, in environmental security discourse, given that even though many of the environmental issues are global ones, the consequences of environmental degradation are usually observed and felt at the local or regional levels.[6]

---

[2] [1] p. 3.

[3] [1] p. 5.

[4] [2] p. 15.

[5] [2] p. 23.

[6] [2] p. 24.

2 Climate Change as a Security Issue

The concept of security has expanded, particularly after the end of the Cold War, in terms of both security objects and security threats. For example, climate change, technological development, demand for resources, growing environmental concern, and the growth of overlapping legal regimes and regulations, are all factors that potentially give rise to security concerns of various states and other actors. While the expansion of the security concept can be seen as a policy response to accommodate those new challenges, its analytical framework is yet to be fully examined in the discipline of international law, which would guide us on how we should conceive those new security threats and address them internationally.[7]

"The core concept of security can be used as a foundation of analytical considerations concerning all aspects (dimensions) of security.

5. The idea of human security requires further scholarly discussion. Otherwise it would become an instrument of ideology and its uses in policy making and in social communication will be very limited."[8]

Much research suggests that while the impacts of climate change on human security will be experienced most in developing countries, human security is at risk for vulnerable populations everywhere [3–5].

"Human security will be progressively threatened as the climate changes (high agreement, robust evidence). Human insecurity almost never has single causes, but instead emerges from the interaction of multiple factors [12.1.2; 12.2]. Climate change is an important factor in threats to human security through (a) undermining livelihoods [12.2], (b) compromising culture and identity [12.3], (c) increasing migration that people would rather have avoided [12.4], and (d) challenging the ability of states to provide the conditions necessary for human security [12.6]" [6].[9]

The framing of climate change as a security issue has been controversial. "*Some countries associate climate change risks with conventional security risks and many countries are concerned about the risks climate change poses to relations between states* (see Sects. 5 and 6)."[10]

3 Social Context and Key Questions

There are a lot of possible societal responses on the adverse changes of the natural environment, and inside and outside the borders these responses all bring up security questions and risks,—inner conflicts, delinquency, war, migration.

---

[7][2] p. 33.

[8][2] p. 15.

[9][6] p. 2.

[10][6] p. 5.

It seems expedient to examine these security risks in terms of the social and political responsibility. The fundamental questions, for which the science and the politics look for answers equally, are:

Who has to take care on the human communities, and to protect them from the crisis situations?

What the individuals and the social communities may do for the prevention of the natural disasters and/or for the treatment of their consequences?

How to prevent the escalation of the security risks?

How it is possible to precede the newer environmental catastrophe caused by the lack of security, by the modern wars?

Which ones are the most endangered countries? What is the context between the climate change and the security risks?

Meanwhile, although industrialized countries are responsible for 60 percent of the greenhouse gas emissions that contribute to climate change, developing countries suffer the "worst and first" effects of climate-related disasters, including droughts [7], floods, and storms, because of their geographical locations. In A Climate of Injustice, J. Timmons Roberts and Bradley Parks analyze the role that inequality between rich and poor nations plays in the negotiation of global climate agreements [8].

Giving a very simplified answer, we may say that those countries and regions are the most endangered and vulnerable ones which face the challenges of the climate change defencelessly, on the one hand, with a weak infrastructure, an one-sided and underdeveloped economy and weak state administration; on the other hand, in which countries the climate change and the demographic changes together may cause political conflicts.

For example the former colonies of which borders have been drawn exclusively in the interest of colonists, which were rivals with each other, and not taking into consideration the natural borders and settlements of different ethnic groups and tribes. These countries are hit by political instability; by fragile economy; by the lack of the middle class and the social cohesion.

Speaking for example about the Sahel region Sahel (Burkina Faso, Chad, Mali, Mauritania and Niger):

Climate change and ecosystem degradation increase the unpredictability of rainfall.

Population growth is among the highest in the world (on average, the population of the Sahel doubles every 25 years). This increases pressure on natural resources and food supply.

Chronic poverty: The Sahel states rank at the bottom of the 2011 UN Human Development Index.

Regional economic disparity (between Sahel countries and coastal countries) *and low resistance to external economic shocks* (e.g. the food price crisis of 2008) contribute significantly to the fragility of the Sahel. As a result, food insecurity in the Sahel is primarily a matter of income and not production.

Weakness of public finances and national institutions in some countries hampers adequate responses to the increasing frequency of crises that affects the region [9].

Speaking about the Arabic countries, the situation may be characterized with the next factors:

"The exclusive nature of many social, political and economic institutions, the *lack of or existence of only limited pluralistic participation in social, economic and political life, and in some cases, the limited respect for cultural diversity* has manifested itself in the form of social discontent and popular uprisings; lack of state acceptance by citizens and the history of exclusion, inequality and neglect for human rights on social cohesion" [10].

Finally, we do not avoid to mention also, that the wars, armed conflicts, the effects of military activities have terrible polluting consequences, and their deleterious effects are expounded on a much wider area, than the size of the territory where really manoeuvres can be localized. On the other hand "Armed conflict disrupts markets and destroys infrastructure, limits education and the development of human capital, causes death and injury to workers, and decreases the ability of individuals, communities and the state to secure credit [11–13]. Conflict thus creates poverty and constrains livelihoods: that in turn increases vulnerability to the impacts of climate change [14][11]; "Armed conflict can decrease the capacity of governments to function effectively (…) A lack of trust in government commitment or capacity to respond, the presence of police or military forces that lack legitimacy, or recent conflict between government and local forces, hampers the ability of these institutions to provide effective relief [15]."[12]

References

1. Czesław Mesjasz: SECURITY AS AN ANALYTICAL CONCEPT. Paper presented at the 5th Pan-European Conference on International Relations, in The Hague, 9–11 September 2004 Cracow University of Economics Cracow, Poland. http://www.afes-press.de/pdf/Hague/Mesjasz_Security_concept.pdf
2. Hitoshi Nasu: The Expanded Conception of Security and International Law: Challenges to the Uncollective Security System The Amsterdam Law ForumVU Universityof Amsterdam
3. Naess, L.O., Norland, I.T., Lafferty, W.M., Aall, C.: Data and Processes linking vulnerability assessment to adaptation decision-making on climate change in Norway. Glob. Environ. Change **16**(2), 221–233 (2006)
4. Leichenko, R.M., O'Brien, K.L.: Environmental change and globalization: double exposures. Oxford University Press (2008)
5. Berrang-Ford, L., Ford, J., Peterson, J.: Are we adapting to climate change? Glob. Environ. Change **21**, 25–33 (2011)
6. UN AND THE CLIMATE CHANGE. Climate Change 2014: Impacts, Adaptation, and Vulnerability. WG II. Intergovernmental Panel on Climate Change, WMO, UNEP. Summary for Policymakers (approved) and Final Draft (accepted), Chap. 12 Human security. http://ipcc-wg2.gov/AR5/images/uploads/WGIIAR5-Chap12_FGDall.pdf
7. Future drouhgts will be shockers study says http://www.nbcnews.com/id/39741525/ns/us_news-environment/t/future-droughts-will-be-shockers-study-says/#.VkNYQrcvddg

[11] [6] p. 18.

[12] [6] p. 19.

8. A Climate of Injustice Global Inequality, North-South Politics, and Climate Policy by J. Timmons Roberts and Bradley Parks, Overview. http://mitpress.mit.edu/sites/default/files/titles/content/9780262681612_sch_0001.pdf
9. http://europa.eu/rapid/press-release_MEMO-12-215_en.htm?locale=en. The European Commission's response to the food crisis and long-term food insecurity in the Sahel region of Africa
10. United Nations Development Program—Region: Arab States. Promoting Social Cohesion in the Arab Region (PSCAR)
11. Stewart, F., Fitzgerald, V. (eds): War and underdevelopment: volume 1, the economic and social consequences of conflict. Oxford University Press, Oxford (2001)
12. Goodhand, J.: Enduring disorder and persistent poverty: a review of the linkages between war and chronic poverty. World Dev. **31**(3), 629–646 (2003). https://econpapers.repec.org/article/eeewdevel/
13. Blattman, C., Miguel, E.: Civil war. J. Econ. Lit. **48**(1), 3–57 (2010). https://doi.org/10.1257/jel.48.1.3. https://chrisblattman.com/documents/research/2010.CivilWar.JEL.pdf
14. Nigel, J.: Livelihoods in a conflict setting. Norsk Geografisk Tidsskrift-Norwegian Journal of Geography **63**(1), 23–34 (2009)
15. Wisner, B.: Risk and the neoliberal state: why post-mitch lessons didn't reduce El Salvador's earthquake losses. Disasters **25**(3), 251–268 (Oct 2001). https://www.researchgate.net/journal/0361-3666_Disasters

Climate Change and Infectious Diseases

Rebecca Hinz, Hagen Frickmann and Andreas Krüger

1 Introduction

The occurrence of several infectious diseases has been linked to climate and weather variability in certain parts of the world. Climate can indirectly enhance the vulnerability for infectious diseases due to malnutrition resulting from climate dependent crop failure or stress, for instance caused by flooding [4]. Direct influences have been observed particularly for infectious diseases transmitted by insect vectors (vector-borne), mainly mosquitoes, contaminated food and water (food- and water-borne) and animals (zoonotic) [1, 2, 5].

1.1 Vector-Borne Infectious Diseases

Survival and reproduction rate of both insect vectors (like mosquitoes, ticks, or sandflies) and pathogens (such as several viruses, bacteria or parasites) depend on certain optimal climatic conditions, particularly temperature, precipitation, humidity, wind and daylight duration. Therefore even the smallest changes of these factors due to extreme weather events or change of season can affect the incidence, transmission

R. Hinz (✉) · H. Frickmann · A. Krüger
Department of Microbiology and Hospital Hygiene (XXI), Tropical Microbiology and Entomology Branch at the Bernhard Nocht Institute for Tropical Medicine, Bundeswehr Hospital Hamburg, Hamburg, Germany
e-mail: hinz@bnitm.de

H. Frickmann
Institute for Medical Microbiology, Virology and Hygiene, University Medicine Rostock, Rostock, Germany

R. Hinz
Institute of Medical Microbiology, Virology and Hygiene, University Medical Center Hamburg-Eppendorf, Hamburg, Germany

M. Palocz-Andresen et al. (eds.), *International Climate Protection*,
https://doi.org/10.1007/978-3-030-03816-8_34

and distribution of vector-borne diseases, suggesting associated potential long-term shifts due to climate change [2, 5].

Beside its impact on survival and emergence of vectors and pathogens, temperature also affects the biting rate of mosquitoes and their affinity to humans and thereby influences the transmission of vector-borne diseases. Furthermore, higher temperatures have demonstrated to decrease the extrinsic incubation period of pathogens for instance responsible for malaria, dengue, or yellow fever within the vector [4, 6–9]. Though, changes of temperature may force the vector to change its geographic distribution, some vectors have shown to be capable of adapting to certain changes of temperature [2, 10].

Precipitation mainly affects the existence of breeding habitats of mosquitoes. Excessive rainfalls may create new larval habitats and increase mosquito populations and transmission of malaria and other diseases or wash out existing breeding sites and reduce vector emergence, associated with decreased transmission of several mosquito-borne diseases [4, 7, 11]. In times of reduced precipitation or drought, the emergence of mosquitoes might decrease due to lack of breeding sites or as well increase due to new breeding habitats provided by former rivers being reduced to smaller pools or an increased range of water storages of households (see Fig. 3) [2, 4].

Strong wind can influence the geographical range of mosquitoes but also decrease their biting opportunities [4].

Malaria, a parasitic disease accounting for 438,000 deaths in 2015 [12], appears to be the vector-borne disease most sensitive to long-term climate change and demonstrates seasonality in highly endemic areas and susceptibility for extreme weather events [2, 11]. The malaria transmitting *Anopheles* mosquitoes require warm moist climate for survival and breed in small water pools in the natural environment. High humidity and formation of new breeding habitats due to excessive monsoon rainfalls or droughts, for instance related with the El Niño Southern Oscillation (ENSO) phenomenon [13], are able to enhance mosquito breeding and survival. In the year after an El Niño event, the malaria epidemic risk can increase around five-fold [2, 7, 14].

Though reduced temperatures and modest warming can increase malaria transmission, severe increases of temperature may as well reduce transmission [5, 15]. Modelling suggests that even only small increases of the global temperature of 2–3 °C might not only increase the number of individuals at risk of malaria by several hundred million, but moreover extend the seasonal duration of malaria in endemic areas. Furthermore, a global rise of temperature by 0.5 °C might increase the occurrence of malaria transmitting mosquitoes up to 30–100% and expand the geographic range of the disease [2].

Rift Valley Fever (RVF) causes significant livestock losses, particularly in sheep, but also in cattle, camels and goats (Fig. 1). Most human infections remain asymptomatic or present with a mild flu-like disease, though lethal cases are reported, particularly in the course of epidemics [16–19].

RVF has first been reported outside Africa in 2000 with cases in the Arabian Peninsula. This epidemiological shift, potentially associated with livestock trade and

Fig. 1 Typical Rift Valley fever hosts (R. Hinz)

climatic factors promoting the emergence of transmitting vectors, provokes concerns of a potential emergence of RVF in Europe [16–18].

Primarily, transmission to domestic animals occurs by mosquitoes, particularly by the *Culex* and *Aedes* genera [16, 18], whereas humans in endemic areas generally become infected via the zoonotic route due to contact with body fluids or tissues of sick animals or associated animal products [16]. However, epidemics affecting humans often result from vector-borne transmission promoted by increased vector emergence in times of above-average rainfall or floods [2, 17–19] within seasonal variability [5] or during the warm phase of the ENSO phenomenon [16]. These findings are applied in forecasting models and for mapping the potential risk of RVF outbreaks in endemic areas [2, 16].

Dengue Fever is the most rapidly spreading mosquito-borne viral disease, frequently transmitted in Asia, the Pacific, the Americas, the Caribbean, and Africa. The first cases of local dengue fever transmission in Europe since the 1920s have been reported in 2010 in France and Croatia, followed by a dengue fever outbreak on the Madeira islands of Portugal with over 2,000 cases and another autochthonous case in 2013 in France [5, 20–22].

Fig. 2 Potential *Aedes* breeding site, flower pot, Bangkok, Thailand (R. Hinz)

Transmission occurs by *Aedes* mosquitoes, which prefer breeding in man-made containers as provided in urban habitats (Figs. 2 and 3). Dengue fever is a flu-like illness. Potentially lethal severe manifestations comprise haemorrhagic fever or shock and require intensive care treatment [20, 21, 23].

The global dynamic expansion of dengue fever is suggested to have multifactorial causes including climate change, virus evolution, urbanisation, global travel and trade [5, 10, 23, 24]. Climate conditions, such as precipitation, temperature and humidity and their variability due to extreme weather events like typhoons, El Niño, droughts or floods have demonstrated an influence on the distribution of the vectors and the incidence of dengue fever [5]. The occurrence of dengue fever and chikungunya in Europe has been linked to the import of *Aedes albopictus* due to global trade of used tires, increased air travel and reduced vector control measures [10, 21, 25, 26].

Many other vector-borne diseases have been linked to certain alterations of weather and climatic conditions due to change of season or extreme weather events, including Japanese encephalitis, Ross River virus disease, West Nile fever, Murray Valley encephalitis, chikungunya, leishmaniasis, Lyme disease, and tick-borne encephalitis [2, 4, 5, 14, 15, 27].

Fig. 3 Potential *Aedes* breeding sites, water filled pots, Kumasi, Ghana (R. Hinz)

1.2 Zoonotic Infectious Diseases

Changes of weather and climatic factors or the occurrence of extreme weather events may have an impact on the distribution and emergence of animal hosts, particularly rodents, and the zoonotic transmission of certain infectious diseases. The size of local rodent populations can increase due to enhanced food availability following heavy rainfalls but also decrease associated with floods, heat, and droughts, however, leading to an enhanced contact of rodents to humans [2, 4, 5, 14]. An increase of rodent populations has furthermore been observed following mild wet winters [13].

Hantavirus is mainly transmitted to humans by the inhalation of dust contaminated with infectious urine or faeces of rodents [20]. An increased incidence of Hantavirus pulmonary syndrome occurring in the Americas has been linked to increased rodent populations due to above-average precipitation, mild temperatures and ENSO related heavy rainfalls and floods [2, 7, 28, 29]. These findings have been used for ENSO related disease risk mapping. But also the occurrence of haemorrhagic fever with renal syndrome (HFRS) due to Hantavirus in Asia and Europe has been influenced by climatic factors, such as temperature, precipitation, and humidity [5, 28, 29].

Plague is caused by the bacterium *Yersinia pestis*. Infection usually occurs vector-borne by the bite of a flea which was infected by a rodent host, but may also occur due to direct contact with contaminated tissues [5, 20]. Plague is endemic in Africa, primarily Madagascar, Asia and the Americas [30].

The emergence of plague has been associated with seasonality and climate variability, mainly temperature, increased precipitation and floods, related to ENSO and

Indian Ocean Dipole (IOD) [5, 29, 30]. Heavy rainfalls are not only capable of increasing the population size of rodents but also of fleas and therefore increase the risk of plague transmission. Droughts on the other hand may also result in an increased risk of plague, as rodents get closer to human habitats searching for food [2, 14].

1.3 Water-Borne and Food-Borne Infectious Diseases

Every year diarrhoeal disease kills approximately 760,000 children under the age of five. Infection occurs due to contaminated water and food or man-to-man transmission due to poor hygiene in times of water shortage [2, 13, 31]. The risk of diarrhoeal disease related to climate change is estimated to rise up to 10% until 2030 in certain regions [2].

Heavy rainfalls and floods, often related to ENSO or monsoon rains, may contaminate water supplies with enteric pathogens, such as *Cryptosporidium* spp., *Giardia duodenalis*, diarrhoea-associated *Escherichia coli*, *Vibrio cholerae*, *Shigella* spp. or *Salmonella enterica* ssp. *enterica* serovars, resulting in an increased risk of diarrhoeal disease [2, 5, 13, 32]. In this scenario, precipitation, floods or even storms may serve as a vehicle for the spread of causative infective agents [2]. However, water shortage due to droughts or low precipitation can be associated with concentration of enteric pathogens in available water supplies and likewise increase the risk of diarrhoeal disease [4].

Temperature has an impact on growth, survival, seasonality and geographical range of several pathogens. Whilst *Campylobacter* spp. appear to prefer lower temperatures, rising temperatures can increase the occurrence of diarrhoeal diseases due to enhanced growth of other pathogens, such as *V. cholerae or Salmonella* spp. Furthermore, higher temperatures are generally associated with increased water consumption, particularly increasing the risk of infection with enteric pathogens in times of safe water shortage due to floods or droughts [2, 4, 5, 13]. ENSO related increased sea surface temperatures associated with enhanced plankton growth, potentially serving as reservoirs of *Vibrio cholerae*, have been linked to cholera outbreaks. Uncommonly high temperatures during the winter associated with El Niño as well as relative humidity increases have also been reported to increase the occurrence of diarrhoeal disease [2, 5, 13, 29].

2 Summary

The emergence of several infectious diseases can somehow be linked to climate, but most likely not exclusively. However, most available data refer to climate or weather variability, extreme weather events or seasonal climate changes, suggesting impacts on infectious diseases due to climate change. Further long-term studies should evaluate the link between climate change and its impact on hosts, vectors, pathogens and the transmission of infectious diseases [3, 4, 10, 27, 32].

References

1. Wei, J., Hansen, A., Zhang, Y., Li, H., Liu, Q., Sun, Y., Xue, S., Zhao, S., Bi, P.: The impact of climate change on infectious disease transmission: perceptions of CDC health professionals in Shanxi Province, China. PLoS ONE **9**, e109476 (2014)
2. World Health Organization: Climate Change and Human Health: Risks and Responses. World Health Organization, Geneva (2003)
3. Suk, J.E., Ebi, K.L., Vose, D., Wint, W., Alexander, N., Mintiens, K., Semenza, J.C.: Indicators for tracking European vulnerabilities to the risks of infectious disease transmission due to climate change. Int. J. Environ. Res. Public Health **11**, 2218–2235 (2014)
4. Wu, X., Lu, Y., Zhou, S., Chen, L., Xu, B.: Impact of climate change on human infectious diseases: empirical evidence and human adaptation. Environ. Int. **86**, 14–23 (2016)
5. Smith, K., Woodward, A., Campbell-Lendrum, D., Chadee, D., Honda, Y., Liu, Q., Olwoch, J., Revich, B., Sauerborn, R.: Human health: impacts, adaptation, and co-benefits. In: Confalonieri, U., Haines, A. (eds.) Climate Change 2014: Impacts, Adaptation, and Vulnerability. Part A: Global and Sectoral Aspects. Cambridge, United Kingdom and New York, NY, USA, pp. 709–754 (2014)
6. Patz, J.A., Olson, S.H.: Malaria risk and temperature: influences from global climate change and local land use practices. Proc. Natl. Acad. Sci. U. S. A. **103**, 5635–5636 (2006)
7. Patz, J.A., Hahn, M.B.: Climate change and human health: a One Health approach. Curr. Top. Microbiol. Immunol. **366**, 141–171 (2013)
8. Lindgren, E., Andersson, Y., Suk, J.E., Sudre, B., Semenza, J.C.: Public health. Monitoring EU emerging infectious disease risk due to climate change. Science **336**, 418–419 (2012)
9. Adelman, Z.N., Anderson, M.A., Wiley, M.R., Murreddu, M.G., Samuel, G.H., Morazzani, E.M., Myles, K.M.: Cooler temperatures destabilize RNA interference and increase susceptibility of disease vector mosquitoes to viral infection. PLoS Negl. Trop. Dis. **7**, e2239 (2013)
10. Krüger, A.: [Aedes species as arbovirus vectors]. Warnsignal Klima, Gesundheitsrisiken. Gefahren für Menschen, Tiere und Pflanzen (2014), pp. 1–7
11. Kovats, R., Bouma, M., Haines, A.: El Niño and Health, Protection of the Human Environment—Task Force on Climate and Health, p. 54, Geneva, 1999
12. World Health Organization: Malaria fact sheet. Date accessed: 31.07.2016, updated: 2016
13. Kovats, R.S., Bouma, M.J., Hajat, S., Worrall, E., Haines, A.: El Nino and health. Lancet **362**, 1481–1489 (2003)
14. Khasnis, A.A., Nettleman, M.D.: Global warming and infectious disease. Arch. Med. Res. **36**, 689–696 (2005)
15. Ebert, B., Fleischer, B.: Global warming and spread of infectious diseases. Bundesgesundheitsblatt Gesundheitsforschung Gesundheitsschutz **48**, 55–62 (2005)
16. Nanyingi, M.O., Munyua, P., Kiama, S.G., Muchemi, G.M., Thumbi, S.M., Bitek, A.O., Bett, B., Muriithi, R.M., Njenga, M.K.: A systematic review of Rift Valley Fever epidemiology 1931–2014. Infect. Ecol. Epidemiol. **5**, 28024 (2015)
17. World Health Organization: Rift Valley fever, Fact sheet N°207, Revised May 2010. Date accessed: 04.08.2016, updated: 2010
18. Mansfield, K.L., Banyard, A.C., McElhinney, L., Johnson, N., Horton, D.L., Hernandez-Triana, L.M., Fooks, A.R.: Rift Valley fever virus: a review of diagnosis and vaccination, and implications for emergence in Europe. Vaccine **33**, 5520–5531 (2015)

19. Lemon, S., Sparling, P., Hamburg, M., Relman, D., Choffnes, E., Mack, A.: Vector-Borne Diseases: Understanding the Environmental, Human Health, and Ecological Connections. Workshop Summary Institute of Medicine, Washington, DC (2008)
20. European Centre for Disease Prevention and Control: Annual epidemiological report: emerging and vector-borne diseases, ECDC surveillance report. Stockholm (2014)
21. World Health Organization: Dengue and severe dengue, Fact sheet. Date accessed: 05.08.2016, updated: 2016
22. Marchand, E., et al.: Autochthonous case of dengue in France, October 2013. Euro. Surveill. **18**, 20661 (2013)
23. Murray, N.E., Quam, M.B., Wilder-Smith, A.: Epidemiology of dengue: past, present and future prospects. Clin. Epidemiol. **5**, 299–309 (2013)
24. Senior, K.: Climate change and infectious disease: a dangerous liaison? Lancet Infect. Dis. **8**, 92–93 (2008)
25. Medlock, J.M., Hansford, K.M., Schaffner, F., Versteirt, V., Hendrickx, G., Zeller, H., Van Bortel, W.: A review of the invasive mosquitoes in Europe: ecology, public health risks, and control options. Vector Borne Zoonotic Dis. **12**, 435–447 (2012)
26. European Centre for Disease Prevention and Control: The climatic suitability for dengue transmission in continental Europe, Technical report, 21 p., Stockholm (2012)
27. Luber, G., et al.: Human Health (Chap. 9). In: Melillo, J., Richmond, T., Yohe, G. (eds.) Climate Change Impacts in the United States: The Third National Climate Assessment. U.S. Global Change Research Program, United States of America, pp. 220–256 (2014)
28. Klempa, B.: Hantaviruses and climate change. Clin. Microbiol. Infect. **15**, 518–523 (2009)
29. Patz, J.A., Campbell-Lendrum, D., Holloway, T., Foley, J.A.: Impact of regional climate change on human health. Nature **438**, 310–317 (2005)
30. Kreppel, K.S., Caminade, C., Telfer, S., Rajerison, M., Rahalison, L., Morse, A., Baylis, M.: A non-stationary relationship between global climate phenomena and human plague incidence in Madagascar. PLoS Negl. Trop. Dis. **8**, e3155 (2014)
31. World Health Organization: Diarrhoeal disease, Fact sheet N°330. Date accessed: 07.08.2016, updated: 2013
32. Patz, J.A., Frumkin, H., Holloway, T., Vimont, D.J., Haines, A.: Climate change: challenges and opportunities for global health. JAMA **312**, 1565–1580 (2014)

Climate Change Impacts on Society and the Economy

Adaptation to Climate Change and Sustainability in Hungary

Mária Szalmáné Csete

1 Adaptation to Climate Change

Global climate change is making appearances with increasing frequency in the scientific and political arenas, and it is presented as one of the key problems of the coming decades. The 5th Assessment Report of the Intergovernmental Panel on Climate Change (IPCC), which is a synthesis of the most current research, states with more certainty than ever before that warming is accelerating, and there is a human factor involved [1]. In contrast, there is still no scientific consensus regarding the statement that warming is at least in part due to human activity [2]. However, most experts agree that, whatever the reason for global warming, it requires countermeasures and adaptation [3], for the continued survival of our species. There are three groups of possible solutions: limiting emissions, adaptation and increasing climate consciousness. The necessity of adaptation is underlined by the following:

The effects of climate change can already be felt [1]
Greenhouse gas (GHG) emissions are not decreasing, their concentration in the atmosphere is still growing [4]
Some of the impacts are now unavoidable [5]
Interventions are costly [5, 6]

Adaptation is the natural response of humanity to external stimuli, and it has been necessary for survival throughout history. This ability is the core of a sustainable society, however, there are considerable uncertainties as to what kind of changes we must prepare for and when these changes will occur. Uncertainty is coded into the nature of the problem and therefore cannot be eliminated completely, but its extent can be limited through research and the development of our forecasting systems. In

M. Szalmáné Csete (✉)
Department of Environmental Economics, Budapest University of Technology and Economics, Budapest, Hungary
e-mail: csete@eik.bme.hu

M. Palocz-Andresen et al. (eds.), *International Climate Protection*,
https://doi.org/10.1007/978-3-030-03816-8_35

addition, the presence of uncertainty cannot be used as a pretense for the lack of action if we are to minimize risks.

1.1 Climate Policy and Adaptation

The global, regional and local interventions intended to tackle the expected impacts of climate change were initiated in the 2000s. The international treaties were mostly focused on limiting emissions, with adaptation slowly gaining ground.

The European Union (EU) started dealing with climate change from the 1980s [7], at first through addressing emissions, adaptation only became a factor in the last decade [8]. The Green and White papers (of 2007 and 2009, respectively) demonstrated an increased focus on adaptation, and were followed up on by the "EU Strategy on adaptation to climate change" in 2013. This document, while not containing concrete interventions, defined the principles of adaptation and the potential directions for future actions [9]. In order to facilitate adaptation and provide access to relevant information, the European Commission (COM) created the Climate-ADAPT platform, which includes data, studies and good practices from countries and cities all over Europe.

The EU also supports local adaptation programs, having created the Covenant of Mayors for local decision makers to aid cities in their efforts to meet climate and energy targets set by the EU. In 2015, the name of the institution changed to the "Covenant of Mayors on Climate and Energy", adding new adaptation targets to the existing mitigation goals. New members must submit a SECAP (Sustainable Energy and Climate Action Plan) instead of the earlier SEAP (Sustainable Energy Action Plan), which includes the adaptation goals set by the city.

Hungary joined the effort for adaptation quite early. The joint research program of the Ministry for Environment and Water and the Hungarian Academy of Sciences, titled VAHAVA (Change—Impact—Response) was one of the first important steps. The program, initiated in 2003, aimed to assess the expected impacts of climate change and identify possible responses, synthesizing scientific inputs from various fields. The program laid the groundwork for climate policy and research in Hungary, while also attempting to bring the issue to the attention of the general public [10]. It is in part thanks to this work that the first National Climate Change Strategy (accepted in 2008) recognized adaptation as a key component in climate policy.

2 Hungarian Adaptation Policy in Practice

The mapping of adaptation practices in Hungary is as yet incomplete, making further analysis and classification problematic. National level initiatives are the most documented. As a member of the EU, Hungary is taking part in its climate adaptation initiatives, and has a climate change strategy (revised version expected to be accepted by Parliament soon at the time of this writing).

On the national level, adaptation analyses available in literature are mostly made for specific sectors, such as tourism [11, 12], transport [13] and energy management, especially in buildings [14].

Hungarian case studies with a regional approach can also be found [15–17]. Studies also exist on settlement level programs, calling attention to their shortcomings and potential for improvement. The common theme in available analyses is a tendency towards monitoring, in other words, the planning and evaluation of adaptation based on indicators [18].

Available data on settlement level practices is scarce. The number of municipalities active in this field can be somewhat roughly approximated through data on membership in organizations dealing with climate change. The two most important such organizations are the Association of Climate-Friendly Municipalities and the Hungarian Climate Alliance. Many municipalities also joined the Covenant of Mayors for Climate and Energy. However, the membership in these organizations does not give a complete picture on „climate-friendly" cities, since membership alone yields no concrete information on interventions that are being implemented, nor does it account for municipalities that are active in the area of climate change, but are not members in any of these organizations [19].

It is even more difficult to gather data on the activities of NGOs and local communities. There is no complete database on the organizations that are dealing with adaptation to climate change, or the activities these organizations perform.

In general, it can be stated adaptation initiatives in Hungary are mostly initiated by the government or local municipalities. These institutions command the necessary financial background and expertise, and can organize activities locally, enabling long-term, proactive planning. However, experiences show that the necessary knowledge and financial tools are only available to larger municipalities and cities. In smaller communities, dealing with climate change is possible if external financial frameworks and tenders are available, or if the commitment of local professional is sufficient to sustain an active community that is prepared to act without external financial support. The opportunities for local action on the part of municipalities may be further limited by the availability of human resources and know-how, making improved access to information crucial. This may be achieved by trainings, sharing information, or providing access to databases such as the National Adaptation Geo-information System (NAGiS).[1]

3 Adaptation and Sustainability on Local Level in Hungary

Social and economic processes have spatial and temporal dimensions, and in the implementation of sustainability in practice, this is compounded by the relationship

[1]More information about NAGiS: http://nater.mfgi.hu.

with the natural environment. Information on the basic characteristics, an understanding of the situation and attitudes is necessary for the stakeholders to recognize synergies, and for the devised coherent strategy to work.

The fields of climate change and sustainability in regional development are characterized by their spatial characteristics, a holistic approach, appearance in multiple dimensions (global, regional, local), and the need to account for relationships between the environment, society and the economy. Therefore, adaptation to climate change cannot be done in isolation, but rather in conjunction and harmony with the local initiatives and programs [20]. Local level documents for adaptation, sustainability and strategic planning have numerous synergies and point towards sustainable urban development. In Hungary, the tools that serve the practical implementation of both sustainability and adaptation include the Local Agenda 21 program, climate strategies and the Environmental Protection program.

Adaptation to climate change is a collaborative undertaking, requiring cooperation on different levels of society, politics and governance. At present, the adaptation strategy is part of the climate strategy, and its objective is to increase resilience to expected impacts, by shifting the emphasis from prevention and control to learning to live with an ever-changing, and sometimes dangerous environment [21].

Even though there is some overlap between the documents presented in Table 1, they cannot be substituted for each other. In the case of local documents, environmental protection programs are prepared pursuant to regulations. Voluntary programs are unlikely to include both a climate strategy (which contains a chapter on adaptation), and a Local Agenda 21 within a given municipality. Based on the available documentation on Local Agenda 21 programs, it seems that they follow the manual released in 2002, and are therefore mostly somewhat stereotyped. Experiences also indicate that these documents sometimes remain strategic in nature and may never be implemented in full, being used only to attain higher scores in tender processes.

Table 1 Local level strategic documents dealing with sustainability and adaptation to climate change in Hungary

	Local Agenda 21	Local climate strategy	Environmental protection program
Legal framework	Defined as optional	None	Act LIII of 1995, §46.
Pillars of sustainability	Economy, society, environment	Economy, society, environment	Environment
Top down vs. bottom up	Top down + Bottom up	Bottom up	Top down
Timeframe	Long term (3 generations)	Long-term, indefinite	Long-term
Mandatory vs. voluntary	Voluntary	Voluntary	Mandatory
Number of settlements that have this document	No data available	Members of the Association of Climate-Friendly Municipalities (33) + a few additional settlements	In theory, all Hungarian municipalities

[a]Total number of Hungarian settlements in 2016: 3,155 (346 town and 2,809 village)

The impacts of climate change hinder and complicate the transition toward sustainability in the preparation for expected impacts (human resources), prevention (increasing the site of green areas, action plans, stockpiling pharmaceuticals, etc.), management and recovery (after the impact has occurred, financial, technical, institutional, etc. conditions). Climate change can also have a fundamental influence on the local quality of life, income, health etc., which make up the basis for a livable city. The core of an adaptation strategy includes making sure that risks can be managed, their consequences minimized, providing a clear framework of responsibilities and providing the necessary financial and technical conditions.

In summary, the livable city is at the center of adaptation and sustainability. The objective of a local adaptation strategy is to provide the socio-economic bedrock for a city that is livable, prevents risks, minimizes damage and is flexible in reacting to climate change. This is achieved by devising an innovative strategic framework that supports the transition towards sustainability.

References

1. IPCC: Climate Change 2014: Synthesis Report. Contribution of Working Groups I, II and III to the Fifth Assessment Report of the Intergovernmental Panel on Climate Change [Core Writing Team, Pachauri, R.K., Meyer, L.A. (eds.)]. IPCC, Geneva, Switzerland, 151 p., 2014
2. Cook, J., Oreskes, N., Doran, P.T., Anderegg, W.R.L., Verheggen, B., Maibach, E.W., Carlton, J.S., Lewandowsky, S., Skuce, A.G., Green, S.A., Nuccitelli, D., Jacobs, P., Richardson, M., Winkler, B., Painting, R., Rice, K.: Consensus on consensus: a synthesis of consensus estimates on human caused global warming. Environ. Res. Lett. **11** (2016)
3. IPCC: Summary for policymakers. In: Climate Change 2014: Impacts, Adaptation, and Vulnerability. Part A: Global and Sectoral Aspects. Contribution of Working Group II to the Fifth Assessment Report of the Intergovernmental Panel on Climate Change (2014)
4. EEA: Atmospheric greenhouse gas concentrations. http://www.eea.europa.eu/data-andmaps/indicators/atmospheric-greenhouse-gas-concentrations-4/assessment (downloaded: 31/05/2016) (2015)
5. ACT: Planning for adaptation to climate change. Guidelines for Municipalities. Adapting to Climate Change in Time. Life Project No LIFE08 ENV/IT/000436, 222 p. http://base-adaptation.eu/sites/default/files/306-guidelinesversionefinale20.pdf (2013)
6. Stern, N.H.: The Economics of Climate Change: The Stern Review. Cambridge University Press, Cambridge (2007)
7. Wurzel, R., & Connelly, J. (eds.): *The European Union as a leader in international climate change politics*. Routledge, 2010
8. Faragó, T.: Nemzetközi klímapolitikai együttműködés. Magyarország részvétele és feladatai. GROTIUS E-KÖNYVTÁR/ 59 2013. http://www.grotius.hu/doc/pub/QZLCSC/2013-06-14_farago_tibor_grotius-ekonyvtar-59.pdf (2013)
9. COM(2013)216: An EU strategy on adaptation to climate change. http://ec.europa.eu/transparency/regdoc/rep/1/2013/EN/1-2013-216-EN-F1-1.Pdf (2013)
10. Faragó, T., Láng, I., Csete, L. (eds.): Climate Change and Hungary: Mitigating the hazard and preparing for the impacts. The VAHAVA Report. Budapest, 124 p. (2010)
11. Csete, M., Szécsi, N.: The role of tourism management in adaptation to climate change—a study of a European inland area with diversified tourism supply. J. Sustain. Tour. **23**(3), 477–496 (2015)
12. Csete, M., Pálvölgyi, T., Szendrő, G.: Assessment of climate change vulnerability of tourism in Hungary. Reg. Environ. Change **13**(1), 1043–1057 (2013)

13. Szendrő, G., Csete, M., Török, Á.: The sectoral adaptive capacity index of hungarian road transport. Priodica Polytech.-Soc. Manag. Sci. **22**(2), 99–106 (2014)
14. Hrabovszky-Horváth, S., Pálvölgyi, T., Csoknyai, T., Talamon, A.: Generalized residential building typology for urban climate change mitigation and adaptation strategies: the case of Hungary. Energy Build. **62**, 475–485 (2013)
15. Csete, M.: Adaptation to climate change related to sustainability at Lake Tisza Hungary. In: IOP Conference Series: Earth and Environmental Science, vol. 6: Paper 392016(1) (2009)
16. Schmidt-Thome, P., Greiving, S. (eds.): European Climate Vulnerabilites and Adaptation: A Spatial Planning Perspective. In: Case study: Climate Change Impacts on the Hungarian, Romanian and Slovak Territories of the Tisza Catchment Area. Oxford: Wiley-Blackwell, 330 p. (2013)
17. Csete, M., Horváth, L.: Sustainability and green development in urban policies and strategies. Appl. Ecol. Environ. Res. **10**(2), 185–194 (2002)
18. Szalmáné Csete, M., Buzási, A.: Managing local adaptation processes in Hungary. Int. J. Manag. Cases (IJMC) **18**(1), 13–22 (2016)
19. Csete, M.: Klímaváltozás és a települések fenntarthatósága. Klíma-21 Füzetek 5, pp. 71–88 (2007)
20. Takács-Sánta, A.: A települési klímaprogramok nemzetközi tapasztalatai – tanulságok a hazai intézkedésekhez. Klíma-21 füzetek, 54. szám, pp. 22–36 (2008)
21. Wamsler, C., Brink, E., Rivera, C.: Planning for climate change in urban areas: from theory to practice. J. Clean. Prod. **50**, 68–81 (2013). https://doi.org/10.1016/j.jclepro.2012.12.008

How Can CO_2 Emission Be Reduced During Food Production?

Éva Erdélyi and Daniella Boksai

1 Cutting Greenhouse Gas Emission, Mitigation Strategy

At the United Nations Framework Convention on Climate Change Conference in Cancun, in November 2010, the Heads of State reached an agreement with the aim of limiting the rise in global temperature to 2 °C relative to preindustrial levels. Earth overshoot day, when we began to use more from nature than our planet can renew in the whole year is getting earlier every year (e.g. 8th August in 2016). It is recognized that long-term future warming is primarily constrained by cumulative anthropogenic greenhouse gas emission that deep cuts in global emission are required, and that action based on equity must be taken to meet this objective. It is necessary to fill the gap with more ambitious mitigation efforts.

There are different environmental indicators, footprints measuring the impacts of human activities on the environment. *Ecological footprint* measures the quantity of biologically productive land and water required to both provide the resources consumed and absorb the waste produced by population or by a single human activity. *Water footprint* measures water use in terms of volume of evaporated and/or pollution water for the entire supply chain per time unit throughout the life cycle. *Carbon footprint* is representing GHG emission generated by the human activities throughout the life cycle, represented in terms of tones of equivalent CO_2. Carbon footprint is used to describe the amount of greenhouse gas (GHG) emissions caused by a particular activity or entity and thus a way for organizations and individuals to assess their contribution to climate change [1].

In this study we were interested in the effects of food production on the environment, so we focus on measuring the CO_2 emission during the life cycle of food

É. Erdélyi (✉)
Budapest Business School, Budapest, Hungary
e-mail: SzaboneErdelyi@uni-bge.hu

D. Boksai
Szent István University, Gödöllő, Hungary

M. Palocz-Andresen et al. (eds.), *International Climate Protection*,
https://doi.org/10.1007/978-3-030-03816-8_36

production processes. Hertwich and Peters [2] quantified GHG emissions associated with the final consumption of goods and services for 73 nations and 14 aggregate world regions, and analysed the contribution of 8 categories. Based on that it is concluded that on the global level 72% of GHG emissions are related to household consumption, food accounts for about 23.5% in average, which is high. Based on that calculation we showed that the average footprint expressed in tons of CO_2 equivalent per year per person is 9,404 [tCO_2e/p] in average, with big variability (CV = 0.78) from min. 0.7 to max. 33.8. The greatest impact products are food, drink, private transport and housing. Impacts of food and services are more significant in developing countries, while mobility and manufactured goods share rise fast with income, and dominate in rich countries. Contribution of food varies from 7 to 67%.

One of achievements in combating climate change is enhancing management systems to reduce carbon emissions and promoting carbon footprint labelling systems for products. The concept of carbon footprint also captures the interest of business, consumers and policy makers. Supplier engagement should be built into the overall project work plan, with roles, responsibilities and is necessary for understanding the product's life cycle and for gathering data.

2 Product Carbon Footprint

Research of the effects of production on the environment is a very complex challenge. In our case study we were looking for what kind of aspects and what types of data are needed for each stage of the methodology to calculate carbon footprint. Carbon dioxide emission of production of food products can be recognized by tracing by calculation from step to step of the products life cycle. The term 'product carbon footprint' refers to the GHG emissions of a product from raw materials through production (or service provision), distribution, consumer use and disposal/recycling. During our research we worked with partners from industry, calculating carbon footprint of some representatives of their product groups, using the standard of PAS 2070 (Assessing the life cycle greenhouse gas emissions of goods and services).

The calculation of carbon footprint has a huge data requirement. Building a process map as a first concrete stage is also a serious procedure. The goal of this stage is to identify all materials, activities and processes that contribute to the chosen product's life cycle. It is necessary to be in contact with the suppliers, to understand the product's life cycle and for gathering data. Supplier engagement should be built into the overall project work plan, with roles, responsibilities and milestones clearly defined and understood. The process map includes all stages and potential emission sources from any activity that also contributes to the delivery or use of the product. Carefully checking boundaries,—what methodology doesn't include (e.g. immaterial emissions sources)—is also very important.

2.1 Methodology of Calculating Carbon Footprint

A British organization called Carbon Trust is a company set up by the British government in 2001 to help UK businesses lower carbon emissions and reduce energy costs. It is taking stages to help consumers better understand the carbon footprint created by their food. Their carbon footprint label clearly marked with the amount of grams of CO_2 created by the product, measures emissions of a product from source to shelf. The carbon footprint is added up from different sources. All the stages of life cycle should be analysed. The Publicly Available Specification (PAS 2050, new version is PAS 2070) commenced in June 2007 at the request of Defra (Department for Environment, Food and Rural Affairs) and the Carbon Trust. It contains BSI (British Standards Institution) Standards Solutions meeting method for measuring the embodied GHG emissions from goods and services and is used as a basis of product carbon footprint calculations, whose most important stages are shown next.

A food product carbon footprint could be gained through analysing the products life cycle from step to step. Collecting data is the most difficult part of the process and needs very important cooperation with producers. It depends on interviews and focuses on the most significant inputs first and identifies their respective inputs, manufacturing processes, storage conditions and transport requirements. The quantification of the total amount of all materials into and out of a process is referred to as 'mass balance'. The mass balance stage provides confirmation that all materials have been fully accounted for and no streams are missing. The equation of the product carbon footprint value is the sum of all materials, energy and waste across all activities in a product's life cycle multiplied by their emission factors. The calculation itself simply involves multiplying the activity data by the appropriate emission factors. Next we introduce the steps of calculations.

Building a process map: The goal of this stage is to identify all materials, activities and processes that contribute to the chosen product's life cycle; includes all stages and potential emission sources from any activity that contributes from source to shelf. Developing a product process map starts by breaking down the selected product to its functional units and focusing on the most significant inputs first, then identifying their respective inputs, manufacturing processes, storage conditions and transport requirements (Table 1, based on PAS 2070). Checking boundaries is an important stage and means that the methodology does not include sources of immaterial emissions (which represent less than 1% of total footprint), human inputs to processes, transport of consumers to retail outlets and animals providing transport.

Data types and collecting data: Two types of data are necessary for calculating the carbon footprint: activity data and emission factors. Activity data refers to all the material and energy amounts involved in the product's life cycle (material inputs and outputs, used energy, transport, etc.). Emission factors provide the link that converts these quantities into the resulting GHG emissions: the amount of greenhouse gases emitted per 'unit' of activity data (e.g. kg GHGs per kg input or per kWh energy used). Activity data and emissions factors can come from either primary or secondary

Table 1 Process map steps for products, 'business-to-consumer'

Raw materials	*Manufacture*	*Distribution/retail*	*Consumer use*	*Disposal/recycling*
All inputs used at any stage in the life cycle. Includes processes related to raw materials.	All activities from collection of raw materials to distribution and all materials produced.	All steps for transport and related storage, retail storage and display.	Energy required during the use phase. .	All steps in disposal: transport, storage, processing; energy required in this process and direct emissions due to it.

sources: primary data refers to direct measurements made internally or by someone else in the supply chain about the specific life cycle of a product. Secondary data refers to external measurements that are not specific to the product, but rather represent an average or general measurement of similar processes or materials (e.g. industry reports or aggregated data from a trade association). Data could be obtained from an emission factor database and through interviews with suppliers.

Mass balance and calculating carbon footprint: The quantification of the total amount of all materials into and out of a process is referred to as 'mass balance'. The mass balance stage provides confirmation that all materials have been fully accounted for and no streams are missing. The equation for product carbon footprint calculation is the sum of all materials, energy and waste (activity data) across all activities in a product's life cycle multiplied by their emission factors, as follows.

3 Case Study for Calculating, and Decreasing CO_2 Emission

We show an example of a food product, e.g. a cereal product named Bulata of Biopont Company with a simple life cycle, which was the first case study of our research. The first stage of calculations is taking farming emissions into consideration (as the production of raw materials, estimated). Second stage of the calculation is based on the transportation of raw materials to the location of production (in the case of this cereal product, mill) and then to the place of manufacture. Packaging is a very important element of product carbon footprint calculation, as well. The "packaging" stage also includes stages of the production (packaging material) and transport (to the packaging location). The next stage of the products life cycle is transportation to storage centre and stores. At the end, we have to take into consideration the emission factors of retail, disposal and landfill decomposition. The result od calculation is given in Table 2. Based on our calculations the carbon footprint of a 200 g package

Table 2 Footprint analysis of Bulata

Raw materials	Farming	Transport	Technology	Retail	Disposal	Total
wheat malt Bulata packaging material	500 (estimated)	4 5.6 4.57 84.199	40 (milling) 35 (malting) 25 (extruding) 53 (packaging)	225	0.0015 0.0223	
Total	500	98.369	153	225	0.0238	976.398

Source Own calculations, given in kilograms per CO_2 emission per tonne product

of this snack product is 0.195 kg CO_2. Compared to a potato chips snack product, it is only the third of its emission.

Quantifying the carbon emission sources will help to understand what impact a company, a product or an organization has on climate change. It helps manage the carbon emissions and make reductions over time, furthermore it helps find and identify areas for reducing emissions, which will often result in cost savings, as well. Transportation is a step of the productions life cycle which has a significant carbon footprint, but could be reduced [3]. It is present in the products life cycle in more steps like seen on the "Map" of the Bulata Products life cycle on Fig. 1.

If we can get closer the production place of the packaging material (Pírtó) it helps reduce the emission, but also if we think of transport with no empty journey on the way back. Calculations show that in that case the emission of transport can be reduced half, which means in this case from 10% of the emission to 5%. Another possible strategy for reducing the carbon footprint is a well-managed technology, retail or farming. Benefit of growing locally and sustainable thinking is very important in reducing emissions, as well.

4 Conclusion and Discussion

In this study we introduce the idea of calculating the carbon footprint of food products and the comparison of the same food with different histories. This thinking and results could help producers and suppliers and also consumers in their environment friendly thinking and decision making. According to a Carbon Trusts study two thirds of consumers are more likely to buy a product if action is being taken to reduce its carbon footprint.

The calculation of the carbon footprint of a product could help companies to understand how products and supply chains are responsible for carbon emissions and help companies identify the most effective ways of reducing it. Showing the possibilities we intend to encourage them to make an effort to minimize the rate of damage to the environment. Labelling could give customers access to better information about the potential impact of climate change of every product they buy and hopefully initiate change as a result, providing benefits for the environment and our

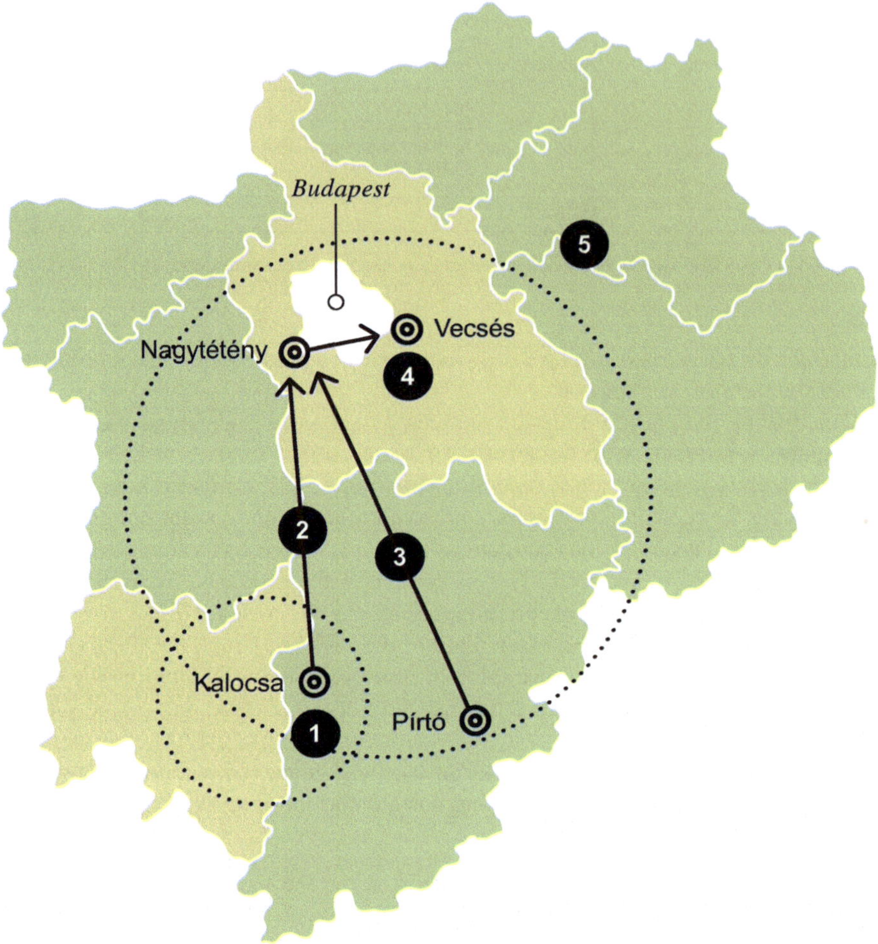

Fig. 1 "Map" of the Bulata Products life cycle: to the production place (1), from the production place to the packaging location (2), from the production place of the packaging material to the packaging place (3), from the packaging place to the storage place (4), to the stores (5)

future living. Comprehensive adoption of financial tools will play an essential role in promoting the development of a low-carbon economy by boosting green investment, increasing employment, saving energy and reducing carbon emissions.

References

1. Schaltegger, S., Csutora, M.: Carbon accounting for sustainability and management. Status quo and challenges. J. Clean. Prod. 1–16 (2012)

2. Hertwich, E.G., Peters, G.P.: Carbon footprint of nations: a global, trade-linked analysis. Environ. Sci. Technol. **43**(16), 6614–6620 (2009)
3. Erdélyi, É., Boksai, D.: Steps towards sustainable food production, a case study for transport, 1. Transilvanian Horticulture and Landscape Studies Conference, Tirgu Mures-Marosvásárhely, p. 24 (2011)
4. PAS 2070: Assessing the life cycle greenhouse gas emissions of goods and services, http://shop.bsigroup.com/en/Browse-by-Sector/Energy–Utilities/PAS-2070/

Part V
Climate and Environmental Protection in the Education and Communication

Deciphering Change in the Alaskan Landscape

Katie Ione Craney

1 Introduction

How do Alaskans' plan for uncertainty when the very land they rely on is changing underneath their feet, before their eyes, and at a rate they may not be able to keep up with? Like art, living in the north provides an essential feedback loop in being human. The land reminds us to pay attention, proceed with caution, check conditions, and plan accordingly. Sometimes we are lucky and narrowly escape catastrophe, other times the land consumes us. The land is still coming out of a geologically recent ice age; evidence is everywhere. In the maritime climate of Southeast Alaska, water—both liquid and solid—has forced its way with the help of gravity to shape bedrock, tear down mountainsides, and carve deep fjords. Earthquakes, landslides, tsunamis, volcanic eruptions, glacial advance, retreat and rebound, have all shown that the earth is very much alive and continues to be reborn. Many of these events go untamed, unmanaged, and unnoticed by most humans since Alaska is so large that many events happen without significant threat to human population centers.

For centuries, the naturally occurring forces of this land have woven together with human habitation. However, the overall population of Alaska has remained low; an estimated 740,000 [1] people live in the largest, most sparsely populated state in the United States. There are roughly 600 acres per person, though more than half of Alaska's population lives in Anchorage, the states largest city.

The commotion of this place has channeled where and how people live, travel, and ultimately survive. With a changing climate, this already dynamic landscape is directly experiencing human-caused climate change at a quickened pace. Alaskans' are watching a once frozen landscape melt and change at an alarming rate without fully understanding the implications of how warmer temperatures will shift current ecosystems and cultural understandings of place.

K. I. Craney (✉)
Haines, AK, USA
e-mail: katieionecraney@gmail.com

M. Palocz-Andresen et al. (eds.), *International Climate Protection*,
https://doi.org/10.1007/978-3-030-03816-8_37

2 Deciphering Change Through Art

As an artist living in Alaska, how I visually decipher what is happening around me is ultimately determined by my immediate surroundings. Living roughly 550 miles south of the Arctic Circle, my home in Southeast Alaska faces similar climatic challenges as those in the far north. As humans, our environment psychologically affects us, whether we are conscious of it or not. The environment we surround ourselves with affects our mental and physical well-being [2]. Environment defined in these terms is broad and can mean anywhere: our homes, a city street, a park, or a designated wilderness area. We are often aware of what brings comfort, make us happy or sad, and what sparks certain memories by the items we choose to surround ourselves with. What about the land we surround ourselves with? What of the visual depictions of the land we are surrounded by? How can art flip the conversation to better understand how human behavior affects the natural world?

Art is vital to helping us experience and reveal our inner perception of massive environmental change. It uncovers how we balance the concept of global change in our deepest emotions and inner world, similar in style to the 19th century landscape romantics who challenged the perception of human dominance of the land, towards humans living with, and in awe of, a landscape.

3 The Art of Observing

The Alaskan landscape surrounding my home plays an essential role in my work. I rely on the land for its tangible qualities and pay close attention to its nuances. Taking the time to look closely at the land allows for greater interpretation of what's at stake, as well as inspiring further investigation and action to preserve it. The naturally occurring forces of this place remind me, as an art maker, to pay close attention, as if the land is speaking and I need to pause long enough to hear the story. Through the art making process, I can take a landscape and break it down like an ecologist would, looking at different layers, stories, and interconnections that create the specific place. This is how I approach much of my research and work including, *Grief Dares Us*, where I experimented with adding and subtracting materials, and eventually deconstructed the original piece. What remained became the substrate to layer the existing materials. The ink drawing on tissue paper is layered with encaustic wax as a way to seal the drawing to the metal while providing tension between malleable and rigid materials. In this paradox of hard and soft is where I find the pliability needed to live fully in the North (Fig. 1).

One example of the paradox of living in the north is the data produced by climate change studies; it has been presented time and time again. In Alaska, we hear the news on our local public radio station, see images from across the state; we see change happening right outside our doors. We talk about thinning Arctic sea ice, melting permafrost, tidewater glaciers grounding, earlier springs, salmon fisheries closing,

Fig. 1 *Grief Dares Us.* Ink, tissue, gauze, and encaustic hot wax on hand-cut scrap metal. (K. I. Craney, 2016)

massive common murre die-off, the large mass of warm water, known as "the blob," in the North Pacific-ocean [3], how yellow-cedar is experiencing climate change-induced dieback [4]; the list goes on. For many, these climate warnings may be heard, but not necessarily listened to [5]. Whether or not Alaskan's are listening, it can be argued that we are physically experiencing change, whether we agree on how or what is causing the change, and are left to emotionally deconstruct the multidimensional phenomena that is happening right outside our front door.

Climate change, for many, may seem too big, remote, or not of immediate personal concern [6]. Columnist for *The Nation* and *The Guardian*, Naomi Klein states that cognitive dissonance is a part of being alive in this moment in history that we look but then look away [7]. If we are not looking, are we turning our back to the climate crisis? There is an argument for visual saturation and being overrun with images of doom, disappearance, and loss. As an artist, these are questions I face in my research

Fig. 2 *You Are a Tender History of Ice*, mixed media on 50 metal plates. (K. I. Craney, 2016)

and process of how to communicate an inner loss of landscape. We can watch the land change, but how it changes us strikes a deeper empathy towards place and being human.

One way I choose to communicate this change and loss is through materials and application, as a majority of my work is on found, hand-cut scrap metal. The juxtaposition between the ephemeral quality of my subject matter and the longevity of human-made metal offer a physical layer to the conceptual loss I'm interested in. For similar reasons, I used repetition of size, color palette, and materials. In the series, *You Are a Tender History of Ice*, (Fig. 2), I created 50 pieces, all 2.5 by 3 in., as a way to tell individual stories that all, inevitably, have a similar outcome.

How does ice, and the loss of it, physically and emotionally affect us? In *Feedback Loop*, (Fig. 3), I researched ice-albedo feedback and the consequences of surface color and reflectivity, particularly how the necessary reflectiveness of ice, its albedo, not only affects pack ice—ice in the Arctic that remains year-round—but everyday life for millions of people around the world. As ice melts, the surrounding water absorbs more sunlight, rather than reflecting it back into space, therefore increasing warming in the ocean, which continues the cycle of melt and absorption of solar energy. The overall albedo of the Arctic decreases due to more open water than ice. This is known as a positive feedback loop [8]. To visually transcribe this loop into art, I chose to work with silver leaf on metal to mimic ice in the form of a silhouette of two hemlock trees outside my front door. The contrasting reflection draws in outside influences of the surrounding environment and can change the composition entirely. The viewer also becomes part of the composition and a part of story through this reflection. By deliberately including the surrounding environment and whoever

Fig. 3 *Feedback Loop*. Silver leaf on hand-cut scrap metal. (K. I. Craney, 2016)

views the work, the viewer becomes an active, rather than passive participant in the story.

4 To See Is to Understand

Seeing establishes our place in the surrounding world and at the same time unsettles it in ways words used to explain it never resolve. A child sees before learning to speak and can therefore relate to the surrounding environment. The late John Berger, art critic, artist, and author, argued that to see is to understand, engage with, and critically assess the work [9]. Contemporary art can be defined in this way, to have the viewer be an active participant in the work rather than a passive observer. As a participant with the art, viewers have the ability to engage in immediate discussion about the work. Contemporary artists can use this communication tool to their advantage to discuss the complexities of climate, and ask viewers of their work to take a closer look not only at the art, but also at the situation that informed the artist and art in the first place, in order to make sense of the larger picture. However, art making can only take us so far and is a small piece in the puzzle of understanding, adapting to, and taking action towards the myriad of issues we are confronted within this physically changing natural and cultural landscape.

Artists who explore the changing landscape provide us with a visual homage of what it is like to be human in the 21st century [10]. By constructing narratives on species loss, glacier melt, temperature change, and cultural adaptations, artists have a unique ability to discuss what is difficult for many to imagine; turning the invisible into visible truth. Art works as a metaphor for the challenges humans are up against. However, art is not a solution to the troublesome problem, rather, it offers a type of solace in a world wrought with unpredictability. Art offers viewers a chance to reflect on the world as they know it and see a different perspective through an artist's interpretation.

5 Conclusion

How artists define, interpret, and represent the current landscape is a way to catalogue and share current and future visual perspectives. Telling the story of the place I live, the challenges Alaska faces, and the outcomes ahead, is only a tiny thread in the larger weave told through art, visual communication, and discourse. My work is a reflection on loss, memory, rebound, and change, while also a window into this particular place. Sharing through art is only one way to catalogue a single perspective of what it is like to observe, partcipate, and engage in a time of massive human-caused change.

Acknowledgements Sections of this chapter have been previously published in *artArctica*, a Finland based art dialog and blog for artists based in, or inspired by, the Arctic, www.artArctica.com; and in an interview with the *United Nations Framework Convention of Climate Change* in partnership with *Julie's Bicycle*, for an #Art4Climate initiative, www.unfccc.int.

References

1. State of Alaska Department of Labor and Workforce Development, Research and Analysis, Juneau, AK (2017). http://live.laborstats.alaska.gov/pop/
2. Wells, N.: How Natural and Built Environments Impact Human Health. Department of Design & Environmental Analysis, Outreach & Extension, Cornell University College of Human Ecology (2017)
3. Rosen, Y.: Alaska Murre Die-Off Led to Reproductive Failure for Survivors. Alaska Dispatch News, May 2017
4. Oakes, L.E., et al.: Conservation in a social-ecological system experiencing climate-induces tree mortality. Biol. Conserv. **192**, 276–285 (2015)
5. Linden, E.: Climate warnings: heard, but not listened to. Yale Climate Connections (2015). https://www.yaleclimateconnections.org/2015/05/climate-warnings-heard-but-not-listened-to/

6. Nicholson-Cole, S.A.: Representing climate change futures: a critique on the use of images for visual communication. Futurescapes **29**(3), 255–273 (2005)
7. Klein, Naomi: This Changes Everything: Capitalism vs The Climate, p. 3. Simon & Schuster Paperbacks, New York (2014)
8. Collier, M.: The Melting Edge: Alaska at the Frontier of Climate Change, p. 15. Alaska Geographic Association (2011)
9. Berger, J., Dibb, M.: Ways of seeing. BBC Enterprises, London (1972)
10. Bunting, M.: The rise of climate-change art. The Guardian (2009). https://www.theguardian.com/artanddesign/2009/dec/02/climate-change-art-earth-rethink

Eco-Themes and Climate Change in Literature

Gábor Tüskés

If we wish to understand the true substance of nature, there is probably no better guide than Humboldt's *Views of Nature* [2]. The collection is written in classic German, and consists of six studies, which have been supplemented by a narrative since the second edition. Humboldt is not concerned with simply presenting factual information, but rather seeks to shed light on the inner workings of the forces of nature and the harmonious cooperation between them, and to put forward a comprehensive concept, the essence of which is the unison between the material world and man's ethical-aesthetic stance. During his journey in America prior to writing his book, Humboldt did not restrict himself to merely studying one discipline, but wanted to gain an overview of the whole of nature. His approach is that of a genuinely curious person who believes in morphological examination; someone who views nature as a single entity, as opposed to the isolating tendencies prevailing in modern scientific research.

One piece of the collection, entitled *Life-force, or the Genius of Rhodes*, Humboldt's only philosophical-allegorical treatise, is of special interest to us [3]. The point of origin is the vitalist teaching on life-forces, which later Humboldt rejected as a consequence of having studied chemistry and physiology; he never, however, completely abandoned the idea about the connections between the organic and inorganic conditions of matter [4]. This organic philosophy is also apparent in his late work, *Cosmos* [5].

Humboldt focuses here on the complexities of the whole, and lays great emphasis on presenting scientific data in a linguistically flawless manner. He begins by describing the universe and ends with depicting mankind. His holistic perspective is apparent from the fact that scientific and historical reasoning follow each other in the chapter discussing human imagination and the effect of nature on man's state of mind. "Contemplating nature," remarks Humboldt, "should be universal, grand and free, unrestricted by considerations of perceived usefulness." The purpose of this is for any image of nature to represent the harmony of nature as a whole, and to be

G. Tüskés (✉)
Magyar Tudományos Akadémia, Budapest, Hungary
e-mail: xviii@iti.mta.hu

M. Palocz-Andresen et al. (eds.), *International Climate Protection*,
https://doi.org/10.1007/978-3-030-03816-8_38

similar to a work of art with regard to the effect it has on the observer. This concept of nature being a living entity has been increasingly dismissed by modern thinking, which is primarily interested in the usefulness of various natural phenomena.

Humboldt's holistic understanding of nature is closely related to Goethe's approach, since both of them were interested in the renewable forces that drive natural processes [6]. According to Humboldt, one such factor is climate, an issue he treats with clearly interest in his works. He not only established close connections between natural phenomena and their order through phytogeography and mineralogy, but also through climatology. During his expedition to America, besides carrying out astronomical and geophysical measurements, he did comprehensive climatological research and recorded his observations about the characteristics of climate. He was one of the first to distinguish between continental and maritime climate, and presented an overview of how plants spread and interact with their environment in relation to changes in temperature.

According to the preface to the second and third edition of *Views of Nature*, his expedition commissioned by the tsar into the Asian territories of Russia, which he embarked on in 1829, had significantly altered—inter alia—people's perception of how climate affected natural processes. The scientific assessment of the journey entitled *Central Asia—Research into mountain ranges and comparative climatology* first appeared in 1843 in French (*Asie Centrale—Recherches sur les chaînes de montagnes et la climatologie comparée*), and a year later in German. The latter was translated by the physician Wilhelm Mahlmann, whose own related findings were included at the end of the second volume in the four charts on temperatures across the two hemispheres [7].

Humboldt systematically studied climate in relation to other natural phenomena. In his methodology he trusted only direct observation and deduction based on factual findings. His research into climatology primarily involved dealing with temperature and the moisture content of the air, while also comparing the climate of Asia, Europe and America and discussing the reason behind isotherm curves. The supplement to the part on climatology contains several observations on the temperature of soil, springs, mines and rivers, together with the issue of underground ice. Humboldt carried out painstakingly thorough studies into the difference in climate between the eastern and western slopes of the Alleghany Range, and likewise explored the effect of deforestation and the expansion of human civilisation on the annual, winter and summer average temperatures. This book has been recognised ever since as one of the fundamental works of comparative climatology.

It is not by chance that the first criticism of the damage inflicted on nature by man first appeared in American literature, and preceded any ecological movement by several decades. One of the first occurrences and classic examples is Herman Melville's *Moby Dick*, which appeared in 1851, eight years after Humboldt's previously mentioned work. This allegorical novel is a treasure trove of various narratives, traditions, mythological, biographical and other motifs, which all reveal the process by which the United States, originally an agrarian state, became an industrialised country through the exploitation of nature. The novel is a polyphony of styles, themes and ideals, much more complex than for us to be able to reduce it to a single underlying idea. However, the majority of interpretations agree that the hunt for the white whale as a metaphor is a large-scale criticism of rational thinking and the change in attitude towards nature [8]. The whale-hunters regard their prey not as a living creature that is a part of its environment but merely as raw material. The hunt, during which all natural, technical and human resources are utilised, symbolises how economic considerations, the desire to discover and the basic concept of conquering are interwoven with each other, thus becoming the vehicle and driving force behind early globalisation.

The real protagonist of the novel is the white whale, which represents the beauty, unpredictability and might of not only the sea but of nature as well—something against which man can measure his own limits. The downfall of its rival, Captain Ahab, who had sworn to kill the whale, and his crew ultimately depicts the fate of mankind and the victory of nature over human folly. While Melville mocks the superficial positivism of the 19th century and criticises the American concepts of acquiring as much as possible and the supremacy of utilitarian rationalism, he also gives an insight into hitherto unknown realms of nature and human conscience. He synthesises the experience of everyday life with the historical teachings of over two thousand years, and at the same time realises that interfering with the order of nature to such an extent will have unforeseeable consequences for humanity.

Poetic interest in nature in the 20th century is primarily determined by the awareness of man belonging to nature yet growing ever more detached from it. Joseph Conrad's short story, *The Heart of Darkness* (1899) presents the story of a trade company's expedition in Central Africa. The various stages of their journey appear as the lost watch-posts of a civilisation led by the "corrupt missionaries of development." Kurtz, the ivory merchant, ruthlessly exploits those living under his command, but eventually ends up falling victim to the wild. The symbolic depiction of the rain forest emphasises the metaphoric nature of the work. The main idea of *The Waste Land* (1922), the famous cycle by T. S. Eliot, is the futility of our modern secular world and the need to interpret it. The barren place, as a symbolic landscape, carries a double meaning in the fifth part: on the one hand, we live in a world deserted by God, yet we must cross the waste land to reach salvation. Eliot employs the "unreal" city enshrouded in fog as the symbol of modern life.

A recent ecocritical interpretation by Geoff Berry of the symbology of light in these two classic texts of the early 20th century uncovered the anxieties they reveal about the way the symbol of light is co-opted on behalf of the modernising project of colonising nature and its technologies (Berry s. a.). Colonisation by Western civilisation is inseparable from the aim of mastering the earth, thereby creating ecological devastation in its wake. Conrad and Eliot, who are aware of the gloomy consequences of this dream, recognise the limits of such an agenda of mastery, and experiment with alternative concepts. As Berry points out, both authors "recognise the disconnectedness of a stance that defines the earth as dead or inert matter, which could be controlled and transformed by technologically advanced cultures, in turn eroding the possibilities of a deeper relationship with the natural world for the modernised urbanite. Furthermore, they both see that we suffer from our alienation from nature, [...] and that this alienation will ultimately prove to be self-[...] defeating."

A similar, though less familiar example of ecocriticism is Alfred Döblin's novel entitled *Mountains, Seas and Giants* (1924). The author's original purpose was to set the natural-mystical tendencies of his earlier novels against the successful U-topia of a technologically developed world. Although the novel does present, for the first time in German, a significant vision about the future of a world governed by science and technology, it conveys a hopeless perspective. The novel revolves around the inability of man to control nature, whereas he can survive only by consciously establishing a harmonious relationship. According to the novel, the continuous development of technology will result in huge overproduction by the 24th century, which will lead to the disintegration of states and extreme urbanisation of the population. With the invention of synthetic food man's alienation from nature will become complete, a result of which will be the appearance of gigantic humans and mutant combinations of man and plants. Mankind will tear itself apart, and only self-sufficient "settlers" will have any chance of survival. In 1932 Döblin published the second, significantly shortened edition of his work, in which he changed his original concept, and instead of nature takes the side of the degenerate giants. The purpose of this latter version is to present the "historical state of modern humanity", which, as it "wishes to encrust itself within the world of machines is choked by the crust, and can only break the crust with great difficulty."

Criticism of technological development at the cost of destroying nature also gained momentum in post 1945 German poetry. For example, the two most distinct features of Ingeborg Bachmann's poetic oeuvre are the connection between nature and history, and the portrayal of the devastation wreaked by mankind upon nature through images and metaphors of a severely damaged environment. It is also images of nature that define much of Günter Eich's poetry. However, while in his early poems nature serves as the sphere of magical experience, his post-world war poems depict a ravaged environment. Questioning development and criticising blind faith in technology are the main motifs of Hans Magnus Enzensberger's poem *The Sinking of the Titanic* (1978). Warning against the exploitation of nature in the age of technology is a recurring theme of Günter Kunert's poetry, apparent in the volumes *Remembering a Planet* (1963) and *The Innocence of Nature* (1966), and elsewhere.

From the 1960s onwards discourse on ecology in America was inspired by works such as Aldous Huxley's *Island* (1962) and Ernest Callenbach's *Ecotopia* (1975). In the latter the author connects the Utopian-ecological system of "stable balance" with the proposal for a welfare state. Inhabitants of Ecotopia lead their lives in accordance with knowledge of natural processes grounded in science and the theoretically attainable stability of the ecosystem. Nature is a closed, self-regulating system, the stability of which is maintained by recycling processes. As opposed to the standardising approaches of earlier utopias, Callenbach argues in favour of maintaining the diversity of lifestyles, ethics and traditions. Economic and social structures are determined by the principle of living in balance with nature. Early examples of ecofeminist Utopias include Marge Piercy's *Woman on the Edge of Time* (1976) and Sally Miller Gearhart's *The Wanderground* (1979). In the 1960s and 1970s Archie Randolph Ammons stressed the importance of a poetic dialogue with nature and the need to relinquish control over the environment. The fundamental message of

Edward Abbey's utopian novel, *Good News* (1980) is that even ecology cannot save the Western world from total destruction [9].

The issue of climate change and global warming, their possible effects and related human-environment interaction entered English-American literature in the late 20th century after sporadic preliminaries; (these include e.g. Jules Verne: *The Purchase of the North Pole*, 1889; Abe Kobo: *Inter Ice Age 4*, 1970, first ed. *Dai-Yon kampyoki*, 1959; J. G. Ballard: *The Wind from Nowhere*, 1961; John Christopher: *The World in Winter*, 1962; James Blish: *We All Die Naked*, 1969; Kurt Vonnegut: *Slapstick or Lonesome No More!* Arthur Herzog: *Heat*, 1977). (The 1930s saw numerous science fiction films centred around apocalyptic, sometimes climatic destruction as metaphorical responses to the widespread economic and political crises of the day.) Some years later German and French literature were also conquered by the theme. All this took place at the same time that the majority of scientists accepted the theory of anthropogenic climate change. The number of related literary works has since risen to several hundred, making it almost impossible to embrace the subject as a whole. For example, the database of the website "Eco-Fiction—Climate change and eco-themes in literature and the arts", a voluntary outreach project run by the small Moon Willow Press, listed in September 2015 326 titles that had been published mainly in the last fifteen years. This number in itself indicates that these books are read by more people than works of non-fiction on climate change, and therefore have to be taken into consideration with regard to environment protection.

Most of the books listed were published by small independent presses, but a few of them were put out by well-known publishing houses, such as Macmillan, Harper Collins, Faber and Faber, Bloomsbury and Penguin Press. The theme has penetrated into popular culture, and besides literature has made its way into films, TV series, visual arts, theatre, pop music and lately video games and comic books (cf. Wikipedia: "Climate change in popular culture"; goethe.de/klima). Dissertations and other academic works on ecological topics in literature have been written since the 1980s, whereas the systematic investigation of novels focusing on the issue of climate change is rather difficult, as these have only attracted the interest of literary studies more recently [10]. Only a few such works have been analysed in detail. In Timothy Clark's *The Cambridge Introduction to Literature and the Environment* the fictional discourse of climate change features as a main topic of the preface [11]. Some universities now offer climate-change fiction courses that deal with both literature and film [12].

Novels dealing with climate change form a sub-category of ecologically oriented fiction (eco-fiction) that focuses on, or strongly alludes to, a world affected by climate change, usually during and after events that take place when climate variations occur, including the "relational-causal" aspects, and the human consequences. Climate fiction, or cli-fi (the term was coined by climate activist and freelance writer Dan Bloom in 2007) is not a new genre, but a theme in science or speculative fiction that has been around for decades and which is constantly evolving with an ever-increasing number of novels and films. Gregers Andersen suggests reserving the term only for works of fiction of various genres that employ the specific scientific paradigm of anthropogenic global warming in their plot [1]. Other terms for climate change novels are solarpunk and anthropocene fiction. Some literary critics speak of cli-fi as a separate genre or a sub-genre of science fiction [13].

The stories are often written with a moral or political imperative to warn society about the human impact on nature. Imaginative narratives about climate change occupy a wide range of genres. Climate themes are commonly found in science fiction, dystopian fiction, literary fiction, speculative fiction and fantasy, but can also be seen in novels dealing with romance, suspense, thriller, adventure and the like. These themes appear in children's literature, juvenile and adult fiction as well. There is also a host of post-apocalyptic adventure novels aimed at young-adult readers. Climate-themed stories can either be earlier science fiction tales based on imaginary ideas about natural or sudden climate changes, or modern stories based on understandings of current, scientifically proven, human-caused climate change. These stories include scenes of the world's most climatically vulnerable regions. Mythological and Biblical themes, the amplification and re-functionalisation of the literary topos *locus terribilis*, apocalyptic scenarios, the imagery of social breakdown, fragments of traditional narratives and popular narrative motifs are frequently used in the plot. Popular motifs include varying degrees of climate-change-induced destruction, the downfall of mankind, war, the struggle of an individual, couple or community as they face the forces of nature, the ruins of society or a new civilisation. Questions are often raised about the responsibility of the individual, society and leaders, the reinterpretations of values, adaptation to new circumstances, change in human character, solidarity and the role of technology. Cli-fi reflects on how humans interact with each other, as well as how non-human nature is to be respected, and preserved [14]. As in other genres of prose, besides inventiveness, authors' understanding of science and the humanities, together with insight into history and psychology, is crucial.

As a form of conceptualisation of environmental danger, cli-fi is one of the hottest reading trends in our times, but climate change is far too complex an issue for a single definitive novel. The aesthetic value of associated works is extremely varied; only a few of them are better than average. Cli-fi novels as influential as George Orwell's *1984* or Aldous Huxley's *Brave New World* are yet to be written. Climate change has not yet inspired works of literature that adequately reflect on this challenge of historic proportions. The international reputation of books is often guaranteed not by their literary quality but by marketing and the climate industry. Many of them have received prestigious literary awards, like for example, Patrick Ness's *The Knife of Never Letting Go* (2008), Ian McEwan's *Solar* (2010) and Paolo Bacigalupi's *The Water Knife* (2015). Several books have been translated into many languages, like for example, Michael Crichton's *State of Fear* (2004), Sophie D. Crockett's *After the Snow* (2012), Karen Dionne's *Freezing Point* (2008), Antti Tuomainen's *The Healer* (2013, first ed. *Parantaja*, 2010), Barbara Kingsolver's *Flight Behaviour* (2012) and Nathaniel Rich's *Odds Against Tomorrow* (2013). A few of them have been adapted for the screen, like for example, David Mitchell's *Cloud Atlas* (2004). Some have received positive reviews in distinguished newspapers, like *Time* magazine, the *Guardian*, *The Financial Times* and *The New York Times*. Many have a separate entry in Wikipedia. Among the authors we can find not only writers from the literary mainstream, but also debutants, journalists and scientists. Time will tell which pieces will lose their relevance or prove to have a lasting effect. The best novels weave together the emotional, rational and moral threads, and provide us with new ways of seeing nature and our place in it.

There is plenty of rhetorical and narrative potential in the representation of the risk global climate change poses. Some cli-fi novels establish new connections between local, national and global forms of awareness by the use of innovative aesthetic forms combining epic, allegory and views of our planet as a whole, employing techniques of montage, fragmentation, collage and zooming. A number of prominent novels have recently depicted climate change or associated phenomena, such as rising sea levels and global warming. These include, for example, Margaret Atwood's *Oryx and Crake* (2003), Will Self's *The Book of Dave* (2005) and Kim Stanley Robinson's *Science in the Capital* trilogy (*Forty Signs of Rain*, 2004; *Fifty Degrees Below*, 2005; *Sixty Days and Counting*, 2007). One possible appeal in representations of climate change lies in the way in which the climatically changed world of a distant future is depicted. A lot of cli-fi novels are highly sophisticated and deeply rooted in science. They focus on the more subtle, early impacts of climate change, on the lives of scientists, activists and others working on climate-related challenges, or on how life will change for ordinary people in the future. Climate change scenarios sometimes merely serve as a background for other themes and situations. Similar to global warming dramas from Hollywood, authors like Michael Crichton dismiss climate change as exaggerated, misleadingly claiming it to be a hoax propagated by environmental activists, and tries to relieve mankind of its responsibility.

In Robinson's aforementioned trilogy, the representation of climate change appears as a complex set of compromises within the generic space of Utopian science fiction [15]. Robinson does not simply establish a future, climatically changed novelty; instead he determines its status in universally challenging and self-reflecting ways. He presents a scenario of an abrupt climate change in the near future; the catastrophe is told from the perspective of a group of scientists and political advisors in Washington DC. The trilogy presents us with no obvious hero, and the action incorporates current scientific research. The imagined paradigm of a society that does not yet exist supplies a cognitive map of what does exist. The moral of the trilogy suggests that only science and politics together can save the day. The overall experimentation and mixing of espionage thriller and political romance is provocative and intentional. Robinson's representation of a world affected by climate change is, as Adeline Johns-Putra pointed out, "a sophisticated revision of generic habit and habitat, a re-imagining of the world-building impulse that characterises science fiction."

The assumption that anthropogenic global warming will, in the future, result in extreme disintegration and human violence on a large scale appears not only in a number of works of popular science, but also in several contemporary climate fictions, like for example, Marcel Theroux's *Far North* (2009), Steven Amsterdam's *Things We Didn't See Coming* (2009) and Helen Simpson's *A Diary of an Interesting Year* (2010). In Simpson's short story, the Biblical narrative of Babel is one of the templates that lends itself to imagining a future climatically changed world [1]. Humanity's progress is set into reverse by its own scientific and technological abilities, as industrialisation first leads to global warming and then sets into motion a world wherein climate refugees, incapable of communicating, are engaged in a constant struggle for survival. This imaginary world faces rather grim prospects, to say the least.

A lot of contemporary cli-fi authors depict a different world affected by climate change, one that we can call the world of the Last Judgment (e.g. Frank Schätzling: *The Swarm*, 2004). However, what is new compared to the ancient imagery of the Last Judgment in contemporary cli-fi is that here it is no longer God, but rather the non-human world, "Nature" itself that acts as both judge and executioner [1]. This notion obviously stems from James Lovelock's controversial Gaia theory, according to which our planet is a self-regulating organism capable of even endangering mankind in order to preserve itself. This idea of the Last Judgment appears in a number of products of Western popular culture, and forms part of a dominating repertoire of narrative templates applied in the imagination of anthropogenic climate change.

For my last example I would like to turn to German science fiction, namely Dirk C. Fleck's novel *Maeva!* (2011), a sequel to his work *The Tahiti Project* (2008). With strong journalistic dimensions, the novel shows the multi-faceted and complicated relationship between local and global actions in a future world affected by global warming [16]. Fleck refers directly to existing political institutions, condemns current renewable energy projects, and presents a nuanced picture of a climatically changed world, in which the expansion of renewable energy cannot serve as the only solution to the problem. According to Fleck a more balanced approach, a life in greater harmony with nature could become part of a solution to the impending environmental crisis. The World Wide Web and digital media play a decisive role in the novel and feature as important tools for ensuring a global perspective. As Antonia Mehnert pointed out, this cli-fi "is meant to serve as a nodal point at which the readers reflect on the narrated future outcomes, and re-consider their own way of life. It thereby contributes to an understanding of climate change as a magnifying glass for the long-term implications of our short-term choices, and as a mirror to re-consider what we really want to achieve for ourselves." The scenarios that Fleck develops explore the diverse environmentalisms and eco-political reactions to escalating crises in a climatically changed world, while at the same time emphasizing the cultural contexts that not only constitute but also constrain these developments.

1 Conclusions

1. Alexander von Humboldt was one of the first to raise awareness of the importance of climate change caused by mankind. The understanding of nature as an organic entity, Humboldt's concept of the balance and interaction of forces within nature, the relationship between man and nature, and regarding science

and the humanities as a single unity are all notions worth taking into consideration in our times, since they might help us establish a mutually favourable relationship with nature and encourage a creative approach to environment protection.

2. Today there are no irrefutable arguments against the theory of anthropogenic climate change. Future discussions are likely to revolve around the options of accommodation, possible preventive measures for the avoidance of a situation even more threatening than the present, and the relationship between growth and climate protection. Ten years ago the Stern Report revealed that global climate protection could be achieved for less than one percent of the world's gross output.
3. Climate change is a cultural change, and climate literacy is more than just understanding climate science. It is about understanding the social, cultural and human dimensions of climate change. It is about grappling with the impacts of climate change on our minds and emotions, and also the ethics of climate change. Thinking and writing about nature and the climate requires all the subtlety, and sophistication writers can muster [17, 18].
4. In the 21st century, ecological humanities have been responsible for showing the way industrialisation has mobilised environmentally damaging practices over recent centuries [19]. In order to understand global warming, we should not only consider its scientific explanations, but also see it within a broader network of discourses and social practices. In good literature especially, we can find innovative approaches to the topic. Literary studies can make a significant contribution to the rapidly evolving debate on climate change and climate protection [20].
5. Once the image of man conquering nature has finally been shattered, it is time to heed the warnings of historical ecology, and listen to what writers and poets say about how man's relationship with nature has changed in the past, about traditional strategies concerned with protecting forests and waters, that is our climate, about current developments in climatic and hydrological conditions, about ruining our environment and the fact that this is irreversible.
6. Besides the scientific, technological and economic approaches to climate change, it is necessary to incorporate the ethical dimension, together with the considerations of social sciences and cultural studies. Interpreting changes and methods of addressing these problems might vary considerably in relation to culture and ideology. The best climate change novels call for a new perspective when tackling global, regional and local questions which do not fit into national politics. Literary studies may explore several unknown aspects of the cultural dynamics of climate change and climate protection.
7. It is widely known that climate change greatly influences the global hydrological cycle and regional sub-cycles. Various works of fiction show that climate change and issues of water supply deserve attention in order to prevent potential disasters and conflicts as a result of water shortage and rise in sea levels.
8. Forecasting and envisaging future anthropogenic global warming cannot solely be a matter for scientific investigation. It is also a culturally constructed phe-

nomenon, as the new is perceived through the cognitive schemes of the old in a process of dialectical adjustment.

9. While the effects of former eco-catastrophes were often clearly visible, the effects of global warming seem hard to grasp. There is a significant knowledge gap between the familiar preoccupations of everyday life and the abstract future of a climatically changed world. Good cli-fi is important because it goes beyond abstract predictions and statistics to show the reality of a possible future, and render visible, personal and tangible that which is as yet unimaginable. For many people it provides stronger motivation to act and to show consideration than looking at scientific data might do [21].
10. Contemporary climate fiction emphasizes the end of the idea that the non-human world is a God-given pond for human mastery, and helps us to go beyond the destructive divide between Nature (written with a capital n) and society. It helps us to perceive the world as a whole, and stimulates an understanding of global interconnectedness and attention to human and non-human, natural and cultural places.
11. Cli-fi helps to explore the reasons for climate change denial and its implications, including wishful thinking, ideology, consumer culture and active lobbying by the fossil fuel industry. It might help to change the reader's outlook as far as global warming is concerned. A further contribution to climate protection may be to persuade mankind to take care of the environment and change its actions now.
12. Good climate fiction is a powerful critical lens: a way to focus on what the future might hold, a way in which we can view global warming and its possible consequences if we do nothing to stop it. It nestles in our minds and urges us to consider other perspectives and adopt new solutions. The best cli-fi novels make us think deeply about the human side of climate change: they raise new questions about what it means not only to survive, but to be human and take responsibility.

Literature Cited

1. Andersen, G.: Cli-fi and the Uncanny. Interdisc. Stud. Lit. Environ. (in print) (2015)
2. von Humboldt, A.: Ansichten der Natur, mit wissenschaftlichen Erläuterungen. Cotta, Tübingen/Stuttgart (1808)
3. von Humboldt, A.: Life-Force, or the Genius of Rhodes. Yearb. Comp. Lit. **58**, 165–168 (2012)
4. Weatherby, L.: Life-force, or the genius of Rhodes. Translator's preface. Yearb. Comp. Lit. **58**, 162–163 (2012)
5. von Humboldt, A.: Kosmos. Entwurf einer physischen Weltbeschreibung, vol. 5. Cotta, Stuttgart/Tübingen (1845–1862)
6. Bratranek, F.T. (ed.): Goethe's Briefwechsel mit den Gebrüdern von Humboldt (1795–1832). Leipzig, Brockhaus, pp. 305–322, 346–357 (1876)
7. von Humboldt, A.: Central-Asien. Untersuchungen über die Gebirgsketten und die vergleichende Klimatologie, vol. 2. C. J. Klemann, Berlin (1844)

8. Tarizzo, D.: Speculative evolution. Darwin, Freud, and the Whale. Yearb. Comp. Lit. **58**, 71–94 (2012)
9. Heller, A., Hölbling, W., Zacharasiewicz, W. (eds.): Utopian Thought in American Literature. Untersuchungen zur literarischen Utopie und Dystopie. Narr, Tübingen (1988)
10. Trexler, A., Johns-Putra, A.: Climate change in literature and literary criticism. Wiley Interdisc. Rev. Clim. Change **2**, 185–200 (2011)
11. Clark, T.: The Cambridge introduction to literature and the environment. Cambridge University Press, Cambridge (2011)
12. Sommer, B. (ed.): Cultural Dynamics of Climate Change and the Environment in Northern America, vol. 3. Brill, Leiden (2015)
13. Mehnert, A.: Things we didn't see coming—riskscapes in climate change fiction. In: Mayer, S., Weik von Mossner, A. (eds.) The Anticipation of Catastrophe. Heidelberg, Winter, pp. 59–78 (2014)
14. Kunkel, B.: Die Erfindung der Klimawandelliteratur. Analyse & Kritik, Nr. 599/18.11.2014. www.akweb.de (2014)
15. Johns-Putra, A.: Ecocriticism, genre, and climate change: reading the utopian vision of Kim Stanley Robinson's science in the capital trilogy. Engl. Stud. **91**, 744–760 (2010)
16. Mehnert, A.: Climate change futures and the imagination of the global. In: 'Maeva!' by Dirk C: Fleck. Eur. J. Lit. Cult. Environ. **4**, 23–35 (2012)
17. North, P.: Introduction. Yearb. Comp. Lit. **58**, 1–4 (2012)
18. Raglon, R., Scholtmeijer, M.: Heading off the trail. Language, literature, and nature's resistance to narrative. In: Lane, R.J. (ed.) Global Literary Theory. An Anthology. Routledge, London/New York, pp. 759–768 (2013)
19. Iovino, S.: Ecocriticism oder: Wenn die Literatur vom Anderen spricht. In: Butzer, G., Zapf, H. (eds.) Theorien der Literatur. Grundlagen und Perspektiven, Bd. 5. Francke, Tübingen/Basel, pp. 205–216 (2013)
20. Vasak, A.: Météorologies. Discours sur le ciel et le climat, des Lumières au romantisme. Champion, Paris (2007)
21. Schneider, B.: Ein Darstellungsproblem klimatischen Wandels? Zur Analyse und Kritik wissenschaftlicher Expertenbilder und ihren Grenzen. Kritische Berichte **3**, 80–90 (2010)
22. Berry, G.: Modernism, Climate Change and Dystopia: An Ecocritical Reading of Light Symbology in Conrad's 'Heart of Darkness' and Eliot's 'The Waste Land'. artsonline.monash.edu.au//colloquy/download/colloquy_issue-twenty-one/berry.pdf (s. a.)
23. Heise, U.: Sense of place and sense of planet: the environmental imagination of the global. Oxford University Press, Oxford; New York (2008)
24. www.goethe.de/klima

Visualization in Climate Modelling

Michael Böttinger and Niklas Röber

1 From Observations to Simulations

Systematic meteorological observations, regularly carried out worldwide since the mid of the nineteenth century, enable us today to analyze the climatic development of the historical past [1]. The latest updates of these observations indicate an increase in the global mean 2 m-temperature by about 1 °C from 1850 to 2015. Observations are needed to better understand the nature of the climate system. However, in order to estimate possible future climate changes, simulations with numerical coupled models of the climate system are the only option we have. Today's complex Earth system models encompass the most important physical, chemical, and biological processes and interactions between the components of the climate system. Since the actual future development of human actions impacting the climate is not known a priori, a variety of climate scenarios is being developed to determine the range of possible future climate changes by performing climate simulations.

2 The Role of DKRZ

The German Climate Computing Center (Deutsches Klimarechenzentrum, DKRZ) is a national high performance computing (HPC) center, founded in 1987 to enable scenario simulations and other climate research with numerical models. Since then, climate models developed at the Max-Planck Institute for Meteorology (MPI-M) have been used for simulations carried out at DKRZ with respect to the German contributions to the IPCC reports. Due to the complexity of the models, their spatial resolution, and the very long periods simulated, climate modeling has always been a

M. Böttinger (✉) · N. Röber
Deutsches Klimarechenzentrum, Hamburg, Germany
e-mail: boettinger@dkrz.de

M. Palocz-Andresen et al. (eds.), *International Climate Protection*,
https://doi.org/10.1007/978-3-030-03816-8_39

compute- and data-intensive discipline: within the fifth phase of the Coupled Model Intercomparison Project, CMIP5 [2], and with respect to the 5th IPCC Report [3], for example, about one Petabyte of model output was produced at DKRZ. Such simulation projects are demanding, and DKRZ supports the scientists in all aspects of their modeling activities to facilitate the efficient execution of the projects. The support activities include optimization of model codes and workflow as well as management and visualization of the vast simulation data.

3 Climate Model Data

Climate model data is multivariate (it consists of various different scalar and/or vector quantities), 3-dimensional, and time-dependent. Although the periods typically simulated span up to hundreds of years, the data needs at least to be available on a 6-hourly basis in order to capture the daily cycle in the analysis, and, consequently, a quite large number of time steps need to be stored within climate simulations. Ensemble techniques, which are commonly used to estimate the internal variability in simulations, add a further dimension to the data.

In the 1990s, most climate models used rectilinear grids for computation and data storage. Today, however, a variety of more complex grids is used in order to make the models more realistic and to enhance their numerical properties. For analysis and visualization of climate model output, curvilinear, rotated, and unstructured (irregular) grids need to be considered as well.

As a result of the long-lasting cooperation and data exchange between international modeling groups contributing to large joint projects like CMIP5, international de facto standards for climate model data and associated metadata have been developed. Particularly, the NetCDF file format [4] is widely used for the exchange of model data; it is machine-independent and, by the use of metadata, self-describing. The metadata contains information on the dimensions, the grid, the quantities contained, the units used, and much more. The NetCDF metadata conventions for climate model output, the CF conventions [5], act as a standard and are regularly adapted to new requirements.

4 Geoscientific Visualization Workflow

Beside statistical analyses, data visualization is the most important tool to gain knowledge from the vast numerical output produced by climate models. The goal of scientific visualization is to transform the information contained in numerical data into images, which humans can much better comprehend than large fields of floating point numbers. In the process of scientific visualization, the amount of information is drastically reduced. Ideally, only the information needed to gain the desired knowledge from the data remains in the final visualizations. The overall

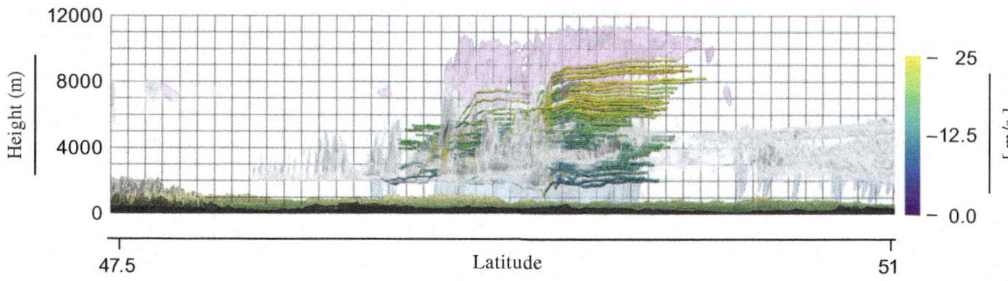

Fig. 1 Vertical slice of a high-resolution cloud-resolving simulation with ICON: cloud water (gray volume rendering), cloud ice (magenta isosurface), rain water (blue isosurface) and wind visualized with ParaView using an orthographic view

visualization workflow, which is often referred to as the *visualization pipeline* [6], consists of three consecutive steps: *filtering*, *mapping*, and *rendering*. Depending on the specific analysis tasks, the *filtering* part can consist of several individual steps such as format conversion, interpolation of the data, smoothing, selection of subsets, dimension reduction, and others. The resulting reduced data set is then mapped to geometric objects, which are finally rendered to images.

Most of the domain-specific visualization systems include functionality needed for basic *filtering steps*. More complex analyses of gridded geoscientific NetCDF data, however, are often better (or more efficiently) computed with separate tools such as cdo [7] or nco [8] prior to visualization. Generally, one has to decide which parts of the visualization pipeline need to be executed interactively. The fewer filtering operations (such as selection, dimension reduction, interpolation, data smoothing, etc.) are part of the interactive visualization pipeline, the shorter are the response times that can be achieved in interactive applications. On the other hand, the more processing steps are performed outside of the visualization application, the more effort is necessary if these parameters need to be modified later in the visualization process.

The *mapping step* refers to the mapping of numerical values to geometries and/or colors. 2D scalar data given on geo-referenced grids is mostly mapped to color-filled grid cells, isolines or filled contours. The result is combined with a map background in order to illustrate the geographic context. 2D methods can also be applied in 3D space, which might be useful to display the spatial location of the 2D slices presented. Beyond that, 3D visualization methods enable the mapping of 3D scalar data to 3D geometric objects such as isosurfaces. Transparency is often used in addition to the color mapping step to exclude unneeded information from the visualization. The mapping of data values to colors/transparency values is called *transfer function*. In the *volume rendering* technique, a transfer function is used for mapping scalar data given on a 3D grid to *voxels* with color and transparency (Fig. 1).

Instead of a simple map background, the 3D orography/bathymetry of the geographic region, eventually colorized by height or by a texture, is used to add geographic context to 3D visualizations (Fig. 2). However, for the majority of analyses published in the field of climate sciences, 2D visualizations are presented,

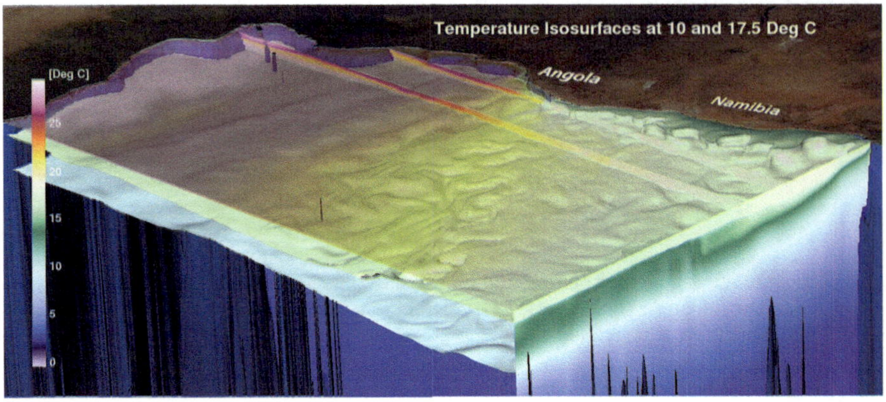

Fig. 2 3D visualization of regional ocean model data using Avizo. Isosurfaces and slices are combined to show the spatial structure of the 3D temperature field

although the original data is mostly 3D (Fig. 2). As an example, the roughly 1500 pages of the last IPCC report [3] do not contain any 3D visualizations of 3D data at all, and only very few 3D visualizations of 2D data can be found ([3], Fig. 1.14).

Vector data such as wind or ocean currents have a magnitude and a direction. Accordingly, directed glyphs such as arrows are used in vector field visualizations. The size and color of the glyphs can be used to encode the magnitudes. Other methods often used for visualizing vector data include streamlines (paths of massless particles for stationary vector fields) and trajectories (pathlines of massless particles for time-dependent data). All these methods can be used in both 2D and 3D; though vector data, especially in the case of 3D visualizations and high resolution models, needs to be thoroughly reduced in order to avoid visual clutter.

Depending on the area of interest and the focus of the visualization, a *map projection* is finally applied to the colorized geometric representations of the data and the geographic context. Typical projections often used are *equidistant cylindrical* (lat-lon), *spherical* (globe), *Mollweide* (Fig. 3) and *polar stereographic*.

Based on the geometries and colors derived before, images are rendered on a screen or exported to files in the final *rendering step*. To allow for publication quality, 2D visualizations are mostly exported using vector graphics formats such as PDF or PostScript. In 3D systems, this is not possible; virtual cameras are used to render pixel images of the view onto the geometries. For 3D rendering at interactive frame rates, powerful GPUs are utilized.

To account for the temporal dimension of the data, 1D visualizations of integral quantities (such as the global mean of the 2 m-temperature) or snapshots of 2D visualizations at different points in time are used within publications in the climate domain. By using time animation, the temporal evolution of spatial patterns can be represented much more intuitively. This applies to both 2D and 3D visualizations.

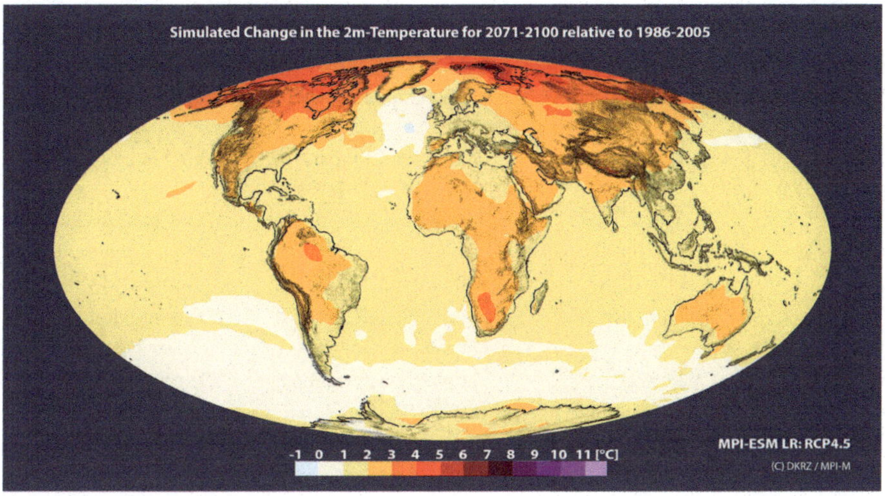

Fig. 3 2D visualization with Avizo using the equal area "Mollweide" projection. Note the shading resulting from rendering a high-resolution 3D relief

5 Visualization Frameworks Used in Climate Research

In 2008, Nocke et al. presented a brief overview of climate visualization, describing techniques and tools used by practitioners [9]. However, due to increased complexity in models and grids, increased resolution, and the widespread use of ensemble techniques, the requirements as well as the availability of suitable visualization software have changed since then. A more recent overview of selected domain-specific and general-purpose tools used by the climate community is given in Table 1. Although limited to meteorological applications, the detailed survey of Rautenhaus et al. [10] nicely presents the current state of techniques and tools used in the weather forecast and climate research communities.

Except for the commercial general purpose systems Avizo and IDL, all visualization solutions listed in Table 1 are freely available, and partially also open source. All of these systems allow the import of NetCDF climate model data, but often only support a subset of the grid types used in the community. All systems have their strengths and weaknesses; therefore, and depending on the visualization task, a variety of frameworks is employed at DKRZ, of which NCL, ParaView and Avizo are the tools used the most. All three systems support various grid types, including the unstructured grid of the ICON model that is jointly developed by the MPI-M and the German Weather Service DWD. While NCL, developed at the National Center for Atmospheric Research (NCAR), is a domain-specific programmable scripting language for data analysis and high quality 2D visualization, ParaView and Avizo are GUI-driven interactive 3D visualization systems with extensions for gridded climate data. Both utilize GPUs and—to some extent—multiple cores, although the scalabil-

Table 1 Selected visualization systems used by climate research practitioners

Type	Name	Location	Properties
Domain-specific	NCL	http://www.ncl.ucar.edu/	2D script-based
	IDV	http://www.unidata.ucar.edu/software/idv/	2D/3D interactive GUI
	Vapor	https://www.vapor.ucar.edu/	3D interactive GUI
	UV-CDAT	http://uvcdat.llnl.gov/	Collection: 2D/3D tools
	GrADS	http://cola.gmu.edu/grads/	2D script-based
	Ferret	http://www.ferret.noaa.gov/Ferret/	2D script-based
	GMT	http://gmt.soest.hawaii.edu/	2D script-based
General-purpose	ParaView	http://www.paraview.org/	3D interactive GUI
	Avizo	https://www.fei.com/software/avizo3d/	3D interactive GUI
	IDL	http://www.harrisgeospatial.com/	2D script-based
	Python/matplotlib	http://matplotlib.org/	2D script-based

ity and flexibility of ParaView is far better with respect to large data and distributed client-server or in situ visualizations.

Compared to Avizo, a powerful commercial application providing domain-specific tools such as map projections and geographic mapping, ParaView additionally offers several techniques for the visual analysis of multivariate data sets [11]. By support of brushing in combination with linked views (e.g. scatter plots, parallel coordinates, 3D views), correlations and dependencies between different quantities can be visually explored [12]. Figure 4 shows a selection of cells with a mean sea surface height of -1.5 m or less. In the parallel coordinates view, the selection of cells—which are located around the Antarctic continent—exhibits a strong correlation between very cold and salty water and low sea surface height, i.e. very dense water that tends to sink—an important aspect of the thermohaline circulation.

Due to an increase in computing capacity and maturing climate models that are able to run at finer spatial and temporal resolutions, data sizes and the complexity of the simulations increase as well. As part of the HD(CP)^2 project, the ICON model was extended to permit large eddy simulations at cloud resolving scales to foster the understanding of clouds and precipitation processes [13] (Fig. 1). The finest resolution of 3D data generated within HD(CP)^2 is sampled on 150 levels with 22.5 million cells per level and is stored at 1 min intervals. Consequently, new strategies are required to handle those very large data sets.

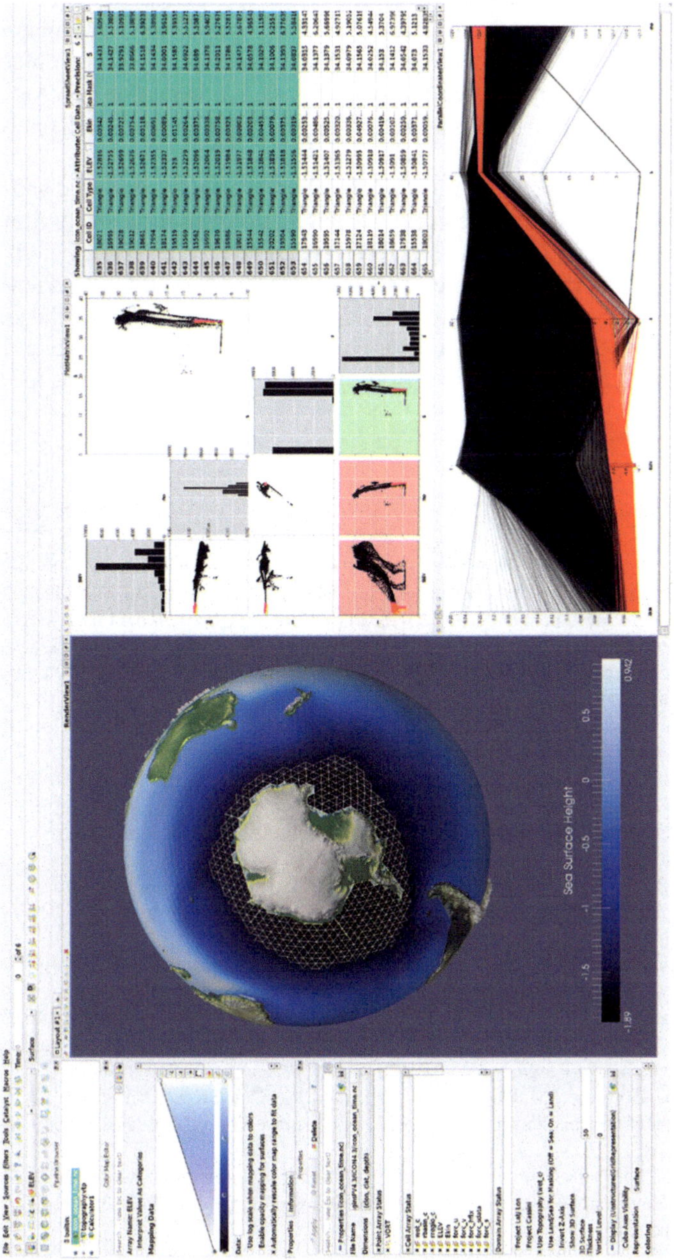

Fig. 4 3D visualization of ICON ocean data. A selection of cells with a sea surface height of −1.5 m or less shows a strong correlation between low temperature, high salinity, high density and low sea surface height

One solution is *in situ compression* and *level-of-detail rendering*. Although the data size stored on disk remains very high, the data is converted into a more manageable form that even allows an interactive exploration. VAPOR, a 3D visualization software also developed at NCAR, provides a wavelet-based data model and interactive level-of-detail rendering for very large rectilinear data sets. A current joint effort of NCAR, DKRZ and the University of Calgary pursues an extension of VAPOR to also allow a level-of-detail-based visualization for unstructured grids such as ICON and MPAS [14, 15].

Another possibility is *in situ visualization*, in which the data is analyzed and visualized "in situ" on HPC systems concurrently to simulation runs instead of writing the full model output to disk for later analyses. Catalyst, an extension to ParaView, allows creating interactive or automated in situ visualizations by connecting ParaView with simulation models using a Catalyst Adaptor [16]. The advantage is that the size of data stored on disk, i.e. images and animations, is vastly reduced. However, if the visualization fails to capture the desired information, the simulation needs to rerun.

References

1. Morice, C.P., Kennedy, J.J., Rayner, N.A., Jones, P.D.: Quantifying uncertainties in global and regional temperature change using an ensemble of observational estimates: the HadCRUT4 data set. J. Geophys. Res. **117**, D08101 (2012). https://doi.org/10.1029/2011JD017187
2. Taylor, K.E., Stouffer, R.J., Meehl, G.A.: An overview of CMIP5 and the experiment design. Bull. Am. Meteorol. Soc. **93**, 485–498 (2012). https://doi.org/10.1175/BAMS-D-11-00094.1
3. Stocker, T.F., Qin, D., Plattner, G.-K., Tignor, M., Allen, S.K., Boschung, J., Nauels, A., Xia, Y., Bex, V., Midgley, P.M. (eds.): Climate Change 2013: The Physical Science Basis. Contribution of Working Group I to the Fifth Assessment Report of the Intergovernmental Panel on Climate Change. Cambridge University Press, Cambridge, United Kingdom and New York, NY, USA, 1535 pp (2013). https://doi.org/10.1017/cbo9781107415324
4. http://www.unidata.ucar.edu/software/netcdf/, https://doi.org/10.5065/d6h70cw6
5. http://cfconventions.org/
6. Aigner, W., Miksch, S., Schumann, H., Tominski, C.: Visualization of Time-Oriented Data, 1st edn. Springer Publishing Company, Incorporated (2013). https://doi.org/10.1007/978-0-85729-079-3
7. CDO 2018: Climate Data Operators. Available at: https://code.mpimet.mpg.de/projects/cdo/
8. NCO 2018: netCDF Operators. Available at: http://nco.sourceforge.net/
9. Nocke, T., Sterzel, T., Böttinger, M., Wrobel, M.: Visualization of Climate and Climate Change Data: An Overview. In: Ehlers, M., et al. (eds.) Digital Earth Summit on Geoinformatics 2008: Tools for Global Change Research, Wichmann, Heidelberg, pp. 226–232 (2008)
10. Rautenhaus, M., Böttinger, M., Siemen, S., Hoffman, R., Kirby, R.M., Mirzargar, M., Röber, N., Westermann, R.: Visualization in meteorology—a survey of techniques and tools for data analysis tasks. IEEE Trans. Vis. Comput. Graphics (TVCG) **24**(12), 3268–3296 (2018)
11. Ayachit, U.: The ParaView Guide: A Parallel Visualization Application. Kitware (2015)
12. Becker, R.A., Cleveland, W.S.: Brushing scatterplots. Technometrics **29**(2), 127–142 (1987)
13. http://hdcp2.zmaw.de
14. Jubair, M.I., Alim, U., Röber, N., Clyne, J., Mahdavi-Amiri, A., Samavati, F.: Multiresolution visualization of digital earth data via hexagonal box-spline wavelets. In: IEEE Conference on Visualization, Chicago, USA, 2015

15. Jubair, M.I., Alim, U., Röber, N., Clyne, J., Mahdavi-Amiri, A.: Icosahedral Maps for a Multiresolution Representation of Earth Data, to appear in International Symposium on Vision, Modeling and Visualization, Bayreuth, Germany, 2016
16. Ayachit, U., Bauer, A., Geveci, B., O'Leary, P., Moreland, K., Fabian, N., Mauldin, J.: ParaView Catalyst: Enabling In Situ Data Analysis and Visualization. In: Proceedings of the First Workshop on In Situ Infrastructures for Enabling Extreme-Scale Analysis and Visualization (ISAV2015). ACM, New York, NY, USA, pp. 25–29 (2015)

Reuniting the Two Moieties of Human Knowledge: The Wisdom at the Intersection of Art and Science

Chantal Bilodeau

1 Introduction

There is a story in the native Tlingit culture of Southeast Alaska that describes how the glacier in Glacier Bay used to be far up the bay:

> It was said you could clearly see up the bay. Through the mountains there you could see the glacier w-a-a-a-a-a-y up the bay; it was only a tiny piece. [...] It couldn't be seen much from the river; it could only be seen from way out [7].

But one day, the glacier started to grow:

> Suddenly people said, "What's wrong with the glacier? It's growing so much!" They used to see it w-a-a-a-a-a-ay up the bay. But now it was near, getting closer [...] It was now moving f-a-a-a-a-a-st. They said [...] the way it was growing was faster than a running dog [7].

According to the story, a young girl from the village broke a taboo by calling the glacier and enticing it with a piece of sockeye dryfish: "Glacier. Here. Here. Here" [7]. In response to her call, the glacier lurched forward, grinding the bottom of the bay, turning the water murky "like diluted milk" and causing "whirlpools [to churn] over to the surface like the tide" [7]. In a short amount of time, the glacier overtook the land forcing Tlingit villagers to abandon their homes and relocate to nearby islands.

Modern science tells us that Glacier Bay "was habitable for many centuries up until 300 years ago, when a final glacial surge would have forced the human habitants to flee their homeland" [6]. Monteith et al. further explain that geological data shows there was a tight window of time in the 1700s where the glacier would have overrun the area, reached its peak and begun to retreat [10]. Though it was expressed differently in the Tlingit story, science confirms that the glacier surged several miles within a generation or two, traveling at an exceedingly quick pace on a geological timescale.

C. Bilodeau (✉)
The Arctic Cycle, New York City, USA
e-mail: chantal1402@me.com

M. Palocz-Andresen et al. (eds.), *International Climate Protection*,
https://doi.org/10.1007/978-3-030-03816-8_40

While science gives us precise data about what happened in Glacier Bay, the Tlingit story encapsulates much more. Through rhythm, imagery and metaphor, we learn about the scope of the event and get observations about its manifestations *and* its social, cultural and emotional impact. The story is, in fact, a blend of scientific information and artistic storytelling. Or, to borrow terms from the Center for Research on Environmental Decisions (CRED), it is a blend of information that targets the brain's analytical processing system, which controls analysis of scientific information, and the experiential processing system, which controls survival behavior and is the source of emotions and instincts [2].

2 When Fact Meets Emotion

Over the last few centuries, rapid advances in science and the increased specialization of knowledge have caused science and art to grow apart. Today, the gap between them has widened to the point where the possibility of meaningful interaction is, to many, unthinkable. For the most part, science is regarded as the more valuable discipline. It helps us understand and predict natural phenomena, makes our lives comfortable, and is supposedly objective. Art, on the other hand, is seen as much more amorphous, highly subjective, and often a luxury that contributes little to day-to-day life.

However, with the increased urgency and unprecedented scope of climate change, which calls for cooperation across physical and intellectual boundaries, there is a movement to reunite the two moieties of human knowledge and explore the rich intersection where facts and emotions meet. This impulse is supported by the fact that decades of climate change communication based on scientific data alone has yielded very little result. We need new strategies.

There is strong evidence from the social sciences that between the brain's two processing systems, the experiential processing system is the stronger motivator for action [8]. The experiential processing system relates current situations to past experiences, often evoking strong feelings and making them dominant in processing (CRED 16). As described by Marx et al.: "A sufficiently vivid description of a situation permits listeners or readers to place themselves in the story, thereby to be influenced by strong positive or negative affect, and to imagine the actions that they would take" [9].

By engaging the brain's experiential processing system through images, music, dance and storytelling, art can trigger reflection, generate empathy and foster new ideas. It can help people imagine different ways of being and relating [11]. It can also, in relation to climate change, make the crisis visible, audible and felt. When we combine art's strengths with the latest scientific insights, we can create experiences that are aesthetically and emotionally satisfying, and have the potential to move people to action.

3 Art in Action

To be effective, artists addressing climate change tend to use strategies that fall into one of five categories: sounding the alarm, celebrating the natural world, making the science visible, envisioning a positive future, and giving agency to communities. Each category has a specific focus, as described below:

Sounding the alarm: The first category stresses the magnitude and urgency of the climate crisis, and illustrates the dire consequences awaiting us if we fail to curb our greenhouse gas emissions. Work tends to be overtly political, sometimes confrontational. The focus is on raising awareness about what is being lost and underlining our collective responsibility.

Celebrating the natural world: Work that falls into this category is often inspired by an indigenous worldview where everything is deeply interconnected. There is an effort to view humans as part of the natural world instead of separate from it, and to promote peaceful values and an attitude of contemplation. The focus is on appreciating and protecting what we have.

Making the science visible: The third category is the one that most directly intersects with climate science. Work translates scientific data into sensory and/or emotional experiences. It offers new ways of looking at and understanding science. The focus is on making science accessible and relevant to people's lives.

Envisioning a positive future: Work in this category is concerned with positive scenarios. It proposes a vision of the world we can all embrace and/or examples of the steps we need to take to get there. The focus is on imagining alternative realities and solutions to our climate crisis.

Giving agency to communities: Maybe the most effective strategy in terms of generating action, artists in this category work in partnership with communities to address local problems. People are active participants in the art-making process. The focus is on empowering communities.

Though the strategies and goals may be different, work in all five categories is directly or indirectly informed by science and offers new perspectives on climate change. Below are two examples from "Making the science visible" and "Envisioning a positive future" where I examine the art/science intersection in more detail. The first example discusses photographer James Balog and his Extreme Ice Survey program. The second example looks at Santiago Muros Cortés's public art/renewable energy project Solar Hourglass, the winning entry in the Land Art Generator Initiative 2014 design competition.

3.1 James Balog: The Memory of the Landscape

James Balog is an American nature photographer who rose to fame in 2012 as the subject of the documentary film *Chasing Ice*. Trained as a geomorphologist, in 2007 Balog founded the ongoing photography program Extreme Ice Survey (EIS) [5].

Determined to find a way to express climate change visually and convinced that "the story [was] in the ice," he rigged a series of still cameras to automatically take one photo per hour of daylight over the course of three years [3]. He secured the cameras to rocks overlooking glaciers in Alaska, Greenland, Iceland and Montana, and waited.

After a few setbacks the cameras delivered, each yielding approximately 8,000 photos per year. Photos are assembled in time-lapse videos, creating stunning visual evidence of the effect of climate change on glaciers. Not only is the ice melting at the edge and are glaciers retreating at an alarming rate, but they are losing significant volume. This evidence is all the more poignant when we understand its implication. As Balog explains after retrieving a memory card from one of the cameras: "This is the memory of the landscape. That landscape is gone. It may never be seen again in the history of civilization" [3].

By changing the timescale at which we look at glaciers, Balog makes visible something that has never been seen before. He makes it possible to watch a scientific process *as it is happening*. It would take several minutes for someone to explain this phenomenon with scientific data and, most likely, the idea would remain abstract. But three years of visual data compressed in a video less than a minute long is understood instantaneously. Faced with Balog's images, we can no longer be passive listeners; we become active witnesses.

EIS also brings attention to another timescale, which tends to exist outside of human consciousness: the geological timescale. We are accustomed to thinking about geological processes as slow unfolding events that happened in the distant past. The time-lapse videos force a different perspective, one where we must reckon with massive geological changes taking place right before our eyes. In Balog's own words: "When a glacier that's been here for 30,000 years or 100,000 years is literally dying in front of [your] eyes, you're very aware that sometimes you go out over the horizon and you don't come back" [3].

Few of us live close enough to glaciers to have a personal relationship with them. The fact that they are melting can seem like a distant problem we don't need to concern ourselves with. Balog bridges that distance by bringing the glaciers to us "in the flesh" and translating the science into evocative images. His work encourages the development of a personal relationship with the glaciers, an understanding of how their health affects us all, and it plants the seed for the desire to protect them.

Finally, in addition to being informed by science, EIS contributes directly to it. Balog has built relationships with scientist to "provide [them] with basic and vitally important information on the mechanics of glacial melting" [1]. Not only is the science made visible through art, but art and science have found a way to be mutually beneficial.

3.2 Land Art Generator Initiative: Beauty in Renewables

The Land Art Generator Initiative (LAGI) strives to create a sustainable world by supporting the design of public art that doubles as a source of renewable energy for the benefit of local communities and environments. The blending of art, architecture, engineering, and other sciences offers a hopeful vision of a future where fossil fuels have been replaced by clean energy, and power plants have become objects of beauty.

One of the programs of LAGI is an international design competition that takes place biennially. Since 2010, the competition has invited artists to design for sites in Dubai/Abu Dhabi (2010), New York City (2012), Copenhagen (2014), and Santa Monica (2016). The design site for the 2014 competition was Refshaleøen, an old shipyard in Copenhagen. First place winner Santiago Muros Cortés from Argentina proposed a giant Solar Hourglass: "Rather than using sand to measure time, the Solar Hourglass uses the power of the sun to electrify hundreds of homes while providing a breathtaking setting for inspiration and relaxation" [4].

The Solar Hourglass is made of two curved white dishes positioned atop each other. The installation resembles a somewhat abstracted hourglass and functions as one, but using a solar beam instead of sand as trickling material. Solar energy is collected in the upper dish, and travels down through a transparent tube to a receiver located under the bottom dish, where it is transformed into electricity.

Though technically a power plant, the Hourglass is a beautiful piece of public art. Envisioned as an urban park, the bottom dish provides a platform where families can sit and eat, where couples can lie on the gentle slope and enjoy views of the Copenhagen harbor, or where musicians can present impromptu performances. The upper dish provides shade from the intense summer sun or shelter from sudden rain showers. At night, the solar beam traveling through the tube goes off and the thin layer of OLED (organic light-emitting diode) that covers the installation lights up on the surface of both [dishes], transforming the Hourglass into a pair of elegantly curved planes that shine off the edge of Refshaleøen (*ibid*).

The Hourglass "aims to send an optimistic message to those who visit it" [4]. In a world where popular media is saturated with negative scenarios, the project offers a refreshingly positive look at the future as well as a visual reminder that there is still time to stop climate change, if we choose to act now. Built from recycled steel and aluminum and creating no emissions or by-products of any kind, the Hourglass adds beauty to the urban environment, enhances the lives of surrounding communities, and can potentially change our relationship to energy by making it beautiful, sustainable, and safe. As a purely engineering project, this power plant would not convey as strong a message as it does now. Its artistic component is what makes it a hopeful icon for what lies ahead if we make the right decisions.

4 Conclusion

Art and science are humans' two fundamental ways of understanding the world. Science provides concrete information about the structure and behavior of the physical and natural world. Art uses the power of imagination to help us shape ideas and feelings, and investigate our place on earth. Historically, the two disciplines have been closely linked but over the past few centuries, they have increasingly grown apart.

Until recently climate change communication has relied on science alone to convey the urgency of the global climate crisis, but this approach has yielded only mild results. Scientific data primarily engages the brain's analytical processing system. In order for communication to be effective, social sciences tell us it must also engage the experiential processing system.

By combining art and science and using strategies from one of five categories, artists can engage both processing systems and have a more affective impact on people. The two examples examined in this article—James Balog's Extreme Ice Survey program, and Santiago Muros Cortés's Solar Hourglass, which won LAGI's 2014 design competition—show that art can help contextualize science, offer new perspectives, and give climate change a voice and a face. It can personalize events that otherwise appear distant in time and space. It can have a positive impact on communities and environments, and create hopeful icons for what a sustainable world may look like.

When the Tlingit created the story about Glacier Bay, they knew instinctively that it had to be vivid enough to allow storytellers to remember it and pass it down generations. Artists today can do the same for climate change. By creating memorable experiences, they can touch people and make climate change much more difficult to ignore.

References

1. About EIS—Why Does EIS Exist? Extreme Ice Survey. Web, 27 May 2016
2. Center for Research on Environmental Decisions: The Psychology of Climate Change Communication: A Guide for Scientists, Journalists, Educators, Political Aides, and the Interested Public. New York, 2009, 15. Print
3. Chasing Ice. Dir. Jeff Orlowski. Featuring James Balog. Submarine Deluxe, 2012. Film
4. Competition—2014 Copenhagen—Winners. Land Art Generator Initiative. Web, 28 May 2016
5. Extreme Ice Survey. Web, 28 May 2016
6. Glacier Bay—Early Peoples. National Park Service. Web, 11 May 2016
7. James, S.: Glacier Bay history. In: Dauenhauer, N.M., Dauenhauer, R., Juneau, A.K. (eds.) Haa Shuká, Our Ancestors: Tlingit Oral Narratives. Sealaska Foundation, pp. 245–259 (1987). Print
8. Marx, S., Shome, D., Weber, E.U.: Analytic vs. experiential processing exemplified through glacial retreat education module. Center for Research on Environmental Decisions, 2006. Web, 29 November 2015

9. Marx, S.M., Weber, E.U., Orlov, B.S., Leiserowitz, A., Krantz, D.H., Roncoli, C., Phillips, J.: Communication and mental processes: experiential and analytic processing of uncertain climate information. Glob. Environ. Change, 2007. **17**, 47–58. Web, 29 November 2015
10. Monteith, D., Connor, C., Streveler, G., Howell, W.: Geology and oral history: complementary views of a former Glacier Bay landscape. In: Piatt, J.F., Gende, S.M. (eds.) Proceedings of the Fourth Glacier Bay Science Symposium, 26–28 October 2004: U.S. Geological Survey Scientific Investigations Report 2007-5047, 2007, pp. 50–53. Web, 17 May 2016
11. Wright, S.: Introduction. The Art of Life: Understanding How Participation in Arts and Culture Can Affect Our Values. Mission Models Money & Common Cause. United Kingdom, 2007, pp. 4–6. Web, 29 November 2015

Environmental Protection in the Contemporary Art

László Sípos

The contamination of the Earth has reached never before seen dimensions. The rising sea level, the lack of drinking water, the leakage of hazardous substances into the soil, the smog and the thinning of the ozone layer are severe problems of our present. Increasing atmospheric concentrations of greenhouse gases will lead to an increase in the temperature of the atmosphere and the global warming rate will accelerate. While the average atmospheric air temperature between 1905 and 2005 increased by 0.74–0.18 °C, between 1990 and 2100 [1], it is likely to increase by 1.1–6.4 °C [2]. Global warming can also be attributed to most natural disasters [3].

The responsibility of the largest CO_2 emitting countries in the process is enormous, yet unfortunately not recognized by the governments of all states. Some also question the reality of global warming. President Donald J. Trump has denied the science of climate change many times in recent years, calling it a "con job" and a "myth," and even suggested the concept was "created by and for the Chinese in order to make U.S. manufacturing non-competitive." As a presidential candidate, Trump said he did not believe climate change is a significant threat, and that he doubted humans contributed to it. "I consider climate change to be not one of our big problems," he said in 2015. After winning the election, Trump said that on his first day in office, he would redirect the billions of dollars the Obama administration pledged to UN climate change programs toward fixing U.S. infrastructure [4]. China another sizeable national emitter, didn't put all its political capital behind the adoption of ambitious carbon-emission capping targets [5]. The 2015 United Nations Climate Change Conference was held in Paris at, the end of 2015. The conference negotiated the Paris Agreement, a global agreement on the reduction of climate change, the text of which represented a consensus of the representatives of the 196 parties attending it [6]. Significant bets on the protection of the environment are among the most critical

L. Sípos (✉)
István Széchenyi Management and Organisation Doctoral School, University of Sopron, Sopron, Hungary
e-mail: sipos.laszlo@yahoo.com

M. Palocz-Andresen et al. (eds.), *International Climate Protection*,
https://doi.org/10.1007/978-3-030-03816-8_41

missions of humanity. We have to seize every tool to avoid further destruction of our environment. Many artists are aware of this.

Fine art has been in close contact with nature since its birth. The first works of art, fertility sculptures and animal models of cave drawings served cultic purposes. In Egyptian art, gods were man-like, animal-shaped, yet anthropomorphic figures. In contrast, in Greek-Roman art, the subjects were the men themselves. Later, the Middle Ages were more focused on the beyond, and with Renaissance, the idea of humanism was reborn. During the Baroque period, man grew closer to nature, as knowledge of science and mathematics grew exponentially. This new knowledge had a direct, as well as indirect impact on the art of the period [7]. The still life, which we know from Antiquity, became a determinative genre of Baroque. Following this period, the thinkers of the Enlightenment focused on meaning, which triggered revolutionary thinking about nature, God, economy, and society. Romanticism first appeared in painting. The painters of this time emphasized the individual creative expression, the variability of nature and emotions, and the randomness of events. Artistic concepts such as plein air, impressionism, Barbizon, and Art Nouveau are all closely related to the ideas about nature that prevailed in the late 19th century. The desire to unite with nature also appears in architecture. An excellent example of this is the biomorphic architecture of Antoni Gaudi and his cathedral Sagrada Familia, still in construction today.

In the beginning of the 20th century, the rise of the avant-garde movement derailed focus away from nature. However, there were exceptions to this, such as Frank Lloyd Wright's Fallingwater completed in 1937, which was constructed over a 30-foot waterfall and designed according to Wright's desire to place the occupants close to the natural environment. The geometric concrete elements of the building are in perfect harmony with the waterfall and the surrounding forest. Thus, he illustrates the harmonious coexistence of artificial human construction and nature. The destruction caused by the two World Wars and the unprecedented industrialisation of the atomic era once again directed attention towards nature, which turned out to be much more fragile than we thought. The preservation of nature became a theme in the fine arts decades ago, and nowadays, artists are increasingly reflecting on climate change, their primary mission being to highlight the importance of environmental protection. In the following, I will mention a few of these artists.

One of the earliest art trends that focused entirely on this theme was called environmental art. It is a range of artistic practices encompassing both historical approaches to nature in art and more recent ecological and politically motivated types of works. Over the past ten years, environmental art has become a focal point of exhibitions around the world as the social and cultural aspects of climate change come to the forefront [8].

A subtype of environmental art is ecological art, which is an art genre and artistic practice that seeks to preserve, remediate and/or vitalize the life forms, resources and the ecology of Earth by applying the principles of ecosystems to living species and their habitats [9]. An example of this are the works of Joseph Beuys, Hans Haacke, Nicolas Uriburu, or Agnes Denes.

The Austrian artist and architect, Friedensreich Hundertwasser also worked in the field of environmental protection. From the early 1950s, he increasingly focused on architecture, advocating for environmentally friendly buildings. Hundertwasser propagated a type of architecture in harmony with nature in his ecological commitment. He campaigned for the preservation of the natural habitat and demanded a life in accordance with the laws of nature. He wrote numerous manifestos, lectured and designed posters in favour of nature protection, arguing to save the oceans and whales, to protect the rainforests, and to fight nuclear power.

Barry Underwood highlighted the importance of tree planting with his photos. His landscapes have been and continue to be altered by ambitious human activities linked to political, social, economic, climactic, and aesthetic forces. He is interested in connections between land use and the interpretation of a landscape as a politically symbolic environment, reflecting human activity and one's self-definition, as well as our values and beliefs [10].

The images of Naziha Mestaouis' work *One heart, One tree* were projected onto monuments in Paris during the United Nations Climate Conference (COP21) in December 2015. Spectators used their smartphones to participate in creating the "forest of light". A heartbeat sensor on the smartphone gathered data that was used to generate a unique digital tree that was then projected onto the building. For each computer-generated tree, a real tree was planted, enabling participants in the art installation to also become participants in reforestation.

On Earth Day in New York in 2011, large photos of forests were projected at Times Square. Nearly 86,000 people sent the word "tree" by SMS to a phone number, which donated $5 for Mexican, Kenyan, Indian and Philippine tree planting. The amount was deducted from the mobile accounts by the cooperating telecommunication companies. The increasing amount of donations was shown on giant screens by CNN, MTV, Reuters and Toshiba.

Along with the disappearance of forests, many artists are also directing attention to the rising sea levels. Jason de Caires Taylor attempts to illustrate the apocalyptic mood of climate change with the four cavalries in the Thames. Statues are dying twice in the tide. The artist portrays horses' heads as oil pumps to criticize our dependence on oil [11]. Banksy wrote "I do not believe in global warming" on the side of the Reagent Canal in Camden, North London in response to the unsuccessful talk of the Copenhagen Climate Summit meeting. The words written in red paint disappear in the water. Another example is the sculpture *Politicians Discussing Global Warming* of Berlin-based artist Isaac Cordal, which is a stark reminder of our collective failure to act on climate change.

Ludovico Einaudi played piano on a floating board drifting across the Arctic Ocean in Svalbard, Norway. The artist's performance is part of the Greenpeace "Save the Arctic" campaign, with the pianist calling attention to global warming. Einaudi's performance is accompanied by the sound of pieces of glacier falling into the sea. Nelo Azevedo's artwork is a collection of hundreds of carved 'ice-men', perched readily side by side on the steps of a custom house in the city of Belfast, a carefully prepared intervention that slowly thawed under the heat of the day. The process of melting ice leads to the disappearance of polar wildlife. Polar bears became the symbol of that. Mark Coreths' art piece was presented in Copenhagen and is sponsored by WWF to create awareness on the human impact on the climate. Once the ice sculpture has melted—which should take a day or ten—a metal skeleton remains that remotely reminds of the metal skeleton of the Terminator—it does provoke one to reflect upon the question if and what species would take the polar bears place if it were to become extinct [12].

In April 2016, a polar bear sculpture was erected in front of the Danish Parliament on a rod indicating that the animal species were in danger. The statue of Jens Galschiøt was named unbearable. This is a word game: unbearable means it is intolerable, but it also includes the bear word, which refers to the animal. The Danish sculptor has also collaborated with WWF.

In the documentary *Racing Extinction*, the Oscar winner Louis Psihoyos projected light shows depicting breath-taking colours of fauna onto New York City's skyscrapers, while warning of the fact that the species are going extinct due to our environmental degradation. Ian Wolters' work was a more than two metres tall monument that featured a steady 'waterfall' of engine oil running across the engraved names of known climate change deniers. On the Piccadilly Circus organised by Action Aid for five days, the British meteorological institute took over power over one of the giant screens every time rain came down. If the institute's measuring instruments detected rain in the neighbourhood, the software automatically launched a short film filmed by an advertising agency projecting the fate of children in Bangladesh due to climate change. In the country, every day, 50 children are killed because of the extreme weather such as floods or storms and the sea level rise. The film also encouraged passers-by to take part in the Climate Hatching a few days later. Luckily, the list of artists fighting for climate change awareness continues.

An increasing number of public and private institutions support the aspirations of nature conservation, as more people are starting to recognize the severity of global warming. The goal of many artists is to send across a message about the importance of environmental protection in hope that it will inspire active change concerning this issue in our society.

Friedensreich Hundertwasser: Columbus landed in India, 1969.

Barry Underwood: Landscape†Light†Installation, 2014.

Naziha Mestaoui: 1 Heart 1 Tree, 2015.

Isaac Cordal: Politicians discussing global warming, 2011.

References

1. IPCC Fourth Assessment Report https://en.wikipedia.org/wiki/IPCC_Fourth_Assessment_Report (last accessed 21st March 2018)
2. https://web.archive.org/web/20070203164304/http://www.ipcc.ch/SPM2feb07.pdf (last accessed 21st March 2018)
3. https://earthobservatory.nasa.gov/Features/RisingCost/rising_cost5.php (last accessed 21st March 2018)
4. https://www.cfr.org/interactives/campaign2016/donald-trump/on-energy-and-climate (last accessed 11th April 2018)
5. http://thedailynewnation.com/news/96538/facing-disastrous-global-warming.html (last accessed 20th April 2018)
6. https://en.wikipedia.org/wiki/2015_United_Nations_Climate_Change_Conference (last accessed 20th April 2018)
7. https://diverdave75.wordpress.com/2011/02/23/science-during-the-baroque-period/ (last accessed 20th April 2018)
8. https://web.archive.org/web/20140201203816/http://greenmuseum.org/generic_content.php?ct_id=306 (last accessed 23rd April 2018)
9. Strelow, H.: Natural Reality: Artistic Positions Between Nature and Culture/Kunstlerische Positionen Zwischen Natur und Kultur. Stuttgart: Ludwig Forum fur Internationale Kunst (1999). (last accessed 18th March 2018)
10. http://www.barryunderwood.com/statement.html (last accessed 8th April 2018)
11. https://www.theguardian.com/artanddesign/2016/feb/02/drowned-world-europe-first-undersea-sculpture-museum-lanzarote-jason-decaires-taylor (last accessed 8th April 2018)
12. https://www.nextnature.net/2009/12/melting-polar-bear-reveals-a-metal-skeleton/ (last accessed 20th April 2018)

Printed by Printforce, the Netherlands